WILDLIFE IN A CHANGING WORLD

An analysis of the 2008 IUCN Red List of Threatened Species™

RED LIST

IUCN

SSC
Species Survival Commission

Lynx

WILDLIFE IN A CHANGING WORLD

An analysis of the 2008 IUCN Red List of Threatened Species™

Edited by Jean-Christophe Vié, Craig Hilton-Taylor and Simon N. Stuart

The designation of geographical entities in this book, and the presentation of the material, do not imply the expressions of any opinion whatsoever on the part of IUCN concerning the legal status of any country, territory, or area, or of its authorities, or concerning the delimitation of its frontiers or boundaries.

The views expressed in this publication do not necessarily reflect those of IUCN.

This publication has been made possible in part by funding from the French Ministry of Foreign and European Affairs.

Published by: IUCN, Gland, Switzerland

Red List logo: © 2008

Copyright: © 2009 International Union for Conservation of Nature and Natural Resources

Reproduction of this publication for educational or other non-commercial purposes is authorized without prior written permission from the copyright holder provided the source is fully acknowledged.

Reproduction of this publication for resale or other commercial purposes is prohibited without prior written permission of the copyright holder.

Citation: Vié, J.-C., Hilton-Taylor, C. and Stuart, S.N. (eds.) (2009). *Wildlife in a Changing World – An Analysis of the 2008 IUCN Red List of Threatened Species.* Gland, Switzerland: IUCN. 180 pp.

ISBN: 978-2-8317-1063-1

Editors: Chief Editor: Jean-Christophe Vié
Editors: Craig Hilton-Taylor and Simon N. Stuart

Cover design: Lynx Edicions, Barcelona, Spain

Cover photo: Iberian Lynx *Lynx pardinus*. © Joe Zammit-Lucia

Layout by: Lynx Edicions, Barcelona, Spain

Produced by: Lynx Edicions, Barcelona, Spain

Printed by: Ingoprint, S.A., Barcelona, Spain
DL: B-31.360-2009

Available from: IUCN (International Union for Conservation of Nature)
Publications Services
Rue Mauverney 28, 1196 Gland, Switzerland
Tel. +41 22 999 0000
Fax +41 22 999 0020
books@iucn.org
www.iucn.org/publications

Lynx Edicions
Montseny, 8. E-08193 Bellaterra, Barcelona (Spain)
Tel. +34 93 594 77 10
Fax: +34 93 592 09 69
lynx@hbw.com
www.lynxeds.com

In the US:
c/o Postal Express & Fulfillment Center, Inc.
265 Sunrise Highway Suite 1 #252
Rockville Centre, NY 11570, USA

This book is printed on paper with FSC certification.

Contents

Foreword

People all over the world are becoming increasingly aware of the growing challenges facing our future and of the vital links between the natural world and human wellbeing. For generations, views on the status of the world's species remained largely speculative and highly focused on large, charismatic mammals but in recent times we are beginning to understand the overall situation of biodiversity far better - from the smallest invertebrates and fungi to the great trees of our forests and the whales of our oceans.

One of the tools that has helped us to "connect-the-dots" is The IUCN Red List of Threatened Species™ – the most comprehensive information source on the global conservation status of the world's plant and animal species. For decades, IUCN has brought together the knowledge of thousands of the world's leading authorities on species conservation through its expert network known as the Species Survival Commission (SSC). Comprised of over 100 taxonomic and thematic Specialist Groups as well as targeted, time-bound Task Forces to tackle contemporary challenges, the SSC continues to keep pace with the emerging issues confronting the conservation of species across the planet. Formed six decades ago, the SSC now comprises almost 8,000 members - their tireless, voluntary efforts help to expand the frontiers of science through their

contribution to Red List assessments – a tangible and enduring demonstration of their passion and commitment to conserving the world's species.

The process of conducting Red List assessments is extremely labour intensive; historically a labour of love delivered through close cooperation and collaboration amongst members of the SSC, staff of the IUCN Species Programme, and other contributing individuals and institutions around the world. The production of this *Analysis of The 2008 Red List* has been no exception and continues in our longstanding tradition.

In this volume, you will find the most up-to-date information on the patterns of species facing extinction in some of the most important ecosystems in the world and the reasons behind their declining status. For managers this information will assist in designing and delivering targeted action to mitigate these threats. From a policy perspective, the Red List offers a progressively more valuable tool. Increasingly it provides the fundamental information needed to deliver indicators for tracking: progress against national obligations under the Convention on Biological Diversity; the conservation status of those species in international trade under Convention on International Trade in Endangered Species; the extent and magnitude of climate change impacts

for reporting through the United Nations Framework Convention on Climate Change or towards refining our knowledge of migratory species apropos the Convention on Migratory Species.

The Red List has grown continuously in terms of its technical strength and breadth, providing a particularly unique and important tool for decision makers. With all species of amphibians, birds, mammals, reef-building corals, freshwater crabs, conifers, cycads and subsets of other taxonomic groups now assessed, the Red List provides an important foundation piece for conservationists by describing the patterns of species conservation status across landscapes and seascapes. The reader will find that in some areas of the world, for example the Mediterranean, our knowledge on the extent, magnitude and causes of threat is even greater for a wider spectrum of species; especially highly threatened endemics. Such information is crucial for conservation planning at the national, regional and global level.

As a result of its continual updating, expansion and deepening of content, we now know better than ever before that the prognosis for species across the Planet is dire. This volume reports new information on freshwater and marine species, which deliver important ecosystem services, including the provisioning of protein to some of the world's poorest communities. These species are now known to be facing extreme threat from overexploitation and habitat loss. The new insights presented here also help us to better understand the most likely differential responses and geographical patterns expected when the effects of global climate change begin to impact the world's most susceptible species. This cutting-edge work will provide predictive abilities to long-range planning and policy development as the effects of climate change are increasingly felt across the globe.

Through the dedicated efforts of thousands of scientists and practitioners, The IUCN Red List has become one of the most authoritative global standards supporting policy and action to conserve species around the world. We hope this *Analysis of The 2008 Red List* will provide you with new information and insights, which will motivate you to actions of unprecedented intensity and commitment on behalf of these fundamental building blocks of life on Earth.

Holly T. Dublin, Chair (2004-08), IUCN Species Survival Commission

Preface

We live in a world with an overload of information bombarding us every day. Most people, wherever they live, know that wildlife – and by 'wildlife' we mean both animals from the smallest insect to the largest mammal, as well as plants – is to some extent 'endangered'. But what is not generally realized is what this really means – how much of our wildlife is threatened, by what, where, what the consequences are likely to be and if it really matters – to us or to our children.

The IUCN Red List of Threatened Species™ tells us the answers to many of these questions. With a long established history, it is the world's most comprehensive information source on the global conservation status of plant and animal species. It is based on an objective system for assessing the risk of extinction of a species. Species listed as Critically Endangered, Endangered or Vulnerable are regarded as threatened and therefore most in need of conservation attention.

However, The IUCN Red List is far more than a register of names and associated threat categories. Underneath the listings is a gold mine of additional information. This includes a rich compendium of information on threats (e.g., climate change or invasive species), on where the species live, and most importantly information on conservation actions that can be used to reduce or prevent extinctions.

This gold mine comprising the extensive database 'underneath' The IUCN Red List also allows us to undertake analyses to determine, for instance, trends in the status of threatened species, the geography of threatened species as well as analyses of different threats and conservation responses. Some of the results of these analyses are presented here.

Every sector, whether it be trade, financial, or health, has its metrics for monitoring trends. For biodiversity The IUCN Red List is that metric. Around 45,000 species have been assessed to-date. This is a tiny fraction (2.7%) of the world's described species (with current estimates of the total number ranging from 5 to 30 million). We now know that nearly one quarter of the world's mammals, nearly one third of amphibians and more than 1 in 8 of all bird species are at risk of extinction. This allows us to come to the stark conclusion that wildlife (the word used in more technical circles is *biodiversity*) is in trouble, and the extent of the current risk of extinction varies between different species groups. For this reason IUCN is increasing the number of conservation assessments of species in the marine and freshwater realms, and for plants and invertebrate groups. Some early findings of this work are presented here.

A frequent reaction to any release of an update to The IUCN Red List is 'Why does it matter?' As the Millennium Ecosystem Assessment of 2005 made clear, biodiversity constitutes and sustains all life processes on the planet. It contributes utilitarian ecosystem 'good and services' as well as cultural, aesthetic and spiritual values and ultimately a sense of identity. It is thus fundamental to human well being. It is increasingly appreciated that biodiversity loss and ecosystem degradation jeopardises human well being. Examples abound from around the whole world – destruction of grazing lands in Ethiopia by invasive species resulting in whole villages being abandoned; the US fruit industry being no longer able to rely on wild pollinators; and fisheries collapsing worldwide, to name but a few.

From all this 'gloom and doom' arises the question – 'What can we do about it?' Less often articulated in public is a further point – 'Is it even worth bothering given that the situation seems so bad?' In some ways we do not apologize for highlighting 'bad news'. IUCN believes that the release of The Red List acts as a clarion call for the drive to tackle the extinction crisis – and without those facts being made clear the world will not react. It is a 'wake up call' and used as such by governments, NGOs, and civil society as a whole to help spread their messages and educate the world about the need to conserve biodiversity.

The Red List release is also an opportunity for us to show that conservation works. In 2008 we were able to report that the Black-footed Ferret *Mustela nigripes* moved from Extinct in the Wild to Endangered after a successful reintroduction by the US Fish and Wildlife Service into eight western states and Mexico from 1991-2008. Similarly, the iconic Wild Horse *Equus ferus* moved from Extinct in the Wild in 1996 to Critically Endangered this year after successful reintroductions started in Mongolia in the early 1990s. The fact that several important conservation planning tools rely on The IUCN Red List means that even the business community is both calling on and relying on this information to minimize their impact on the world's biodiversity.

Every time it is released The IUCN Red List gets increasing amounts of publicity. This is because it is trusted – not only by the media but by governments, NGOs, businesses and the general public. At the basis of this trust is the remarkable partnership of the world's leading species scientists – the IUCN Species Survival Commission, the IUCN Red List Partnership (including BirdLife International, Conservation International, NatureServe and the Zoological Society of London), together with the IUCN Species Programme which manages, processes and publishes The Red List. It is therefore a product of IUCN's triple helix of members, commission members and secretariat. It is important to recognize and pay tribute to not only the individual authors of the papers in this volume, but all those who contribute their expertise and data, often on a voluntary basis.

In 2002 the most of the world's governments (those who have ratified the Convention on Biological Diversity) set a target to try to begin to arrest the damage to the world's wildlife. It states 'To achieve by 2010 a significant reduction of the current rate of biodiversity loss at the global, regional and national level as a contribution to poverty alleviation and to the benefit of all life on earth.'

As we approach 2010, the world is beginning to assess to what extent this rather technical sounding target has been achieved. As can be seen from the findings presented in this volume we are facing the stark conclusion that the target will not be met.

As the world begins to appraise this situation in the run up to 2010, the International Year of Biodiversity, it is becoming clear that the prognosis for the future of humankind on this planet is tied up with a need to move from a situation which could be described as a patchwork of conservation successes to a whole new approach to biodiversity conservation by all sectors of society. Over the last few years the world has woken up to the threat of climate change. The same now needs to happen in relation to biodiversity conservation. The two are inextricably linked of course given that the destruction of biodiversity contributes to climate change by releasing carbon from forests, wetlands, grasslands and peatlands for example, and its conservation offers solutions to the climate problem as well as to humanity's general well being.

We are hearing a great deal about the economic 'credit crunch'. What we face also in the natural world is a 'credit crunch for biodiversity'. As the world wakes up to its failure to achieve the '2010 target' it is to be hoped that this publication and the ongoing work to produce and update The IUCN Red List of Threatened Species can contribute to a paradigm shift in our efforts to place true and realistic values on our wildlife. We need to set – and then reach - new ambitious targets to value and conserve the fundamental riches of our life support systems, and the wildlife and people that depend on them.

Julia Marton-Lefèvre, Director General, IUCN
Jane Smart, Director, Biodiversity Conservation Group and Head, Species Programme, IUCN

Acknowledgements

General

The IUCN Red List of Threatened Species™ is compiled and produced by the IUCN Species Programme based on contributions from a network of thousands of scientific experts around the world. These include members of the IUCN Species Survival Commission Specialist Groups, Red List Partners (currently Conservation International, BirdLife International, NatureServe and the Zoological Society of London), and many others including experts from universities, museums, research institutes and non-governmental organizations.

Wildlife in a changing world – An analysis of the 2008 IUCN Red List of Threatened Species was made possible thanks to the support of the French Ministry of Foreign and European Affairs.

Compilation and production of The IUCN Red List of Threatened Species™ would not be possible without the financial support of many donors. IUCN would like to thank all the donors who have generously provided funds to support this work, and in particular would like to acknowledge the ongoing financial support from The Rufford Maurice Laing Foundation that enables the production of the annual Red List updates.

Other major donors to the Red List assessment process include the Moore Family Foundation; the Gordon and Betty Moore Foundation; the Critical Ecosystem Partnership Fund; the European Commission; the Esmée Fairburn Foundation; the French Ministry for Foreign Affairs (DgCiD – Direction générale de la Coopération internationale et du Développement); and the MAVA Foundation for Nature Conservation (MAVA Stiftung für Naturschutz / Fondation pour la Protection de la Nature). The 2010 Biodiversity Indicators Partnership and TRAFFIC International supported the analysis of species used for food and medicine. Further details about the specific contributions of these and other donors are included under the acknowledgement sections below for the individual chapters.

To improve and expand the Red List assessment process, further development of the tools used is required. In order to support the new developments an IUCN Red List Corporate Support Group has been established. IUCN would like to acknowledge those organizations that have become members of the support group: Chevron, Electricité de France, Holcim, Oracle, Statoil, and Shell.

The editors would like to thank all authors and contributors for their production of the various parts of the publication as well as Vineet Katariya, Susannah O'Hanlon for producing all the maps, Mark Denil for designing the map layouts, and Kevin Smith for doing all graphics. Thanks are also due to Lynne Labanne who assisted with the design of the publication as well as its distribution, James Ragle and Claire Santer who compiled the long list of assessors and evaluators, Mike Hoffmann and Neil Cox who played a key role in the coordination of global assessments, Leah Collett who helped with data processing and Jez Bird who provided much help with queries on the bird data. Nieves García and Annabelle Cuttelod supervised the Spanish translations of the factsheets summarizing each chapter. Jean-Christophe Vié supervised the French translations. These one-page factsheets are available at: www.iucn.org/redlist.

Freshwater biodiversity: a hidden resource under threat

IUCN would like to acknowledge those donors whose financial contributions support freshwater biodiversity assessments; European Commission (EC); North of England Zoological Society (Chester Zoo); Conservation International; Esmée Fairbairn Foundation; Global Environment Facility through the 2010 Biodiversity Indicators Partnership; IUCN Water and Nature Initiative (WANI); Junta de Andalucia; MAVA Foundation (through the IUCN Mediterranean Regional Office for Cooperation); The Netherlands Ministry of Foreign Affairs (DGIS); Rufford Maurice Laing Foundation; Spanish Ministry of Environment and Wetlands International.

We would also like to acknowledge all the scientific experts from around the world that have underpinned the freshwater species assessments. Without these enthusiastic and dedicated individuals, this analysis would not be possible.

E. Abban, A. Abdelhamid, O. Akinsola, E. Akinyi Odhiambo, M. Ali, A. Amadi, G. Ameka, K. Amoako, M. Angliss, C. Appleton, F.Y.K. Attipoe, A. Awaiss, M. Baninzi, R. Barbieri, J. Bayona, C. Bigirimana, R. Bills, E. Bizuru, N.G. Bogutskaya, J.-P. Boudot, L. Boulos, T. Bousso, R. Brummett, J. Cambray, J.A.G. Carmona, M. Cheek, A. Chilala, S. Chimatiro, V. Clausnitzer, F-L. Clotilde-Ba, W. Coetzer, A.J. Crivelli, N. Cumberlidge, B. Curtis, L. Da Costa, M. Dagou Diop, M. Dakki, F. Daniels, S. Daniels, F.S. Darboe, G.M. De Bélair, F. De Moor, L. De Vos, C. Dening Touokong, M. Diedhiou, E. Dieme-Amting, K.-D. Dijkstra, B. Elvira, J. Engelbrecht, M. Entsua-Mensah, F. Erk'akan, S.A.F. Ferreira, W. Foden, J. Freyhof, M. Ghamizi, J.-P. Ghogue, N. Gichuki, P. Giorgio Bianco, M. Goren, P. Grillas, W. Hagemeijer, C. Hamallah Diagana, G. Howard, H. Howege, A. Ibala Zamba, D. Impson, S. Issa Sylla, H. James, D. Johnson, A. Jørgensen, M. Jovic, V. Kalkman, A. Kane, C. Kapasa, A. Karatash, E. Kaunda, J. Kazembe, J. Kipping, J. Kisakye, M. Kottelat, M. Kraiem, T.K. Kristensen, J. Kruger, F. Krupp, R. Kyambadde, P. Laleye, C. Lange, P.V. Loiselle, T. Lowe, M. Madeleine Manga, P. Mafabi, Z. Magombo, A. Mahamane, P. Makocho, V. Mamonekene, B. Marshall, G. McGregor Reid, A. McIvor, F. Médail, E. Michel, T. Moelants, N. Mollel, K.A. Monney, R. Monsembula, M. Mrakovcic, Mubbala, A. Muhweezi, M. Mutekanga, B. Mwangi, K. Naidoo, S. Ndey Bibuya Ifuta, A. Ndiaye, G. Ndiritu, B. Ndodet, B. Ngatunga, C. Ngereza, F. Niang Diop, A. Nicanor Mbe Tawe, F. Nicayenzi, G. Ntakimazi, M. Nyaligu, T. Nzabi, L. Nzeyimana, P. Ochieng Mbeke, S.S. Ogbogu, B. Olaosebikan, J.M. Onana, O. Opoye Itoua, L. Ouédraogo, D. Oyugi, L. Potter, V. Pouomogne, M. Povz, V. Prié, L. Rhazi, E. Riservato, J. Sa, B. Sambou, B. Samraoui, M. Samways, W. Schneider, K. Schütte, M. Seddon, M. Séga Diop, E. Sieben, J. Simaika, J. Šinžar-Sekulić, P.H. Skelton, J. Snoeks, G. Soliman, J. Somua

a

b

Partners and sponsors: a. Red List Partners, b. Publication sponsor, c. Major Red List sponsors, and d. Red List Corporate Group.

c

d

Amakye, D. Soumaré Ndiaye, J.S. Sparks, A-S. Stensgaard, M. Stiassny, F. Suhling, E. Swartz, S. Tchibozo, P. Tchouto, S. Terry, D. Twedole, T. Twongo, D. van Damme, E. Vela, J. Victor, K. West, and F. Wicker.

Status of the world's marine species

Corals
We thank Tom Haas and the New Hampshire Charitable Foundation, Conservation International, Esmée Fairbairn Foundation, and the Royal Caribbean Cruises Ocean Fund for their generous support of the IUCN Coral Red List Assessment. We also thank the following for help and support: Moonyeen Alava, Jonathan Baillie, Deb Bass, Hector Reyes Bonilla, Thomas Brooks, Hubert Froyalde, Peter Glynn, Scott Henderson, Cleve Hickman, Michael Hoffmann, Danwei Huang, Vineet Katariya, David Knight, Federico Lopez, Roger McManus, Mike Palomar, Caroline Pollock, Rodolfo Quicho, Jonnell Sanciangco, Michael Smith, Muhammad Syahrir, Romeo Trono, Mariana Vera, Dana Zebrowski, Charles Darwin Foundation, Conservation International Philippines, Darwin Initiative, First Philippine Conservation Incorporated, FirstGen Incorporated, Gordon and Betty Moore Foundation, Walton Family Foundation, and the Zoological Society of London.

Groupers
We thank the Society for the Conservation of Reef Fish Aggregations, the David and Lucile Packard Foundation, the University of Hong Kong, the US State Department and IUCN for supporting work on groupers. Members of the Groupers & Wrasses Specialist Group who have worked on assessments over the years are: Phil Heemstra, Howard Choat, Liu Min, Michel Kulbicki, David Pollard, Barry Russell, Beatrice Padovani, Melita Samoilys, Annadel Cabanban, Pat Colin, Matt Craig, Luiz Rocha, Kevin Rhodes, William Cheung, Rob Myers, Being Yeeting, Athila Andrade, Dave Cook, Andy Cornish, Patrice Francour, Mauricio Hostim Silva, Chris Koenig, Graciela Garcia-Moliner, Kwang-Tsao Shao and Sean Fennessy, with excellent technical support from Rachel Wong.

Marine turtles
We would like to thank all those members of the Marine Turtle Specialist Group who have given so selflessly of their time to carry out the detailed research and analysis required to conduct the assessments for each of the turtle species, and all of those who have shared their data so that the assessors could produce the most accurate of assessments. We are grateful to the Marine Turtle Specialist Group (MTSG) Assessment Steering Committee and its Chair for all the hard work and careful independent review of each assessment. We also acknowledge all of the host institutions for MTSG members, which allow us to invest time, energy and dedication into meeting research and conservation goals for these valuable species.

Sharks
We thank Conservation International, the David and Lucile Packard Foundation, Defra the US State Department, IUCN, Pew Lenfest Ocean Program, Marine Conservation Biology Institute and numerous other funders of the SSG's various Red List workshops over the past five years for their generous support. Full details are provided on the Sponsors page of the SSG's website. We thank all members of the Shark Specialist Group and invited regional and international experts who have contributed to assessments for their valuable time and enthusiastic commitment to species conservation, without which this work would not be possible. We also thank Caroline Pollock, Craig Hilton-Taylor and the Global Marine Species Assessment team for help and support.

Broadening the coverage of biodiversity assessments
We greatly acknowledge the funding support of the following organizations: Esmée Fairbairn Foundation, Rufford Maurice Laing Foundation, Global Environment Facility through the 2010 Biodiversity Indicators Partnership, The North of England Zoological Society (Chester Zoo), the Gordon and Betty Moore Foundation through Conservation International, Natural Environment Research Council and the Centre for Population Biology, Imperial College London, and The Fishmongers' Company.

Core team
Jon Bielby, Anne Chenery, Zoe Cokeliss, Blythe Jopling, Sarah Lewis, Paul Lintott, Nicola Lipczynski, Hannah Peck, Gary Powney, Jennifer Sears, Kate Sullivan, Oliver Wearn, Penny Wilson, Sally Wren, and Tara Zamin.

IUCN Specialist Groups and Species Programme
BirdLife International (Stuart Butchart, Ali Stattersfield); Crocodile Specialist Group (Tom Dacey); Freshwater Biodiversity Unit (Will Darwall, Anna McIvor, Kevin Smith); Freshwater Fish Specialist Group (Gordon McGregor Reid); Global Mammal Assessment (Janice Chanson, Mike Hoffman, Jan Schipper); Global Marine Species Assessment (Kent Carpenter, Suzanne Livingstone, Beth Polidoro); Global Reptile Assessment (Janice Chanson, Neil Cox, Simon Stuart); Groupers and Wrasses Specialist Group (Yvonne Sadovy); Mediterranean Red List Assessment (Annabelle Cuttelod); Marine Turtle Specialist Group (Milani Chaloupka); Odonata Specialist Group (Viola Clausnitzer, Vincent Kalkman, Frank Suhling); Red List Unit (Craig Hilton-Taylor, Vineet Katarya, Caroline Pollock); Syngnathid Red List Authority (Amanda Vincent, Heather Koldaway, Sarah Bartnik, Eve Robinson and Sian Morgan Shark Specialist Group (Sarah Fowler, Claudine Gibson, Sarah Valenti); Tortoise and Freshwater Turtle Specialist Group (Anders Rhodin, Peter Paul van Dijk).

Contributors

REPTILES
Klaus Adolphs, Cesar Aguilar, Allen Allison, Natalia Ananjeva, Steve Anderson, Sergio Augusto A. Morato, Mark Auliya, Christopher Austin, Sherif Baha el Din, Raoul Bain, Aaron Bauer, Daniel Bennett, Don Broadley, Sharon Brooks, Rafe Brown, Juan Camilo Arredondo, Ashok Captain, Angus Carpenter, Fernando Castro, David Chapple, José Rogelio Cedeño-Vázquez, B.C. Choudhury, Diego F. Cisneros-Heredia, Lázaro Cotayo, Harold Cogger, Gabriel C. Costa, Teresa Cristina Sauer Avila-Pires, Pierre-Andre Crochet, Brian Crother, Felix Cruz, Ranjit Daniels, Neil Das, Ignacio de la Riva, Kevin de Queiroz, Anslem de Silva, Maria del Rosario Castandea, Lutz Dirksen, Jim R. Dixon, Tiffany M. Doan, Paul Doughty, Dirk Embert, Robert E. Espinoza, Richard Etheridge, Andre Felipe barreto Lima, Xie Feng, Lee Fitzgerald, Fred Franca, Leonardo

Francisco Stahnke, Tony Gamble, Miguel A. García, Juan Elías García-Pérez, Maren Gaulke, Phillipe Geniez, Stephen Goldberg, David Gower, Eli Greenbaum, Lee Grismer, Michael Guinea, Jakob Hallermann, Kelly Hare, Mike Harvey, Harold Heatwole, S. Blair Hedges, Neil Heideman, Robert Henderson, Rod Hitchmough, Karim V. D. Hodge, Paul Horner, Barry Hughes, Mark Hutchinson, Ivan Ineich, Bob Inger, Richard Jenkins, Tony Jewell, Ulrich Joger, Hinrich Kaiser, Dave Kizirian, Paul Kornacker, Axel Kwet, Enrique La Marca, William Lamar, Malcolm Largen, Michael Lau, Matthew LeBreton, Edgar Lehr, Kuang-Yang Lue, César Luis Barrio-Amorós, Luca Luiselli, Vimoksalehi Lukoschek, Mikael Lundberg, Robert Macy, Ulrich Manthey, Jean Mariaux, Otavio Marques, Marcio Martins, Brad Maryan, Nixon Matthews, Gregory Mayer, Werner Mayer, Colin McCarthy, Randy McCranie, Michele Menegon, Sanjay Molur, Tami Mott, Hidetoshi Ota, Jose Ottenwalder, Theodore Papenfuss, Fred Parker, Olivier Pauwels, Tony Phelps, Eric Pianka, Steven Platt, Paulino Ponce-Campos, Robert Powell, Raju Radder, Arne Rasmussen, Chris Raxworthy, Bob Reynolds, Gilson Rivas, Mark-Oliver Rödel, Lourdes Rodríguez Schettino, Nelson Rufino de Albuquerue, Ross Sadlier, Hermann Schleich, Andreas Schmitz, Muhamad Sharif Khan, Glenn Shea, Richard Shine, Roberto Soberón, Ruchira Somaweera, Steve Spawls, Peter Stafford, Bryan Stuart, Rob Stuebing, Gerry Swan, Sam Sweet, Manoel Alonso Tabet, Roberto Ramos Targarona, John Thorbjarnsson, Colin Tilbury, Peter Tolson, Sam Turvey, Johan van Rooijen, Monique van Sluys, Alvaro Velasco, Miguel Vences, Milan Veselý, Gernot Vogel, Milan Vogrin, Raju Vyas, Fabiano Waldez, Van Wallach, Bryon Wilson, Larry Wilson, Kaiya Zhou, and George Zug.

TORTOISES AND FRESHWATER TURTLES
Patrick J. Baker III, Alexandre Batistella, Bill Branch, Russell Burke, Olga Victoria, Castaño Mora, Tomas Diagne, Ken Dodd, Sean Doody, Michael Dorcas, David Emmett, Kevin Enge, Alejandro Fallabrino, Arthur Georges, Justin Gerlach, Shi Haitao, Magaretha Hofmeyr, John Iverson, Michael Lau, Dwight Lawson, Luca Luiselli, William Magnusson, Sebastien Metrailler, Steven Platt, Peter Pritchard, Willem Roosenburg, Tracy Tuberville, Sabine Vinke, Thomas Vinke, and Richard Vogt.

FRESHWATER FISH
Aaron Jenkins, Nina Boguskaya, Will Darwall, Rema Devi, Roberto Esser dos Reis, Tan Heok Hui, Fang Kullander, Philippe Laleye, Flavio Lima, Topis Macbeath, Gordon McGregor Reid, and Jos Snoeks.

DRAGONFLIES
Viola Clausnitzer, Vincent J. Kalkman, Matjaz Bedjanic, Klaas-Douwe B. Dijkstra, Rory Dow, John Hawking, Haruki Karube, Elena Malikova, Dennis Paulson, Kai Schütte, Frank Suhling, Reagan Joseph Villanueva, Natalia Ellenrieder, and Keith Wilson.

CRABS
Fernando Alvarez, Felix Y.K. Attipoe, Martha R. Campos, France-Lyse Clotilde-Ba, Neil Cumberlidge, Savel R. Daniels, Lara J. Esser, Celio Magalhaes, Anna McIvor, Tohru Naruse, Peter K.L. Ng, Mary B. Seddon, and Darren C.J. Yeo.

Species susceptibility to climate change impacts

IUCN's project on species susceptibility to climate change is funded by the John D. and Catherine T. MacArthur Foundation, with contributions from the Indianapolis Zoo. We thank the Centre for Population Biology, Imperial College London for workshop sponsorship and we are grateful to the species experts who participated in the identification of climate change vulnerability traits, namely: Resit Akçakaya, Rob Alkamade, Jon Bielby, Neil Brummitt, Simon Butler, Mar Cabeza, Ben Collen, Keith Crandall, Nick Dulvy, Rob Ewers, Rich Grenyer, Craig Hilton-Taylor, Sarah Holbrook, Joaquin Hortal, Kate Jones, David Keith, Zoe Macavoy, Rob Marchant, Tom Meagher, J.B. Mihoub, David Obura, Shyama Pagad, Paul Pierce-Kelly, Jeff Price, John Reynolds, Ana Rodrigues, Andy Sheppard, and Stephen Williams. Cagan Sekercioglu, Stephen Garnett, Oliver Kruger, David Hole, Brian Huntley, Steve Willis, Rhys Green, Arvind Panjabi, Rob Clay, Ian Burfield, Greg Butcher, Petr Vorisek, Terry Rich, Paul Donald, Bruce Young, Andrew Baker, Peter Harrison, Carden Wallace and Charlie Veron contributed essential species data. Andy Symes, Tristram Allinson and Joe Wood played a vital role in compiling bird data, and Janice Chanson assisted with the compilation of amphibian data. We thank Conservation International for covering the cost of Alex Gutsche's time, and are grateful to members of the SSC network and our partner organizations, particularly BirdLife International, for their valuable contributions.

The Mediterranean: a biodiversity hotspot under threat

Assessing species for the IUCN Red List of Threatened Species relies on the willingness of dedicated experts to contribute and pool their collective knowledge, thus allowing the most reliable judgments of a species' status to be made. Without their enthusiastic commitment to species conservation, this kind of regional overview would not be possible.

We would therefore like to thank the following people, asking for forgiveness from anyone whose name is inadvertently omitted or misspelled:

For Mediterranean amphibians and reptiles
Peter Paul van Dijk for producing the draft species assessments for the tortoises and freshwater turtles, and the following people who gave their time and valuable expertise to evaluate all of the assessments: Mr Rastko Ajtic, Sherif Baha El Din, Wolfgang Böhme, Marc Cheylan, Claudia Corti, Jelka Crnobrnja Isailovic, Pierre-André Crochet, Ahmad Mohammed Mousa Disi, Philippe Geniez, El Mouden El Hassan, Juan Antonio Camiñas Hernández, Souad Hraoui-Bloquet, Ulrich Joger, Petros Lymberakis, Rafael Márquez, Jose Antonio Mateo Miras, Jose Luis Mons Checa, Saïd Nouira, Carmen Díaz Paniagua, Valentín Pérez Mellado, Juan Manuel Pleguezuelos, Paulo Sá-Sousa, Riyad Sadek, Murat Sevinc, Tahar Slimani, C. Varol Tok, Ishmail Ugurtas, Milan Vogrin and Yehudah Werner. We would also like to thank David Knox and Peter Paul van Dijk for assisting with workshop facilitation and subsequent editing of the data.

For the Birds
BirdLife International and its partners for providing the bird species assessments and in particular Stuart Butchart and Jez Bird, for their support and analysis of further data.

For the Mediterranean cartilaginous fish
All of the IUCN SSC Shark Specialist Group (SSG) Mediterranean members

and invited regional and international experts who participated at the San Marino workshop; Marco Affronte, Irene Bianchi, Mohamed Nejmeddine Bradai, Simona Clò, Rui Paul Coelho, Francesco Ferretti, Javier Guallart, Ferid Haka, Nils-Roar Hareide, Farid Hemida, Cecilia Mancusi, Imène Meliane, Gabriel Morey, Manal Nader, Guiseppe Notarbartolo di Sciara, Persefoni Megalofounou, Titian Schembri, Fabrizio Serena, Alen Soldo, Fausto Tinti, Nicola Unçaro, Marino Vacchi, Ramón Bonfil, Nick Dulvy, Ian Fergusson, Sarah Fowler, Charlotte Mogensen and Ransom Myers. Particular gratitude is expressed to Imène Meliane and Ameer Abdulla of the IUCN Global Marine Programme; and Helen Temple of the IUCN Red List Unit for reviewing this document and especially to Sarah Fowler IUCN SSC SSG Co-chair, for her continual support.

We gratefully acknowledge Leonard Compagno and Fabrizio Serena for help in compiling the regional checklist for this report, and Sarah Ashworth, Sarah Valenti and Adel Heenan for all the work they have undertaken in contributing to reviewing and editing species assessments. We would also like to thank Peter Kyne for extremely helpful discussions. Finally, we would like to thank Alejandro Sancho Rafel for providing the illustrations.

For the cetaceans

All of the IUCN SSC Cetacean Specialist Group (SSG) Meciterranean members and invited regional and international experts who participated at the Monaco workshop; Alex Aguilar, Alexei Birkun, Jr., Ana Cañadas, Greg Donovan, Caterina Maria Fortuna Alexandros Frantzis, Stefania Gaspari, Philip Hammond, Ada Natoli, Giuseppe Notarbartolo di Sciara, William F. Perrin, Randall R. Reeves, Renaud de Stephanis, as well as Marie-Christine Grillo and the ACCOBAMS Secretariat for their collaboration and support.

For the Mediterranean freshwater crabs and crayfish

Francesca Gherardi for producing the preliminary assessments for crayfish and Neil Cumberlidge for the assessments of freshwater crabs.

Assessors and evaluators

Producing this analysis would not be possible were it not for the extraordinary enthusiasm, dedication and willingness of many people around the world who contribute an enormous amount of time and effort to supply Red List assessments and the supporting documentation required for The IUCN Red List. In particular, we must acknowledge all the SSC Specialist Group Chairs, Red List Authority focal points, Specialist Group members, and the many field scientists who have been involved in contributing to The IUCN Red List. In the list below we have tried to highlight all the individuals whose contributions appear on the 2008 IUCN Red List of Threatened Species™. With such a long list of names, it is highly probable that we have inadvertently forgotten someone or spelt names incorrectly; please forgive us.

The Editors thank:

Abba, A., Abbott, J.C., Abbott, T., Abdel Rahman, E., Abe, H., Abramov, A., Abrar, M., Agreu-Grobois, A., Abril, V.V., Abu Baker M.A., Abuzinada, A.H., Acero, A., Acevedo Rodríguez, A., Acevedo, M., Acosta A., Acosta, G., Acosta-Galvis, A., Acuña E., Adams, M., Adams, W.F., Addoor, S.N.R., Adema, F., Adoor, S., Aeby, G., Afuang, L., Agarwal, I., Agoo, E.M.G., Aguilar, A., Aguilera, M., Aguirre Leon, G., Aguirre, L., Agwanda, B., Ahmad Disi, G.D., Ahmad Khan, J., Ahmad, N., Ahmed Khan, J., Ahmed, M.F., Ajtic, R., Ajtic, R., Akinyi, E., Al Dosary, M., Al Habhani, H.M., Al Khaldi, A.M., Al Mutairi, M.S., Al Nuaimi, A.S.M., Alberth Rojas, C., Alberts, A., Albornoz, R., Aicala, A., Alcala, E., Alcalde, J.T., Alcaraz, D., AL-Eisawi, D.M.H., Aempath, M., Alfonso, G.L., Alford, R., Algarra Ávila, J.A., Alkon, P.U., Allen, G., Allen, G.R, Allet, M., Allison, A., Almada-Villela, P., Almandáriz, A., Almeida, D., Almeida, Z., Almendáriz, A., Alonso, A.M., Alonso, R., Altrichter, M., Alvarado, S.O., Alvares, R., Alvarez Castaneda, S.T., Alvarez, F., Avarez, R., Álvarez, S., Alvarez, S.J., Álvarez-Castañeda, S., Alves, P.C., Alviola, P., Amanzo, J., Ambal, G., Amézquita, A., Amiet, J.-L., Amir, O.G., Amori, A., Amori, G., Amorim, A.F., Amorós, C.L.E., Amoroso, V.B., Amr, Z., Anacleto, T., Anandanarayanan, Ananjeva, N., Ancrenaz, M., Andayani, N., Andelt, W., Anderson, E.F., Anderson, M., Anderson, P.K., Anderson, R., Anderson, R.P., Anderson S., Andrade, G., Andrainarivo, C., Andreone, F., Andrews, H.V., Andriafidison, D., Andriaholinirina,

V.N., Andrianjakarivelo, V., Andrianjakavelo, V., Ángeles Ortiz, M., Angelici, F.M., Angerbjörn, A., Angulo, A., Anstis, M., Anthony, B., Anwarul Islam, Md., Ao, M., Aparicio Rojo, J.M., Aplin, K., Appleton, B., Appleton, C., Aquino, L., Araújo, M.L.G., Arboleda, I., Ardila-Robayo, M.C., Areces-Mallea, A.E., Arfelli, C.A., Argolo, A.J.S., Ario, A., Ariunbold, J., Arizabal, W., Arnaud, G., Arntzen, J.W., Aronson, R., Arrendondo, A.G., Arrigoni, P.V., Arroyo, S., Arroyo-Cabrales, J., Arumugam, R., Arzabe, C., Asa, C., Asber, M., Ashenafi, Z.T., Ashton, P., Asmat, G.S.M., Assi, A., Assogbadjo, A., Astua de Moraes, D., Atkinson, R.P.D., Attic, R., Attipoe, F.Y.K., Augerot, X., Aulagnier, M., Aulagnier, S., Aune, K., Aurioles, D., Austin, C., Avermid, D., Averyanov, L., Avila Villegas, H., Avirmed, D., Azeraoul, A., Azevedo-Ramos, C., Azlan, A., Azlan, J., Azlan, M.J., Baard, E., Babieri, R., Babik, W., Bachman, S., Bachraz, V., Bahir, M.M., Baigún, R., Bailey M., Baillie, J., Bain, R., Baird, R., Baker, C.S., Baker, J., Baker, L.R., Baldi, R., Baldisseri, F., Baldo, D., Baldwin, R., Balestra, A.D., Balete, D., Balfour, D., Ballesteros, F., Balletto, E., Balmforth, Z., Baloyan, S., Bambaradeniya, C., Ban, N.T., Bañares, A., Bandeira, S., Bangoura, M.A., Banks, P., Banks, S., Bannister, J.L., Bantel, C.G., Baorong, G., Baral, H.S., Barashkova, A., Barbanti, M., Barbieri, R., Barker, A., Barker, A.S., Barlow, J., Barnett, A.A., Barnett, L.A.K., Barney, L., Barquez, R., Barrantes, U., Barratt, P., Barratt, P.J., Barrera, G.S., Barriga, P., Barrio, J., Barrio-Amorós, C., Barroso, G.M., Barry, R., Bartels, P., Bartnik, S., Barua, M., Bass, D., Basso, N., Bastos, R., Basu, D., Batbold, J., Bates, P., Batin, G., Batsaikhan, A., Batsaikhan, N., Bauer, H., Baum, J., Baxter, R., Bayona, J., Bayona, J.D.R., Bazante, G., Beachy, C., Beamer, D., Bearder, S., Bearzi, G., Beasley, I., Beauvais, G.P., Beccaceci, M.D., Beck, H., Beckmann, J., Bedjanic, M., Beebee, T., Beentje, H.J., Beerli, P., Beever, E.A., Begg, C., Begg, K., Behler, J., Beier, M., Beja, P., Bekoff, M., Belant, J., Belbachir, F., Bell, B., Bell, T., Bellarry, C.L., Bellingham, P., Bello, J., Benavente, A., Benavides, G., Benda, P., Benedikt Schmidt, A., Benishay, J.M., Bennett, D., Bennett, M.B., Benshemesh, J., Benzie, J., Berducou, C., Bergallo, H., Bergl, R.A., Bergmans, W., Berlin, E., Bernal, M., Bernal, N., Bernal, R., Bernard, R., Bérnils,

R.S., Bertolino, S., Bertoluci, J., Bertoncini, A., Bertozzi, M., Best, P.B., Bestelmeyer, S., Beyer, A., Bhat, G.K., Bhatnagar, Y.V., Bhatta, G., Bhatta, T., Bhattacharyya, T., Bhuddhe, G.D., Bhupathy, S., Bianchi, I., Bianco, G., Bianco, P.G., Bibiloni, G., Bickford, D., Bidau, C., Biggins, D., Bigirimana, C., Biju, S., Biju, S.D., Bila-Isia, I., Bilgin, C., Bills, R., Bird, J., Birkinshaw, C., Birkun Jr., A.A., Birstein, V., Bishop, P., Bist, S.S., Bisther, M., Biswas, B.K., Bizzarro, J.J., Bjørge, A., Black, P., Black, P.A., Blair Hedges, R.J., Blair, D., Blakenmore, R.J., Blanc, J., Blanca, G., Bleisch, B., Bleisch, W., Blois, J., Blom, A., Blomquist, S., Bloomer, P., Blotto, B., Boada, C., Boeadi, Boesch, C., Bogan, A.E., Bogarin, D., Bogutskaya, N., Bohlen, J., Böhme, W., Bohs, L., Boisserie, J-R., Boistel, R., Boitani, L., Bolaños, F., Bolívar, W., Bolten, A., Bonaccorso, F., Bonaccorso, F.A., Bonai, B.S., Bonfil, R., Bonham, K., Bonifaz, C., Bonvicino, C., Boonratana, R., Borah, M.M., Boratynski, A., Bordoloi, S., Borges-Najosa, D., Born, E.W., Born, M., Borroto, R., Bosch, J., Bossuyt, F., Bosworth, B., Botello, J.C., Boubli, J.P., Boubli, J.-P., Bouchet, P., Boudot, J.-P., Bour, R., Bóveda-Penalba, A.J., Bowler, M., Bowles, M., Boyd, L., Boyer, A.F., Boza, E., Bozdogan, M., Bradaï, M.N., Bradford, D., Bradley Martin, E., Brahim, K., Branch, W.R., Brandão, R., Brandle, R., Brandon, A., Branstetter, S., Brash, J., Brash, J.M., Brasileiro, C., Braswell, A., Braulik, G., Braulik, G.T., Breed, W., Breitenmoser, U., Breitenmoser-Wursten, C., Brereton, R., Brescia, F., Breuil, M., Brickle, N.W., Bridson, D., Brit, D., Brito, D., Brockelman, W., Broderick, A.C., Bronner, G., Brooke, A., Broome, L., Broughton, D.A., Brown, D., Brown, D., Brown, D.S., Brown, J., Brown, M., Brown, P., Brown, R., Brownell Jr., R.L., Bruce, B., Bruckner, A., Bruegmann, M.M., Brugiere, D., Brule, T., Bucal, D., Buckner, S., Buckner, S.D., Buden, D., Buhlmann, K., Buitrón, X., Bukhnikashvili, A., Bumrungsri, S., Burbidge, A., Burdin, A., Burfield, I., Burger, M., Burgess, G.H., Burgess, G.M., Burgoyne, P.M., Burkanov, V., Burke, A., Burneo, S., Burnett, S., Burrows, J., Burton, F.J., Burton, J., Bury, B., Buskirk, J.R., Bustamante, M.R., Butchart, S., Butterworth, D.S., Butynski, T.M., Buuveibaatar, V., Byers, J., Bygrave, P., Cabanban, A., Cabello, J., Cabezudo, B., Cable, S., Cadena, A., Cadi, A., Cadle, J.,

Cailliet, G.M., Cairns-Wicks, R., Cajas, J., Cajas, J.O., Caldas, J.P., Calderón Mandujano, R., Calderon, E., Caldwell, I., Callaghan, D., Camancho, J., Camara L., L., Camarda, I., Cambi, V., Cambray, J., Camhi, M., Campagna, C., Campbell, J.A., Camperio-Ciani, A., Canseco-Márquez, L., Cantley, R., Canty, P., Capper, D., Caramaschi, U., Carauta, J.P.P., Caraway, V., Carbajal, R., Carbyn, L., Cardiff, S., Cardiff, S.G., Cardozo, V., Carey, C., Caringal, A.M., Cariño, A., Cariño, A.B., Carino, P., Carlisle, A.B., Carlson, J., Carlson, J.K., Carlstrom, A., Carmona, J., Carnaval, A.C., Carpenter, K., Carqué Álamo, E., Carr, J.L., Carranza, S., Carrick, F., Carrington, C.M.S., Carrión Vilches, M.Á, Carron, G., Carter, R.L., Carter-Holmes, S., Carvalho, L.d'A.F., Cashatt, E., Caso, A., Casper, B.M., Cassinello, J., Castañeda, F., Castellanos, A., Casteñeda, F., Castillo, A., Castro, A.L.F., Castro, F., Castro, O., Castro-Arellano, I., Castroviejo-Fisher, S., Catenazzi, A., Catling, P., Catzeflis, F., Causado, J., Cavalcanti, T.B., Cavallini, P., Cavanagh, R.D., Ceballos, N., Cedeño, A., Cerón, C., Cervantes, F.A., Céspedez, J., Chakraborty, S., Chalise, M., Chaloupka, M., Chalukian, S., Chan Tak-Chuen, T., Chan, B., Chan, B.P.L., Chan, S., Chan, T., Chanard, T., Changyuan, Y., Channing, A., Chanson, J., Chaparro, J.C., Chaparro-Auza, J.C., Chapman, R., Chapman, R.E., Chardonnet, P., Charvet-Almeida, P., Chaudhry, A.A., Chauhan, N.P.S., Chaves, G., Cheek, M., Chemnick, J., Chen, X.-Y., Cheng, L., Cheylan, M., Chhangani, A., Chiang, P.J., Chiaramonte, G., Chiaramonte, G.E., Chiarello, A., Childerhouse, S., Chin, A., Chiozza, F., Chippindale, P., Chiramonte, G., Chiramonte, G.E., Chiriboga, A., Chitaukali, W., Choat, H., Choat, J.H., Chou Wenhao, Choudhury, A., Choudhury, B.C., Christodoulou, C.S., Christodoulou, S., Christoff, A., Christofides, Y., Chua, L.S.L., Chuaynkern, Y., Chuen, N.W., Chundawat, R.J., Chundawat, R.S., Chundaway, R.A., Chung, R.C.K., Church, D., Cianfrani, C., Cilliers, S., Cisneros-Heredia, D., Clapham, P.J., Clark, D., Clark, J.L., Clark, T.B., Clarke, C., Clarke, D., Clarke, G.P., Clarke, J., Clarke, M., Clarke, M.W., Clarke, S., Clausnitzer, V., Cliff, G., Clò, S., Cloete, E., Clotilde-Ba, F.-L., Clubbe, C., Coelho, R., Coetzee, N., Cogal, D., Cogalniceanu, D., Cogger, H., Cole, N., Collen, B., Colli, G., Collins, J., Collins, K.,

Collins, T., Coloma, L.A., Compagno, L.J.V., Connolly, R., Conran, J.G., Conrath, C., Conroy, J., Constantino, E., Contreras-Balderas, S., Conyers, J., Cook, S., Cook, S.F., Cooke, J., Cooper, N., Cootes, J., Copley, P., Coppois, G., Corbett, L.K., Cornejo, F., Cornejo, X., Cornish, A., Cornish, A.S., Coroiu, C., Coroiu, I., Correia, J.P.S., Cortés, E., Cortés, J., Cortés-Ortiz, L., Cortez, C., Corti, C., Corti, M., Corti, P., Cossios, D., Cossíos, E.D., Costa, L., Costa, P., Cotayo, L., Cotterill, F.P.D., Cotterill, F.W., Cotton, E., Courtenay, O., Covert, B., Cowie, R., Cowie, R.H., Cox, D., Cox, N., Coxe, S., Craig, C., Craig, M., Crandall, K., Craul, M., Craven, P., Crawford-Cabral, J., Creel, S., Crespo, E.A., Crespo, M.B., Cribb, P., Cribb, P.J., Crider, D., Crivelli, A., Crivelli, A.J., Crochet, P.-A., Crochet, P-A., Crombie, R., Cronk, Q.C.B., Crosby, M., Crouse, D., Crump, M., Cruz, G., Cruz-Aldan, E., Csiba, L., Csorba, G., Cuarón, A.C., Cuarón, A.D., Cuéllar, E., Cumberlidge, N., Cumming, D.H.M., Cunningham, M., Curtis, B., Custodio, C., Cuzin, F., Cypher, B.L., D'Anatro, A., Da Costa, L., da Cruz, C.A.G., da Fonseca, G.A.B., Dacey, T., Dagit, D.D., Dahal, B., Dalebout, M.L., Dalponte, J., Daniel, B.A., Daniels, R., Daniels, S., Daniels, S.R., Darbyshire, I., Darman, Y., Darria, J., Darwall, W., Das, I., Das, J., Datong, Y., Dávalos, L., Davenport, T., Davenport, T.R.B., David, J., Davidson, C., Davis, G.S., Dawson, S.E., Dawson, S.M., Day, M., de A. Goonatilake, W.I.L.D.P.T.S., de Almeida, M.P., de Bustos, S., de Carvalho, M.R., de Carvalho, M.R., de Grammont, P.C., de Granville, J.J, de Iongh, H., De Jong, Y., De Jong, Y.A., de la Riva, I., de la Sancha, N., de la Torre, S., de Lange, P.J., De Leon, J., De Luca, D., de Montmollin, B., de Oliveira, L.F., de Oliveira, M.M., de Oliveira, T., de Silva, A., de Silva, P.K., de Smet, K., de Thoisy, B., de Tores, P., De Vega, C., de Villalobos, A., de Villiers, A., de Vogel, E.F., de Wilde, W.J.J.O., de Winter, A.J., Decher, J., Degani, G., Deka, P., Dekker, W., Delbeek, J.C., Delgado, C., Delgado, T., D'Elia, G., Delphey, P., DeLuycker, A.M., Denham, J., Denny, M., Denoël, M., Derocher, A., Derocher, A.E., Desai, A., Deutsch, C.J., DeVantier, L., d'Huart, J.P., Di Bernardo, M., Di Fiore, A., Di Giácomo, E., di Tada, I., Díaz, A., Diaz, A.G., Diaz, G.Q., Díaz, L., Díaz, L.M., Diaz, M., Díaz-Paéz, H., Di-Bernardo, J.C., Dicht, R.F., Dickman, C., Diesmos, A.,

Diesmos, M.L., Dieterlen, F., Dijkstra, K.-D.B., Din, S B.E., Ding-Qi, R., Dinh Thong, V., Dipper, F., Disi, A., Disi, M., Ditchfield, R., Dittus, W., Dixon, R., Do Tuoc, Doadrio, I., Dobbs, K., Dodd, K., Doggart, N., Dold, T., Dolino, C., Dollar, L., Domeier, M., Domingo, A., Domning, D., Donaire, F., Donaire-Barroso, D., Donaldson, J.S., Donaldson, T.J., Dong, S., Donnelly, M., Donnelly, N., Donovan, G.P., Dorjderem, S., Dormeier, M., Doroff, A., Doughty, P., Doumbouya, F., Dowl, J.L., Dowler, R., Down, T., Downer, C., Dransfield, J., Dressler, R.L., Drew, C., Drewes, R., Driessen, M., Drioli, M., Driscoll, C., Drummond, L O., Du Puy, D., Du, L.-N., Duarte, J.M.B , Dublin, H.T., Duckworth, J.W., Ducrocq, M., Dudley, S., Dudley, S.F.J., Duellman, W., Duellman, W.E., Duffy, C., Duffy, C.A.J., Dujsebayeva, T., Duke, S., Dulco, M.E., Dulvy, N., Dunham, A., Dunn, A., Dunnum, J., Dupain, J., Duplaix, N., Durant, P., Durant, S., Durate, J.M.B., Durbin, J., Durbin, L.S., Dutta, S., Dutton, P., Duvall, C., Duya, A., Duya, M., Duya, M.R., Duya, P., Eamkamon, T., Easa, P.S., East, R., Eastwood, A., Ebert, D., Ebert, D.A., Eddowes, P.J., Edgar, G., Edgar, P., Edwards, A., Eeley, H., Ehardt, C., Ehardt, T., Ezirik, E., Eken, G., Eklund, A-M., Ekué, M.R.M., El Din, S.B., El Hassan, M., El Mouden, H., Eldredge, L.G., Elias, P., Ellis, C.M., Ellis, E., Ellis, J., Ellis, J.E., Ellis, M., Ellis, S., Elron, E., Elsey, R., Elvira, B., Emberton, K.C., Emmons, L., Emslie, R., Endrody-Younga, S., Engelbrecht, J., English, K., Enlbrecht, J., Enssle, J., Erdmann, M., Erk'akan, F., Ermi, Z., Espino Castellanos, L.A., Espinoza, C., Esselstyn, J., Esser, L., Essig, F.B., Estupinan, R.A., Eterovick, P.C., Etuge, M., Evans, G., Evans, R., Evans, T., Everett, B., Fa, J., Fabregat Llueca, C., Fagundes, L., Fagundes, V., Fahr, J., Fahr, J., Fairclough, D., Faivovich, J., Faria, D., Faria, V., Farías, V., Farjon, A., Farmer, K.H., Fasola, L., Faulhaber, C.A., Faulkes, C., Feh, C., Feio, R., Feistner, A., Felix, T., Fellers, G., Fellers, G.M., Fellowes, J., Feng, X., Fenner, D., Fennessy, J., Fennessy, S., Ferguson, M., Fergusson, I., Fergusson, I.K., Fergusson, R., Fernades, M., Fernandes, M., Fernández Jiménez, S., Fernández-Badillo, E., Fernandez-Duque, E., Fernanco, P., Fernando, S., Ferrari, S.F., Ferreira, B., Ferreira, D.C., Ferreyra, N., Festa-Bianchet, M., Fil, Y., Findlay, K.P., Finet, Y., Finnie, D., Firsov, G.A., Fisher, D., Fitz Maurice, B., Fitz

Maurice, W.A., FitzGibbon, C., Flaherty, A., Flammang, B.E., Flander, M., Flannery, T., Florence, J., Florens, D., Flores-Villela, O., Flueck, W., Foerester, C., Foggi, B., Folkerts, G., Fonseca, G.d., Font Garcia, J., Forbes, T., Forchhammer, M., Ford, J., Fordham, S., Formas, J., Formas, R., Formozov, N., Forney, K., Foster, G., Foster, S.J., Foufopoulos, J., Foulkes, J., Fouquet, A., Fournaraki, C., Fowler, S., Fraga Arguimbau, P., Fragoso, J., Francis, C., Francis, M., Francis, M.P., Francis, N., Francisco Jiménez, J., Franco, F.L., Franco, M., Frantzis, A., Fredriksson, G., Freire, E.M.X., Freire-Fierro, A., Frere, E., Frest, T., Fretey, T., Frey, J., Freyhof, J., Frias-Martin, A., Fridlender, A., Friedland, K.D., Friend, T., Frodin, D., Fronst, A., Frost A., Frost, C., Frost, D., Frost, D.R., Fruth B., Fu Likuo, Fuenmayor, Q., Fuertes-Ramos, O., Fuller, D., Fuller, T.K., Funes, M., Furey, N., Furtado, M., Furuichi, T., Gadea, J.C.-S., Gadig, O.B.F., Gadsden, H., Gafny, S., Gaikhorst, G., Galat, G., Galat-Luong, A., Gales, N., Galicia Herbada, D., Gallina Tessaro, S., Galline, S., Gamisans, J., Gang, L., Gankhuyag, P., Ganzhorn, J., Garayzar, C.V., Garbutt, N., García Aguayo, A., Garcia de Souza, G., Garcia, G., Garcia, H., García, I., García, J.C., Garcia, J.E., Garcia, J.J., Garcia, M., Garcia, P., Garcia-Moliner G., García-París, M., García-Pérez, J.E., García-Vasco, D., Gardner, A., Gardner, M., Gardner, R., Garfi, G., Garner, T., Garrido, A., Garshelis, D.L., Garske, L., Gartshore, M., Gascon, C., Gasgoigne, A., Gasith, A., Gates, C., Gatti, S., Gaubert, P., Gaucher, P., Geberemedhin, B., Gee, G., Geffen, E., Geiger, D., Geise, L., Geissler P., Geissmann, T., Geist, V., Gelatt, T, Gem, E., Gemmill, C., Gen, P., Geniez, P., Geraldine, V., Gerber, G., Gerber, L., Gerlach, J., Gerson, H., Gese, E.M., Gesti Perich, J., Ghazanfar, S.A., Gianguzzi, L., Giannatos, G., Giannatos, T., Giareta, A.A., Gibbons, W., Gibson, C., Gibson, C.G., Gibson, R., Gil, A.Q., Gil E.R., Gil, Y., Gill, R., Gillespie, G., Gilroy, J., Gimar, B., Gimenez Dixon, M., Ginsberg, J.R., Gippoliti, S., Giri, V., Giusti, F., Gizejewski, Z., Glaw, F., Gledhill, D., Glowacinski, Z., Gober, P., Goedeke, T., Goettsch, B., Golamco, Jr., A., Golden, C., Golding, J., Goldman, C., Goldman, K.J., Goldstein, I., Goldsworthy, S., Gomez, H , Gómez, L.D., Gomez, R., Gomez-Campo, C., Gómez-Hinostrosa, C., Gómez-Laverde, M., Gondek, A.,

Gongora, J., Gonzales, J.C., González, B., Gonzalez, E., González, J., González, M., Gonzalez, P., Gonzalez, S., González, S., Gonzalez, T., González-Espinosa, M., Gonzalez-Maya, J., Gonzalez-Maya, J.F., Gonzalez-Porter, G.P., Goodman, D., Goodman, S., Goodman, S.M., Goossens, B., Gorcon, G., Gordon, I., Gordonia, I., Goren, M., Gosline, G., Gottelli, D., Gour-Broome, V.A., Gower, D., Grach, C., Grady, J. Graham, K., Graham, K.J., Graham, R., Graham, R.T., Granjon, L., Grant, T., Grassman, L., Grasso, R., Gray, A., Gray, M., Greenbaum, E., Greengrass, E., Gregorin, A.D., Griffin, M., Griffiths, H.I., Griffiths M., Griffiths, O., Griffiths, R., Grismer, L., Groenedijk, J., Groves, C.P., Groves. M., Grubb, P., Grubbs, D., Grubbs, D.R., Grubbs, R.D., Gruber, S., Gu Huiqing, Guadagnin, D.L., Guallart, J., Guanfu, W., Guayasamin, J., Güemes, J., Guevara M.A., Guineo, G., Gumal, M., Gunatilake, S., Gunawardena, M., Güner, A., Gunn, A., Günther, R., Gururaja, K.V., Gutierrez, B., Gutiérrez, L., Guttiérrez, E.E., Guttman, A., Guzmán, A., Guzmán, H., Ha, D.S., Haaker, P., Haase, M., Habib, B., Hack, M., Hack, M.A., Hadfield, M., Hadjikyriakou, G., Hadjisterkotis, E., Hadway, L., Haevermans, T., Hafner, D.J., Hahn, N., Hai Yin, W., Haines, W.P., Haitao, S., Hajkova, P., Hall, J., Hall, L., Hallingbäck, T., Hämäläinen, M., Hamer, M., Hamilton, R., Hamilton, S., Hammer, S., Hammerson, G., Hammerson, G.A., Hammond, J., Hammond, P.S., Hampson, K., Han, K.H., Han, S., Hanken, J., Hanski, I., Hanssens, M., Happold, D., Happold, M., Hardersen, S., Harding, L., Hardy, J , Hare, J., Hareide, N., Härkönen, T., Harmelin-Vivien, M., Harris, R., Harris, R.B., Harrison, J., Harrison, J., Hart, J , Hart, J.F., Hart, T., Harvey, M., Hashimoto, C., Hashimoto, T., Hawking, J., Hawkins, A.F.A., Hawkins, C.E., Hawthorne, W , Haxhiu, I., Hayes, W.K., Haynes, J., Hays, D., Haywood, M., Heaney, L., Hearn, A., Heckel, J.-O., Héctor, M., Heddle, M., Heddle, M.L., Hedges, B., Hedges, S., Heemstra, P.C., Heenan, A., Hefner, R., Heide-Jørgensen, M.P., Helgen, H., Helgen, K., Heller, J., Henderson, A., Henderson, R., Henschel, P., Henttonen, H., Herbert, D., Herbert, D.G., Herbert, J., Herbinger, I., Herman, A., Hermann, P., Herrández González, A., Hernández Luís, A., Fernández, C., Hernández, H.M., Hernandez, M., Hernández-Bermejo, E., Herndon, A.P.,

Hero, J.-M., Herrera, M.I., Herrera-Molina, F., Herrero, J., Herrero. J., C., Herrington, R., Herrmann, H.-W., Hersteinsson, P., Hess, W.J., Heupel, M., Heupel, M.R., Heyer, R., Heymann, E., Hicham, M., Hickman, C., Hicks, C., Hiep, N.T., Hierro, M.J., Highton, R., Hill, K.D., Hill, R., Hilton-Taylor, C., Hines, H., Hinton, G., Hinton, G.B., Hinton, G.S., Hobbelink, M.E., Hoces, D., Hodgetts, N., Hodgson, G., Hoeck, H., Hoeksema, B., Hoeksema, E., Hofer, A., Hoffmann, M., Hofmeyr, G., Hogan, Z.S., Hohmann, G., Holden, J., Holdsworth, M., Holekamp, K.E., Hollingsworth, B., Holtzhausen, H., Holtzhausen, H.A., Hon, J., Honer, O., Honess, P., Hong-Wa, C., Hoock, A., Hoogmoed, M., Horner, P., Horodysky, A.Z., Horrocks, J., Horsup, A., Hoskin, C., Hounsome, M., Howard, J., Howell, K., Howeth, J., Hozbor, N., Hrabar, H., Hraoui-Bloquet, S., Htun, S., Hua, Z.Z., Huacaz, D., Huang, D., Huber, D., Hudson, R., Hughes, C., Huibin, Z., Huiqing, G., Huitson, A., Human, B., Human, B.A., Humle, T., Hunt, K.D., Hunter, C., Hunter, L., Huntsman, G., Hurley, M., Hurter, J., Hussain, S.A., Hutson, A.M., Hutson, T., Hutterer, R., Huveneers, C., Ibanez, M., Ibáñez, R., Ibéné, B., Ibisch, P.L., Icochea, J., Idriz, H., Ilambu, O., Iliffe, T.M., Impson, D., Incháustegui, S., Indrawan, M., Ineich, I., Inger, R., Ingle, N., Iñi, Inoue, K., Insall, D., Irwin, K., Isailovic, J.C., Isfendiyaroglu, S., Is-haquou Daouda, H., Ishchenko, V., Ishihara, H., Ishii, N., Iskandar, D., Isnan, W., István Kiss, M.P., Iverson, J., Iverson, J.B., Izawa, M., J Lee, J., Jackson, D., Jackson, J., Jackson, P., Jackson, R., Jacobs, D., Jaeger, J., Jaén Molina, R., Jaffré, T., Jakubowsky, G., James, D., James, R., Jansen, M., Jaramillo, C., Jaramillo, T., Jaramillo-Legorreta, A., Jaya, J.P, Jayat, J., Jayat, J.P, Jdeidi, S., Jdeidi, T., Jeanmonod, D., Jefferson, T.A., Jehle, R., Jenkins, P., Jenkins, R., Jenkins, R.K.B., Jennings, A., Jennings, A.P., Jennings, R., Jensen, J., Jerusalinsky, L., Jhala, Y.V., Jiang Zhigang, Jiang, Y.-E., Jianping, J., Jiddawi, N., Jimena San Martín, M., Jiménez Martínez, J.F., Jiménez, J., Jiménez, J.E., Joger, U., Joglar, R., Johan, O., Johnsingh, A.J.T., Johnson, D., Johnson, K., Johnston, C.H., Jones, A., Jones, C., Jones, G., Jones, M., Jones, T., Jordan, M., Jørgensen, P., Jørgensen, P.M., Jorgensen, S., Joshua, J., Jowkar, H., Juncá, F., Jungfer, K.-H., Juškaitis, R., Juste, J., Kaariye, X.Y., Kadis, C., Kahwa, D., Kalkman, C., Kalkman, V., Kalkman, V.J., Kamenya, S., Kanchanasaka, B., Kaneko, Y., Kappeler, P., Karanth, U., Karatas, A., Karatash, A., Karczmarski, L., Karsen, S., Kasembe, J., Kasuya, T., Katende, A.B., Kaufman, L., Kauhala, K., Kaunda, E., Kawada, S., Kawanishi, K., Kays, R., Kazembe, J., Kebede, F., Kefelioglu, H., Keirulff, M.C.M., Keith, M., Kelly, B.T., Kelly, D.L., Kelly, M., Kelt, D., Kemper, C., Kemper, K., Kendrick, A.J., Kerbis Peterhans, J., Kerle, A., Kessner, V., Keuroghlian, A., Kevan, P.G., Khac Quyet, L., Khalikov, R., Khan, J., Khan, M.K.M., Khan, M.S., Khan, W., Khat Quyet, L., Khawa, D., Khonsue, W., Khorozyan, I., Kierulff, C., Kierulff, C.M., Kierulff, M.C.M., Kiesling, R., Kilburn, R.N., Kiliç, T., King, S.R.B., Kingdon, J., Kingston, N., Kingston, T., Kipping, J., Kis, I., Kiss, I., Kittle, A., Kiwi, L.K., Klingel, H., Knapp, C.R., Knowlton, F., Kochummen, K.M., Kock, D., Koenig, C., Koenig, S., Kofoky, A.F., Köhler, G., Köhler, J., Kohorn, L., Kok, P., Konstant, B., Koprowski, J., Kotas, J.E., Kottelat, M., Kouadio, A., Kovacs, K., Kovács, T., Kovács, T., Kozlowski, A., Kranz, A., Krasinska, M., Krasinski, Z.A., Kraus, F., Kreb, D., Krecsák, L., Krentz, S., Kristensen, T., Kristensen, T.K., Krose, M., Kryger, U., Kryštufek B., A.A., Kryštufek, B., Kuangyang, L., Kubicki, B., Kuchling, G., Kukherjee, S., Kulbicki, M., Kulka, D., Kulka, D.W., Kullander, F., Kumar Chhangani, A., Kumar, A., Kumar, N.S., Kumar, S., Kümpel, N., Kunte, K., Künzel, T., Kupfer, A., Kurczewski, F.E., Kurniati, H., Kurt, F., Kuzmin, S., Kwet, A., Kyambadde, R., Kyne, P., Kyne, P.K., Kyne, P.M., Kypriotakis, Z., La Mantia, A., Labat, H., Labat, J.N., Lacher, T., LaClaire, L., Laegaard, S., Lafrance, P., Lahiri Choudhury, D.K., Laidre, K., Lajmanovich, R., Laker, J., Lakim, M., Lamarque, F., Lamilla, J., Lamilla, J.M., Lammertink, M., Lamónaca, A.F., Lamoreux, J., Lanfranco, E., Lang, B., Lange, C., Lange, C.N., Langguth, A., Langone, J., Lanjouw, A., Lanza, B., Lara-Ruiz, P., Largen, M., Lascelles, B., Last, P., Last, P.R., Lastica, E., Latroú, G., Lau, M.W.N., Laurenson, M.K., Lavilla, E., Lavin, P., Lavin-Murcio, P., Lavrenchenko, L., Lawes, M.J., Lawson, D., Lawson, D.A., Lazell, J., Lea, J., Lea, R., Leach, G., Leandro, L., Leary, T., Leasor, H., Lecis, R., Lee, B., Lee, J., Lehmann, T., Lehn, C., Lehr, E., Leipelt, K.G., Leite, M.R.P., Leite, Y., Leite-Pitman, R., Leiva, A.M., Lemckert, F., Lemine Ould Sidi, M., Lenain, D.M., Lenin, J., León, N.P.L., Leong Tzi Ming, Léon-Yánez, S., Lescure, J., Leslie, D., Lessa, E., Lessa, R., Leus, K., Leuteritz, T., Lew, D., Lewis, D., Lewis, R., Lewison, R., Lhagvasuren, B., Li Zhenyu, Li, S.Y., Liang, F., Libois, R., Licandeo, R.R., Lichtenstein, G., Licuanan, A., Lidicker Jr., W.Z., Lim, B., Lim, T.W., Lima, A., Lima, R., Limpus, C., Linder, J., Lindquist, E., Lindsey, P., Lindstrom, A., Link, A., Linkie, M., Linzey, A.V., Lippold, L., Lips, K., Lira-Torres, I., Lisney, T.J., List, R., Lister, A., Litt, A., Littlejohn, M., Litvaitis, J., Litvinov, F., Liu Zhengyu, Livingstone, S., Lizana, M., Lizcano, D., Lizcano, D.J., Lkhagvasuren, D., Llamozas, S., Lloyd, P., Loader, S., Loc, P.K., Lockyear, J., Loiselle, P., Loman, J., Loman, J., Long, B., Long, M.A., Loots, S., Lopez Arevalo, H., Lopez Gonzalez, C., López Jiménez, N., Lopez Luna, M.A., López Udias, S., López, M., Lopez-Gonzalez, C., Lopez-Luna, M.A., Lorenzen, E., Lorenzo, C., Lorica, R., Lorica, R.P., Lorite, J., Lötters, S., Louis Jr., E., Loureiro, M., Lourie, S., Lovari, S., Lovell, E., Loveridge, A.J., Lovett, J., Loving, J., Low, B., Lowrey, C., Lowrie, A., Lowry, L., Lowry, P., Lowry, P.P., Loy, A., Lu, S.Y., Lu, W., Lü, Z., Lucherini, M., Ludovic, R., Lue, K., Luiselli, L., Luke, Q., Lumsden, L., Luna-Mora, V.F., Lunde, D., Lunn, N., Lunney, D., Luque Moreno, P., Luque, P., Lüthy, A.D., Luu, N.D.T., Lyenga, A., Lymberakis, P., Lynam, A., Lynam, A.J., Lynam, T., Lynch, J., Lynch, J.D., Maas, B., MacCulloch, R., Macdonald, A.A., Macdonald, D., Macdonald, D.W., Mace, G.M., Maciel, N., MacKinnon, J., MacPhee, R., Madhyastha, N.A., Madulid, D., Madulid, D.A., Maeda, K., Maeda-Martinez, A.M., Maffei, L., Magalhaes, C., Magin, C., Magombo, Z., Magombo, Z.L.K., Maharadatunamsi, D., Maharadatunkamsi, D., Maharandatunkamsi, D., Maheswaran, G., Mahony, M., Mahood, S., Mailosa, A., Mailosi, A., Maisels, F., Makocho, P., Makris, C., Malakar, M.C., Malcolm, J.R., Maldonado-Silva, R.A., Mallari, A., Mallon, D., Mallon, D.P., Malombe, I., Malonza, P., Manamendra-Arachchi, K., Mancina, C., Mancina, C.A., Mancini, P., Mancusi, C., Maneyro, R., Manganelli, G., Manh Ha, N., Mann, T., Mannheimer, C., Mannullang, B., Mansur, M.C.D., Mantilla, H., Mantuano, M., Manullang, B., Manzanares, J.M., Manzanilla, J., Mar, I., Maran, J., Maran, T., Marca, E.L., Maree, S., Marijnissen, S.,

Marijnissen, S.A.E., Marinhio-Filho, J., Marinho, F., Marinho-Filho, J., Marino, J., Marker, L., Markezich, A.L., Marks, M. Marler, T., Marmontel, M., Marques, O.A.V., Marquez, F., Marquez, R., Marrero Gómez, M.V., Marrero Rodríguez, A., Marsden, A.D., Marsden, D., Marsh, H., Marsh, L., Marsh, L.K., Marshall Mattson, K., Marshall, A., Marshall, A.D., Marshall, B.E., Marshall, L.J., Martin, R., Martínez Lirola, M.J., Martínez Rodríguez, J., Martinez, J., Martinez, J.L., Martínez-Solano, I., Martínková, N., Martins, M., Martins, M.B., Martins, P., Martin-Smith, K., Marty, C., Martyr, D., Martyr, D.J., Maryanto, I., Masafumi Matsui, B.S., Maskey, T., Maslova, I., Mason, T., Masoud, T.S., Massa, A., Masseti, M., Masseti, T., Mateo Miras, J.A., Mateo, J., Mathew, R., Mathieu Denoël, M.S., Matillano, J., Matola, S., Matson, J., Matsui, M., Matthee, C., Matthee, C.A., Mattoccia, M., Mauchamp, A., Maude, G., Maunder, M., Mauremootoo, J., Mauric, A., Mawson, P., Maxwell, A., May, S., Mayol, J., Mayoral García-Berlanga, O., Mazibuko, L., Mazzoleni, R., Mbeiza Mutekanga, N., McAdam, J.H., McAllister, D., McAuley, R., McCallum, H., McCarthy, T., McCord, M., McCord, M.E., McCormack, C., McCracken, S.F., McCranie, J., McCranie, J.R., McCranie, R., McCreery, K., McDonald, K., McDonald, R., McDougall, P.T., McDowall R.M., McEachran, J.D., McGinnity, D., McGraw, S., McGuinness, C.A., McIvor, A., McKay, J., McKenzie, G., McKenzie, N., McKinnon, J., McKnight, M., McLellan, B.N., McLeod, D., McNutt, J.W., McShea, B., Mead, J.G., Means, B., Measey, J., Mech, L.D., Medecilo, M.P., Medellín, R., Medhi, R., Medici, P., Medina, A., Medina, C., Medina, E., Medina, G., Medina-Vogel, G., Meegaskumbura, M., Meegaskumbura, S., Mehlman, P., Meijaard, E., Meinig, H., Mejia-Falla, P.A., Mellado, V.P., Menard, N., Mendelson III, J., Mendelson, J., Mendes, S.L., Mendoza Qoijano, F., Menegon, M., Menkhorst, P., Menon, V., Menzies, J., Merino, M.L., Merino-Viteri, A., Meritt, D., Meritt, M., Merker, S., Merizanidou, D., Mesa Coello, R., Meyer, A., Meyer, E., Meyers, D., Miaud, C., Mickeburgh, S., Mignucci-Giannoni, A., Mijares, A., Mikkelsen, P., Milan Vogrin, J.L., Milan Vogrin, M., Mildenstein, T., Millar, A.J.K., Miller, A., Miller, A.G., Miller, B., Miller, D.J., Miller, J., Miller, K.A., Miller, S., Mills, G., Mills,

M.G.L., Milne, D., Milner-Gulland, E.J., Ming, L.T., Minter, L., Minton, G., Miquelle, D., Mira, A., Miranda, F., Miras, J.A.M., M-Sook, M., Mitani, J.C., Mitchell, J., Mitchell, N., Mitra, S., Mitré, M., Mitsain, G., Mittermeier, R.A., Mix, H., Moehlman, P.D., Moehrenschlager, A., Mogollón, H., Mohammed, O.B., Moler, P., Molinari, J., Molur, S., Monadjem, A., Monkhzul, T., Monkhzul, T.S., Montenegro, O.L., Montesinos, D., Montúfar, R., Monzini, J., Mooney, N., Moore, J., Moore, L., Mora Vicente, S., Moraes, M., Morales, A., Morales, A.L., Morales, M., Morales-Jiménez, A.L., Morato, S.A.A., Moravec, J., Moreira Fernandes, F., Moreira, G., Moreno Saiz, J.C., Moreno, P., Morey, G., Morey, S., Morgan, A., Morgan, A.J., Morgan B., Morgan, B.J., Morgan, D.B., Morgan, S., Morgan, S.K., Morris, K., Morrison, C., Mortimer, J.A., Moseby, C., Moseby, K., Moss, K., Mostafa Feeroz, M., Mora Poveda, J.F., Motokawa, M., Mouden E.H.E., Mouna, M., Mouni, A., Moura, R., Moura-Leite, J.C., Mousa Disi, A.M., Moya, C., Moyer, D., Moyer, D.C., Mrasovcic, M., Msuya, C., Muddapa, D., Mueller, H., Mueses-Cisneros, J.J., Mugisha, A., Mukherjee, S., Mulawwa, M., Muller, H., Müller, O., Mumpuni, Munis, M., Munks, S., Munny, P., Muños, A., Muñoz, A., Muñoz, L.J.P., Muñoz-Alonso, A., Murdoch, J., Muriel, P., Murray, D., Murugaiyan, P., Mus, M., Musick, J., Musick, J.A., Musser, G., Mustari, A.H., Muths, E., Mycock, S.G., Myers, R.A., Mylonas, M., Nabhitabhata, J., Nader, I., Nader I., Nadler, T., Naeer, P.O., Naggs, F., Nagorsen, D.W., Nagy, Z., Nakamura, M., Nakaya, K., Nakazono, A., Nambou, M., Nameer, P., Nameer, P.O., Namora, R.C., Napoli, M., Naranjo, E., Narayan, G., Naruse, T., Narvaes, P., Nascimento, L.B., Natakmazi, G., Natalia Ananjeva, N., Navarrete, H., Navarro, F., Navas, D., Navas, P., Naveda, A., Navia, A.F., Ndiritu, G.G., Ndjele, M.B., Ndunda, M., Neer, J.A., Neill, D., Neira, D., Nekaris, A., Nel, J.A.J., Nel, R., Nellis, D., Nelson, C., Nelson, K., Nerz, J., Nettmann, H.K., New, T.R., Newby, J., Newell, D., Newton, A., Newton, P., Ng Wai Chuen, Ng, P., Ng, P.K.L., Ngereza, C., Nghia, N.H., Ngoc Thanh, V., Nguyen Duc To Luu, Nguyen Tien Hieo, Nguyen Van Nhuan, Nicayeniz, F., Nichols, G., Nicolalde, F., Nijman, V., Nikolic, T., Nistri, A., Nixon, K., Nixon, S., Noblet, J.F., Noblick, L., Nogales, F., Norman, B., Noronha, M.N., Noss, A.,

Notarbartolo di Sciara, G., Nouira, M.S., Nouira, S., Novarino, W., Novaro, A.J., Novellie, P., Nowell, K., Ntakimazi, G., Núñez, H., Núñez, J., Nusalawo, M., Nussbaum, F., Nyambayar, B., Nyberg, D., Nyhus, P., Nyström, P., Nzeyimana, L., O'Corry-Crowe, G., Oakwood, M., Oates, J., Oates, J.F., Oates, M.R., Obradovitch, M., Obura, C., Obura, D., Ochavillo, D., Ochoa, J., Odhiambo, E.A., O'Donnell, C., O'Donovan, D., Odum, R.A., Oetinger, M.I., Ogi, M., Ogielska, M., Ogrodo, A., Ogrodowczyk, A., Oguge, N., Ohdachi, S.D., Ohler, A., Ojeda Land, E., Ojeda, R., Ojeda., R., Oleas, N., Olech, W., Olgun, K., Oliveira, M.V., Oliveira, R.B., Oliveira, S.N., Oliver, W., Olivieri, G., Olson, A., Olson, L., Olsson, A., Omasombo, V., Ong, P., Ong, R.G., Oommen, O.V., Ordoñez Delgado, L., Orlov, A., Orlov, N., Ortiz, J.C., Oscrno-Muñoz, M., Otgonbaatar, N., Ott, J., Ottenwalder, J., Oval de la Rosa, J.P., Ovaska, K., Ovsyanikov, N., Pacheco, V., Packer, C., Packer, K., Padhye, A., Padial, J., Padovani Ferrera, B., Page, W., Paglia, A., Paguntalan, L.M., Painter, C., Paisley, S., Palacios, E., Palazzi, S., Palden, J., Palis, J., Palmeirim, J., Palmeirim, P., Palomares, F., Pamaong, R., Pamaong-Jose, R., Pan, F.J., Pangulatan, L.M., Pangunlatan, L.M., Paniagua, C.D., Pannell, C.M., Papenfuss, T., Paradis, G., Farauka, F.M., Pardina, U., Pardinas, U., Parent, C., Parker, F., Parnaby, H., Parr M., Parra-Olea, G., Parris, M., Pasolini, P., Passamani, M., Pasta, S., Patel, E., Paton, A.J., Pattanavibool, A., Patterson, B., Patton, J., Patton, J.L., Paul L., Paul, S., Paulson, D., Paunovic, M., Pauwels, O., Pavan, D., Paxton, J., Paxton, J.R., Payan, E., Payne, J., Pearce-Kelly, P., Pearl, C., Pearson, D., Peckover, R., Pecralli, G., Pedraza, S., Pedregosa, M., Pedregosa, S., Pedro Beja, F.A., Pedro Beja, S.K., Pedrono, M., Peet, N., Peeters, P., Peguy, T., Pei, K.J-C., Peixoto, A.L., Peixoto, O.L., Peñas, J., Pennay, M., Per Nyström, B.A., Perälä, J., Percequillo, A., Percequillo, A.R., Percequillo, C., Percequillo, M., Perdomo, A., Pereira, J., Pereira, J.P., Perera, A., Peres, M.B., Pérez Latorre, A.V., Perez, A.M., Pérez, J.C.W., Pérez, J.M., Perez, M., Perez, N., Perez, S., Pérez, S., Perez, V., Pérez-Mellado, V., Pergams, O., Perieras, A., Perkin, A., Perold, S.M., Perret, J., Perrin, M., Perrin, W.F., Perzanowski, K., Peters, S., Pethiyagoda, R., Petrovic, F., Pfab, M.F., Phan Ke Loc,

Pheeha, S., Phillips, C., Phillips, D.M., Phillips, M.K., Phiri, P.S.M., Pickersgill, M., Pierce, S.J., Piercy, A., Piercy, A.N., Pilcher, N., Pilgrim, J., Pillans, R., Pillay, D., Pimenta, B., Pimley, E., Piñeda, C., Pineda, J., Pineda, W., Pinilla, M.P.R., Pino, J., Pino, J.L., Pinto, L.P., Piovezan, U., Pipeng, L., Pires Costa, A., Pires O'Brien, J., Pires-Costa, L., Pita, R., Pitman, N., Pitman, R., Pitman, R.L., Platt, S., Pleguezuelos, J., Ploss, J., Plotkin, P., Plötner, J., Plowman, A., Plumptre, A.J., Pogonoski, J., Pogonoski, J.J., Pokheral, C.P., Pokryszko, B., Polechla, P., Polhemus, D.A., Polidoro, B., Politano, E., Pollard, B., Pollard, B.J., Pollard, D., Pollard, D.A., Pollock, C.M., Pombal, J., Pomilla, C., Pompert, J., Ponce-Campos, P., Ponder, W., Ponder, W.F., Pople, R., Porini, G., Porley, R.D., Potsch de Carvalho-e-Silva, S., Pounds, A., Pourkazemi, M., Povz, M., Powell, J., Powell, J.A., Powell, R., Poyarkov, A., Poynton, J., Pradhan, M.S., Prado, D., Prados, J., Precht, B., Preece, R.C., Price, D., Prina, A., Princee, A., Princee, F., Printes, R.C., Pritchard, P.C.H., Priyono, A., Ptolemy, J., Pucek, Z., Pudyatmoko, S., Puig, S., Puky, M., Punt, A., Puntriano, C.A., Purchase, N., Puschendorf, R., Qarqas, M., Qin, H.-N., Queirolo, D., Queiroz, H.L., Quero, H.J., Querouil, S., Quibilan, M., Quierolo, D., Quijano, S.M., Quintana, C., Quintero Díaz, G., Quintero Díaz, G.E., Qureshi, Q., Rabarivola, F., Rabarivola, J.C., Rabearivelo, A., Rabibisoa, N.H.C., Rabiei, A., Racey, P., Rachlow, J., Rada, M., Raffaelli, J., Raherisehena, M., Rahmani, A.R., Rainho, A., Rajamani, N., Rajeriarison, C., Rakotoarivelo, A.R., Rakotondravony, D., Rakotosamimanana, B., Rakotosamimanana, J.C., Ram, M., Ramala, S.P., Ramanamanjato, J.B., Ramayla, S., Ramiarinjanahary, H., Ramilo, E., Ramirez-Marcial, N., Rand, P.S., Randall, D., Randi, E., Randriamahazo, H., Randriamanantsoa, H.M., Randrianantoandro, C., Randrianasolo, A., Randrianjafy, V., Randrianjohany, E., Randriantafika, F.M., Rangel Cordero, H., Ranivo, J., Ranker, T., Rao, R.J., Rasamimanana, H., Rasamimanana, R., Rashid, S.M.A., Rasmussen, G., Rastegar-Pouyani, N., Rathbun, G., Ratimomanarivo, F., Ratnayeke, S., Ratrimomanarivo, F.H., Ratsi, H., Rattanawat Chaiyarat, Ravichandran, M.S., Ravino, J., Rawson, B., Raxworthy, C., Ray, J., Ray, P.,

Rayaleh, H.A., Razafimahatratra, E., Razafimanahaka, H.J., Read, J., Reading, R., Reardon, M., Reardon, M.B., Reardon, T., Reboton, C., Reed, J., Reeves, R., Reichle, S., Reid, F., Reid, J.W., Reid, R., Reidl, P.M., Reilly, S.B., Reinartz, G., Reis, M., Reizl, J.C., Renjifo, J.M., Rentz, D.C.F., Retallick, R., Reuling, M., Rey, J., Reyes-Bonilla, H., Reyna, R., Reyra-Hurtado, R., Reynolds III, J.E., Reynolds, J.C., Reynolds, R., Reynolds, V., Rhind, S., Rhodes, K., Rhodin, A.G.J., Rice, C., Richard-Hansen, C., Richards, G., Richards, J., Richards, N., Richards, S., Richards, Z., Richardson, M., Richter, S., Riddle, H., Riga, F., Rigaux, P., Rincon, G., Rios-López, N., Rioux Paquette, S., Ripken, T., Rischer, H., Riservato, E., Rita Larrucea, J., Ritchie, E., Rivalta, V., Rivas, B., Rivas, P., Rivas-Pava, P., Rivera, F., Riyad Sadek, S.H.-B., Robbins, M., Robbins, R., Robbrecht, E., Roberton, S., Roberts, C., Roberts, D., Roberts, D., Robertson, P., Robichaud, W.G., Robinson, L., Robinson, T., Rocha, C.F.d., Rocha, L., Rodden, M., Rödel, M.-O., Rodrigues, F., Rodrigues, M.T., Rodrigues, W.A., Rodriguez, A., Rodriguez, B., Rodríguez, L., Rodríguez-Luna, E., Rodriquez, J.C., Roemer, G.W., Rogers, A., Rogers, S., Rohwer, J.G., Rojas, W., Rojas-Bracho, L., Romano, A., Romero Malpica, F.J., Romero, M., Romero-Saltos, H., Romoleroux, K., Ron, S., Rookmaaker, K., Roos, A., Roos, C., Rorabaugh, J., Rosa, R.S., Rosell-Ambal, G., Rosenbaum, H., Rosmarino, N., Ross, J., Ross, J.P., Ross, S., Rossi, R.V., Rossiter, S., Roth, B., Roth, L., Roux, J.P., Rovero, F., Rovito, S., Roy, D., Ruanco, G., Rübel, A., Rubenstein, D., Rubenstein, D.I., Rueda, A.R., Rueda, L., Rueda-Almonacid, J.V., Ruedas, L., Ruggerone, G., Ruiz, M., Ruiz-Olmo, J., Rumiz, D.I., Runcie, M., Runstrom, A., Rushforth, K., Russell, B., Rutty, R., Ryan, S., Ryan, T., Rylands, A.B., Sadek, R., Sadek, R.A., Sadovy, Y., Saeki, M., Sáenz Goñalons, L., Safina, C., Sagar Baral, H., Saha, S.S., Sahlén, G., Salas, A., Salas, L., Saleh, M., Salim, A., Saltz, D., Salvador, A., Salvia, H., Samarawickrama, P., Samba Kumar, N., Sami Amr, Z., Samiya, R., Samoilys, M., Sampaio, C., Sampaio, E., Samudio, R., Samways, M., San Martín, J., San Martín, J.M., San Martin, M.J., Sánchez Gómez, P., Sanchez Rojas, G., Sánchez, B., Sánchez, J., Sánchez, J.M., Sanchez, R., Sanderson, J., Sandiford, M., Sano, A.,

Santana, F.M., Santiago, S., Santiana, J., Santos Motta, F., Santos, G., Santos, S.S.D., Santos-Barrera, G., Sanyal, P., Sarig Gafny, A., Sarkar, S.K., Sarker, S.U., Sarmiudo, R., Sarmudio, R., Sarti Martinez, A.L., Sasaki, H., Sá-Sousa, P., Sato, K., Savage, A., Savage, J., Saw, L.G., Sazima, I., Sbordoni, V., Schabetsberger, R., Schaller, G.B., Scheidegger, C., Schembri, P.J., Scherlis, J., Schilthuizen, M., Schiøtz, A., Schipper, J., Schliebe, S., Schlitter, D., Schmidt, B., Schmidt, P.A., Schmitz, A., Schneider, W., Schnell, D., Scholtz, S., Schraml, E., Schreiber, A., Schulte, R., Schulz, M., Schwaner, T., Schwitzer, C., Schwitzer, N., Scott, D., Scott, E., Scott, M.D., Scott, N., Scott, P., Scott-Shaw, R., Sebastian, T., Secchi, E., Secchi, E.R., Sectionov, Sedberry, G., Seddon, M.B., Sedlock, J., Sefass, T., Segalla, M.V., Seisay, M., Self-Sullivan, C., Selvi, F., Semesi, S., Semiadi, G., Seminoff, J.A., Señaris, C., Sengupta, S., Sepulveda, M., Sequin, E., Serena, F., Serena, M., Serena, S., Séret, B., Seri, L., Serra, J.M., Serrano, M., Servheen, C., Seryodkin, I., Sevinç, M., Seychelles, N.P.T.o., Seydack, A., Shaffer, B., Shah, N., Shank, C., Shanker, K., Shar, S., Sharif Khan, M., Sharifi, M., Sharma, J., Sharma, R.K., Shedden, A., Sheftel, B., Shekelle, M., Shenbrot, G., Shepard, D., Sheppard, A., Sheppard, C., Sherbrooke, W., Sherley, G., Sherrill-Mix, S.A., Shi Haitao, Shoemaker, A., Sholz, S., Shoshani, H., Shrestha, N., Shrestha, T.K., Shuk Man, C., Shunqing, L., Sidi, N., Sidiyasa, K., Siegel, R., Siex, K., Siler, C., Siliwal, M., Sillero-Zubiri, C., Silva Jn., J., Silva Jr, J.S., Silva, C., Silva, G., Silva, N.M.F., Silva, S.P.d.C.e., Silvano, D., Simaika, J., Simaika, J.P., Simkins, G., Simons, M., Simpfendorfer, C., Simpfendorfer, C.A., Sinaga, J., Sinaga, U., Sinanga, U., Sindaco, R., Singadan, R., Singh, L.A.K., Singh, M., Singleton, I., Sinha, A., Sinha, R.K., Sinisterra Santana, J., Sinsch, U., Situ Yingyi, A., Situ, A., Siu, S., Skelton, P., Sket, B., Skog, L.E., Skopets, M., Skuk, G., Slack-Smith, S., Slimani, T., Sliwa, A., Slooten, E., Sluys, M.V., Smale, M., Smale, M.J., Smith, A., Smith, A.T., Smith, B., Smith, B.D., Smith, C., Smith, E., Smith, G., Smith, J., Smith, J.A., Smith, K., Smith, R., Smith, R.K., Smith, S.E., Smith, W.D., Snell, H., Snelson Jr., F.F., Snelson Jr., F.S., Snelson, F., Snoeks, J., Sobel, J., Soberón, R.R., Söderström, L., Sogbohossou, E., Solari, S., Soldo, A.,

Solem, A., Soliano, P., Solís, F., Song J.-Y., Sonké, B., Soriano, P., Soto, J., Soto, J.M.R., Sotomayor, M., Sousa, M.C., Southwell, C., Sovada, M., Soy, J., Sozen, M., Sparks, J.S., Sparreboom, M., Spector, S., Spelman, L., Spence, C., Spironello. W.R., Spitzenberger, F., Sredl, M., Srinivasulu, C., St. Louis, A., St. Pierre, R., Stamm, C., Stanisic, J., Stankovic, S., Starmühlner, F., Start, T., Stauffer, F., Steel, L., Steffek, J., Stehmann, M., Stehmann, M.F.W., Stehmann, S., Steinmetz, R., Steinmitz, R., Stenberg, C., Stensgaard, A.S., Stephenson, P.J., Sterijovski, B., Sternberg, G., Stevens, D., Stevens, J., Stevens, J.D., Stevens, P.F., Stevenson, D.W., Stevenson, P., Stiassny, M., Stiassny, M.L.J., Stier, S., Stöck, M., Stokes, E.J., Strahan, R., Strahm, W., Strauss, M., Streicher, U., Struhsaker, S., Struhsaker, T., Stuart, B., Stuart, C., Stuart, S.N., Stuart, T., Stubbe, G., Stübbe, M., Stuebing, R., Stuppy, W., Suárez Mejía, J.A., Subirá, R., Sugardjito, J., Sugimura, K., Suhling, F., Suin, L., Sukhchuluun, G., Sukumaran, J., Sumardja, E., Sumardja, M.K.M, Sun, W., Sunarto, S., Sunderland-Groves, J., Sundström, L.F., Sunyer, J., Superina, M., Supriatna, J., Suprin, B., Surprenant, C., Suyanto, A., Suyanto, I., Swan, S., Swartz, E., Syahrir, M., Symes, A., Tabao, M., Tabaranza, B., Taber, A., Tabet, M.A., Taggart, D., Tahar, S., Talavera, S., Tallents, L., Talukdar, B.K., Talukdar, B.N., Tan, B., Tan, B.C., Tandang, D.N., Tandy, M., Tannerfeldt, M., Tapia, F., Targarona, R.R., Tarkhnishvili, D., Tatayah, V., Tattersfield, P., Tavares, V., Taylor, A., Taylor, B., Taylor, B.L., Taylor, J., Taylor, N.P., Taylor, P., Taylor, P.J., Taylor, S., Teclai, R., Teixeira de Mello, F., Tejedo M., Tejedor, A., Telfer, W., Telles, A.M., Temple, H., Teta, P., Tezoo, V., Thalmann, U., Thanh Hai, D., Thapa, J., Theischinger, G., Thirakhupt, K., Thomas, P., Thomas, R., Thompson, D., Thompson, F.G., Thompson, J., Thomson, B., Thomspon, D., Thorbjarnarson, J., Thouless, C., Thulin, M., Thun, S., Ticul Alvarez, S., Tikhonov, A., Tilson, R., Timberlake, J., Timm, B., Timm, R., Timm, T., Timmins, R. Timmins, R.J., Timmins, T., Ting, N., Ting,

T., Tinnin, D., Tinsley, R., Tinti, F., Tirira, D., Tu, D., Tizard, R.J., Tocher, M., Tok, V., Tokida, K., Tokita, K., Toledo, L.F., Tolson, P., Tomiyama, K., Tooze, Z.J., Toral, E., Torres, D.A., Torres, R.B., Torrijos, I.A., Touk, D., Tous, P., Townsend, J., Traeholt, C., Tran Quang Phuong, Treloar, M.A., Trembley, R., Trillmich, F., Trinnie, F.I., Trocchi, V., Troìa, A., Truong, N.Q., Tsogbadrakh, M., Tsytsulina, K., Tuniyev, B., Turak, E., Turner, A., Turvey, S., Tuthill, J., Tutin, C., Tutin, C.E.G., Tweddle, D., Twongo, T.K., Tye, A., Tyson, M., Ubaldo, D., Úbeda, C., Ugurtas, I.H., Ulloa Ulloa, C., Ungaro, N., Uozumi, Y., Urbán, J., Urbani, B., Urdiales Perales, N., Usukhjargal, D., Uzzell, T., Vagelli, A., Valderrama, C., Valdespino, C., Valdez, R., Valencia, R., Valente, M.C.M., Valenti, S., Valenzuela, J.C., Valenzuela-Galván, D., Valesco, M., Valezco, P., Vallan, D., van der Elst, R., van der Straeten, E., van Dijk, P.P., van Gruissen, J., van Jaarsveld, A., van Lavieren, E., van Manen, F., van Rompaey, H., van Rompaey, J., van Schaik, C., van Strien, N.J., van Swaay, C., van Swaay, C.A.M., van Weenen, J., van Welzen, P.C., Vana, J., Vanitharani, J., Varela, D., Vargas, I., Vargas, J., Varman, R., Varty, N., Vaslin, M., Vasquez Díaz, J., Vasquez, C., Vasudevan, K., Vaz, A.M.S., Vázquez Díaz, J., Vázquez, E., Vázquez, E., Vázquez, R.C., Vázquez-Domínguez, E., Veiga, L.M., Velasco, A., Velazco, P., Velez-Espino, L.A., Velez-Liendo, X., Velilla, M., Veloso, A., Velosoa, J., Vences, M., Venegas, P., Venkataraman, A., Venturella, G., Vera Pérez, J.B., Vera, M., Verdade, V., Vermeer, J., Vermeulen, J., Veron, G., Vicens Fandos, J., Victor, J., Victor, J.E., Vié, J.-C., Vieira, E., Vieites, D., Vijayakumar, S.P., Vila, A., Villalba, L., Villamil, C., Vincent, A., Vincent, A.C.J., Viney, D.E., Vivar, E., Vivero, J.L., Vizcaino, S., Vogel, P., Vogliotti, A., Vogrin, M., Vogt, R.C., Vohralík, V., Vololomboahangy, R., von Ax, B., von Arx, M., von Cosel, R., von Elenrieder, N., Vonesh, J., Vooren, C.M., Vörös, J., Vovides, A., Vovides, A.P., Vreven, E., Vyas, R., Wabnitz, C., Wacher, T., Wade, P., Wager, R., Wagner, A., Wai, H., Wake, D., Wake, M., Waldemarin, H.F.,

Waldman, B., Waldren, S., Walker, P., Walker, P., Walker, R., Walker, T.I., Wallace, R.B., Wallance, R., Wallays, H., Walsh, P.D., Walston, J., Walstono, J., Wang Ying-Xiang, Wang, D., Wang, J.Y., Wang, S., Wang, X., Wang, Y., Wanzenböck, J., Ward, D., Werguez, D., Warner, J., Warren, M.S., Watanabe, K., Waters, S., Watling, D., Watson, A., Watson, M., Wayne, A., Wayne, R.K., Webb, R., Weber, M., Wecksler, M., Weil, E., Weinberg, P., Weksler, M., Wells, R.S., Wells, S., Welsh, H., Wenge, Z., Wenhao, C., Werner, R., Werner, Y., Werner, Y., Wheeler, J., Whistler, A., Whitaker Jr., J.O., Whitaker, N., Whitaker, R., White, L., White, W., White, W.T., Whittaker, D., Whorisky, F., Wibisono, H.T, Wich, S., Wich, S.A., Wickramasinghe, D., Widmann, P., Wiesel, I., Wiewandt, T., Wigginton, M.J., Wiig, Ø., Wikelski, M., Wikramanayake, E., Wild, E., Wildermuth, H., Wiles, G., Wilhelmi, F., Wilkinson, M., Williams, D.F., Williams, R., Williams, R.S.R., Williams, S., Williams, S.A., Williamson, E.A., Wilson, B., Wilson, K., Wilson, L., Wilson, M.L., Wilting, A., Win Ko Ko, U., Wingate, D., Winter, J., Wintner, S.P., Witsuba, A., Wogan, G., Woinarski, J., Wolseley, P.A., Wong, G., Wong, S., Wood, E., Wood, J., Wood, K.R., Woodman, N., Woodroffe, R., Woods, C.M.C., Cotterill, F.P.D., Woolley, P., Wozencraft, C., Wranik, W., Wright, D., Wright, P., Wuster, W., Xia, W., Xiang Qiaoping, Xie Feng, Xiuling, W., Xuan Canh, L., Xuelong, J., Yaakob, N., Yahr, R., Yahya, S., Yamada, F., Yambun, P., Yánez-Muñoz, M., Yang, B., Yang, J., Yang, J.-X., Yang, S.Y., Yanling, S., Yano, K., Yapa, W., Yensen, E., Yeo, D., Yigit, N., Ying-xiang, W., Yohannes, F., Yokohata, Y., Yom-Tov, Y., Yongcheng, L., Yongzu, Z., Yonzon, P., Young, B., Young, B.E., Young, J., Yoxon, G., Yoxon, P., Yuan, Y.C., Yuezhao, W., Yustian, I., Zagorodniuk, I., Zambrano, L., Zapfack, L., Zappi, D.C., Zaw, T., Zeballos, H., Zeballos, N., Zeballos, Z., Zemanova, B., Zerbini, A.N., Zhao Ermi, Zhigang, Y., Zhou, K., Ziaie, H., Ziegler, T., Zielinski, J., Zima, J., Zimmermann, W., Zoerner, S., Zona, S., Zortea, M., Zorzi, G., Zug, G., and Zweifel, R.

The IUCN Red List: a key conservation tool

Jean-Christophe Vié, Craig Hilton-Taylor, Caroline M. Pollock, James Ragle, Jane Smart, Simon N. Stuart and Rashila Tong

Biodiversity loss is one of the world's most pressing crises with many species declining to critically low levels and with significant numbers going extinct. At the same time there is growing awareness of how biodiversity supports human livelihoods. Governments and civil society have responded to this challenge by setting clear conservation targets, such as the Convention on Biological Diversity's 2010 target to reduce the current rate of biodiversity loss. In this context, *The IUCN Red List of Threatened Species™* (hereafter The IUCN Red List) is a clarion call to action in the drive to tackle the extinction crisis, providing essential information on the state of, and trends in, wild species.

A highly respected source of information

The IUCN Red List Categories and Criteria are widely accepted as the most objective and authoritative system available for assessing the global risk of extinction for species (De Grammont and Cuarón 2006, Lamoreux *et al.* 2003, Mace *et al.* 2008, Rodrigues *et al.* 2006). The IUCN Red List itself is the world's most comprehensive information source on the global conservation status of plant and animal species; it is updated annually and is freely available online at www.iucnredlist.org (Figure 1). It is based on an objective system allowing assignment of any species (except micro-organisms) to one of eight Red List Categories based on whether they meet criteria linked to population trend, size and structure and geographic range (Mace *et al.* 2008).

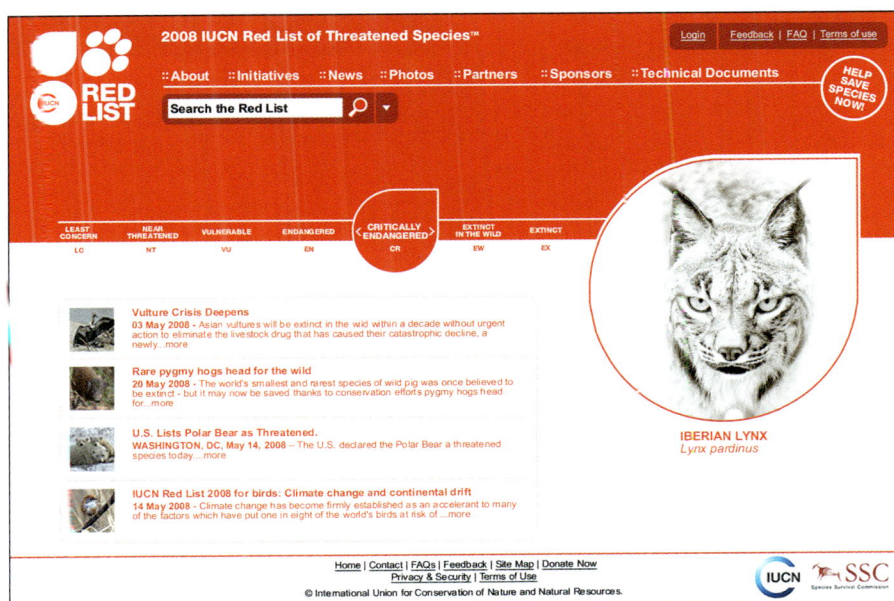

Figure 1. *The IUCN Red List can be viewed in its entirety on www.iucnredlist.org.*

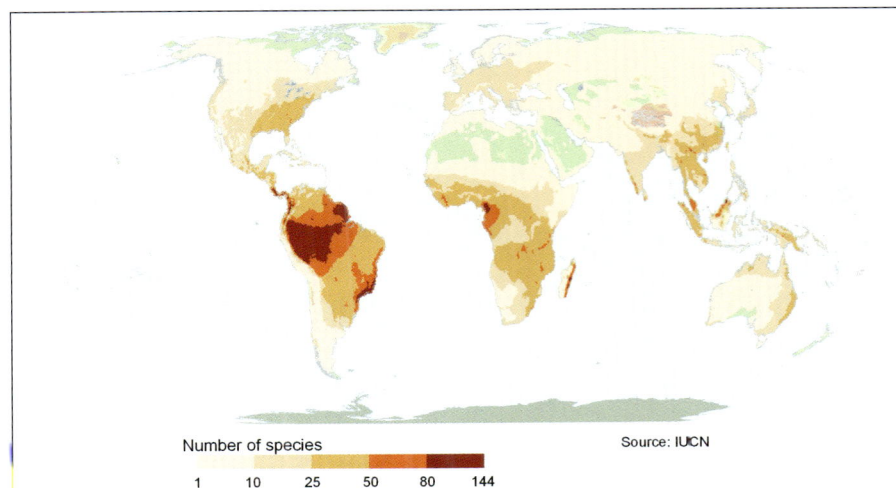

Figure 2. *Red List data allows detailed analysis of biodiversity at various scales across the globe.*

Far more than a list

One of The IUCN Red List's main purposes is to highlight those species that are facing a high risk of global extinction. However, it is not just a register of names and associated threat categories. The real power and utility of The IUCN Red List is in what lies beneath: a rich, expert-driven compendium of information on species' ecological requirements, geographic distributions and threats that arms us with the knowledge on what the challenges to nature are, where they are operating, and how to combat them.

A wealth of information about threatened and non-threatened species

The IUCN Red List is not limited to just providing a threat categorization. For an increasing number of species, be they threatened or not, it now provides extensive information covering taxonomy (classification of species), conservation status, geographic distribution, habitat requirements, biology, threats, population, utilization, and conservation actions. Spatial distribution maps are also becoming available for an increasing number of species (almost 20,000 species on The 2008 IUCN Red List have maps). All this information allows scientists to undertake detailed analyses of biodiversity across the globe (Figure 2).

Only about 2.7% of the world's estimated 1.8 million described species have been

The IUCN Red List includes threatened and non-threatened species such as the Vulnerable Shoebill Balaeniceps rex *and the Least Concern Guianan Cock-of-the-rock* Rupicola rupicola.
© Jean-Christophe Vié

assessed for The IUCN Red List so far; therefore the number of reported threatened species is much less than the true number at serious risk of extinction. The IUCN Red List is, nevertheless, by far the most complete global list of such species available.

Species: the cornerstone of biodiversity

Species provide us with essential services: not only food, fuel, clothes and medicine, but also purification of water and air, prevention of soil erosion, regulation of climate, pollination of crops, and many more. They also provide a vital resource for economic activities (such as tourism, fisheries and forestry), as well as having significant cultural, aesthetic and spiritual values. Consequently the loss of species diminishes the quality of our lives and our basic economic security.

Species are the building blocks of biodiversity and provide us with essential services. Barracudas Sphyraena sp. *in Guinea Bissau and Cork Oaks* Quercus suber *in Portugal.* © Jean-Christophe Vié

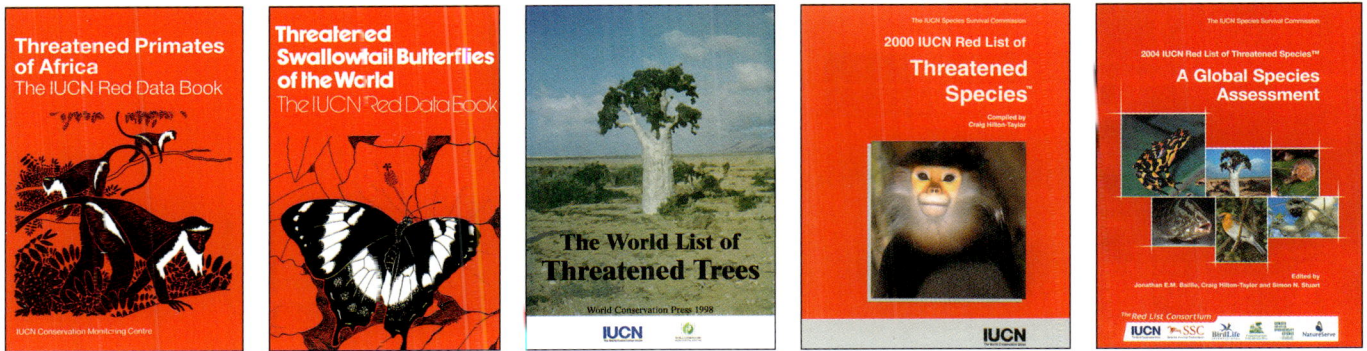

Figure 3. Some examples of past Red List publications.

Species are easier to identify and categorize than ecosystems, and they are easier to measure than genes. They provide the most useful, and useable, indicators of biodiversity status and loss. Species have been extensively studied for more than two centuries and there is an impressive amount of information dispersed around the world, that once compiled and standardized, can be used for developing strategies to tackle the current extinction crisis.

A long and successful history

The IUCN Red List is well established and has a long history. It began in the 1960s with the production of the first Red Data Books (Fitter and Fitter 1987). The concept of the Red Data Book, registers of wildlife assigned categories of threat, is generally credited to Sir Peter Scott when he became Chair of the then IUCN Survival Service Commission in 1963, with the first two volumes (on mammals and birds) published in 1966.

Since the 1960s The IUCN Red List has evolved from multiple lists and books dedicated to animal groups or plants into a unique comprehensive compendium of conservation-related information now too large to publish as a book (Figure 3). However it can be viewed in its entirety on a website managed and maintained by the IUCN Species Programme. It is updated once a year and is freely available to all users of the World Wide Web

Identifying, documenting and monitoring trends

By assessing the threat status of species, The IUCN Red List has two goals: (i) to identify and document those species most in need of conservation attention if global extinction rates are to be reduced; and (ii) to provide a global index of the state of change of biodiversity. The first of these goals refers to the "traditional" role of The IUCN Red List, which is to identify particular species at risk of extinction. However, the second goal represents a more recent radical departure, as it focuses on using the data in the Red List for multi-species analyses in order to identify and monitor trends in species status.

To achieve these goals the Red List aims to (i) establish a baseline from which to monitor the change in status of species; (ii) provide a global context for the establishment of conservation priorities at the local level; and (iii) monitor, on a continuing basis, the status of a representative selection of species (as

Polypedates fastigo – a Critically Endangered amphibian from Sri Lanka. © Don Church

Zanzibar Red Colobus Procolobus kirkii – an Endangered species endemic to Zanzibar island. © Jean-Christophe Vié

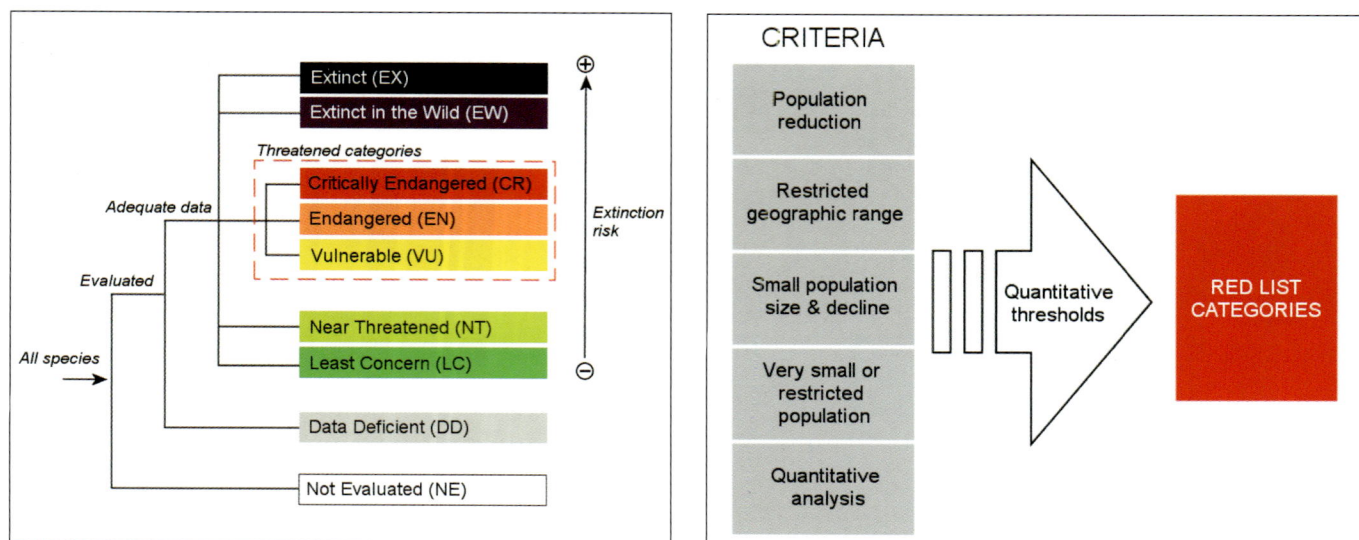

Figure 4. *Structure of the Red List Categories and the five Red List Criteria.*

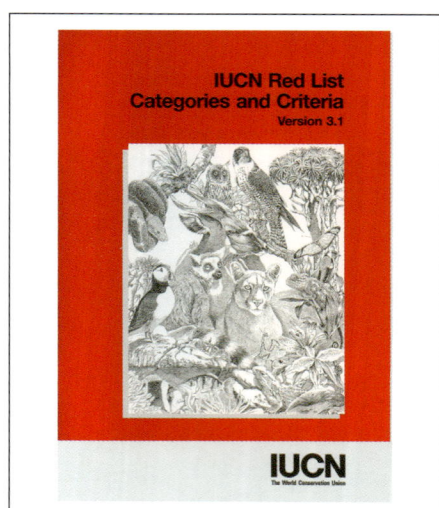

biodiversity indicators) that cover all the major ecosystems of the world.

The high profile, standards and scientific integrity of The IUCN Red List are maintained in the following ways: (i) the scientific aspects underpinning The IUCN Red List are regularly published in the scientific literature (Butchart *et al.* 2004; 2007; Colvyan *et al.* 1999; Mace *et al.* 2008); (ii) the assessment process is clear and transparent; (iii) the listings of species are based on consistent use of the Red List Categories and Criteria and are open to challenge and correction; (iv) all assessments are appropriately documented and supported by the best scientific information available; (v) the data are freely available through the World Wide Web to all potential users; (vi) The IUCN Red List is updated regularly (annually at present) but not all species are reassessed with each update – many assessments

simply roll-over from the previous edition; and (vii) analyses of its findings are regularly published, approximately every four to five years, usually at the time of the World Conservation Congress (Hilton-Taylor 2000; Baillie *et al.* 2004; Vié *et al.* this volume).

From expert judgment to robust criteria

The first Red List Criteria were adopted in 1994 (IUCN 1994) after a wide consultative process involving hundreds of scientists. The IUCN Red List Categories and Criteria were revised in 2001 (IUCN 2001). They currently include nine categories and five quantitative criteria (Figure 4). The *Guidelines for Using The IUCN Red List Categories and Criteria* (http://www.iucn. org/redlist) have been developed and are updated on a regular basis; they provide detailed guidance on how to apply the categories and criteria and aim at providing solutions to specific technical issues to ensure that assessments are conducted in a standardized way across various plant and animal groups.

The IUCN Red List Categories and Criteria are the world's most widely used system for gauging the extinction risk faced by species. Each species assessed is assigned to one of the following categories: Extinct, Extinct in the Wild, Critically Endangered, Endangered, Vulnerable, Near Threatened, Least Concern and Data Deficient, based on a series of quantitative criteria linked to

population trend, population size and structure, and geographic range. Species classified as Vulnerable, Endangered and Critically Endangered are regarded as 'threatened'. The IUCN Red List Criteria were developed following extensive consultation and testing, and involved experts familiar with a very wide variety of species from across the world, and can be used to assess the conservation status of any species, apart from microorganisms.

The Red List Criteria were developed for use at the global scale when the entire range of a species is considered. They can be applied at any regional scale, provided the guidelines for application at regional levels (IUCN 2003) are used, but they may not be appropriate at very small scales.

Working in partnership

The IUCN Red List is compiled and produced by the IUCN Species Programme based on contributions from a network of thousands of scientific experts around the world. These include members of the IUCN Species Survival Commission Specialist Groups, Red List partners (currently Conservation International, BirdLife International, NatureServe and the Zoological Society of London), and many others including experts from universities, museums, research institutes and non-governmental organizations. Assessments can be done by anyone and submitted to IUCN for consideration. Assessments are impartial and are developed and approved based on their scientific merits without consideration of their policy implications. This approach allows for an independent, robust process, requiring rigorous peer-review of all the data. Assessments are periodically updated to ensure that current information is available to users. The IUCN Red List is therefore a synthesis of the best available species knowledge from the world's foremost scientists. Only after the data have been through the peer review process can they be included in The IUCN Red List.

An effort has also been made to work in partnership with other organizations to agree for example, on standard classification schemes and a common language for threats and conservation measures (Salafsky et al. 2008)

A complex and rigorous process

The IUCN Species Programme plays the lead role in helping to fund, convene and facilitate the assessment workshops which drive much of the data gathering and review process for the Red List. It has expanded its staff to facilitate the coordination of assessments. This has allowed the information to grow significantly in recent years, particularly in terms of the number and type of species being assessed, and in the improved richness of the collected data. It has also permitted a significant increase in the quality and consistency of the

Blacktip Reef Sharks Carcharhinus melanopterus *– Near Threatened – on Aldabra atoll, a World Heritage Site in the Seychelles. All 1,045 species of sharks and rays have been assessed.* © Jerker Tamelander

assessments within and across groups of organisms.

Since 2000, a significant effort has been made to increase the number of assessments through assessing entire taxonomic groups, as BirdLife International has done for birds since 1988. This led to the establishment of a central Red List Unit and the establishment of global assessment teams within the IUCN Species Programme. In particular, a Biodiversity Assessment Unit established in partnership with Conservation International is coordinating the work on mammals, reptiles, amphibians and marine species. Other IUCN units are coordinating global freshwater biodiversity and regional species assessments. These units play a key role in running the assessment processes, and also in finding the necessary resources to mobilize the experts' knowledge and bring assessments to completion.

The Species Survival Commission (SSC) currently has 80 Red List Authorities which work very closely with the Species Programme, especially in identifying the leading experts to contribute to assessments, and conducting evaluations of the data as part of the peer-review process. Many of the Red List Authorities are part of SSC Specialist Groups, and some are also within the Red List Partner organizations.

From the field to The IUCN Red List

All species assessments are based on data currently available for the species (or subspecies, population) across its entire

Asian Wild Ass Equus hemionus *– Endangered.* © Jean-Christophe Vié

global range. Assessors take full account of past and present literature (published and grey) and other reliable sources of information relating to the species. For subspecies, variety or subpopulation assessments, a species-level assessment is also carried out.

All submitted assessments are evaluated by at least two qualified reviewers, in most cases assigned by the Red List Authorities. The evaluation process is similar to the peer review process used by scientific journals in deciding which manuscripts to accept for publication.

A sophisticated information management system

IUCN has developed the Species Information Service (SIS), an information management tool to collect, manage, process, and report data – to the point of publication on The IUCN Red List. The SIS allows the contributors to participate in the Red List assessment work more easily than was the case in the past. In addition, through improved data exploration capabilities on The IUCN Red List website, SIS is making the world's most accurate, up-to-date information on species, their distribution and conservation status accessible with flexible, easy-to-use tools to support sound environmental decision-making.

Almost 45,000 listed species

The number of species assessed as threatened keeps increasing every year (Figure 5). By 2008, 44,837 species have been assessed; at least 38% of these have been classified as threatened and 804 classified as Extinct. The documented number of threatened species and extinctions is only the tip of the iceberg, as this number depends on the overall number of assessed species; in addition 5,561 species classified as Data Deficient are possibly threatened (Hilton-Taylor et al. this volume). The number of Extinct species is also a very conservative estimate given that for a species to be listed as Extinct requires exhaustive surveys to have been undertaken in all known or likely habitats throughout its historical range, at appropriate times and over a timeframe appropriate to its life cycle and life form (IUCN 2001). Species that are likely to be Extinct but for which additional surveys might be necessary to eliminate any doubt, are classified in the Critically Endangered Category with a "Possibly Extinct" flag (Butchart et al. 2006).

Comprehensive assessments of every known species of mammal, bird, amphibian, shark, reef-building coral, cycad and conifer have been conducted. There are ongoing efforts to complete assessments of all reptiles, all fishes,

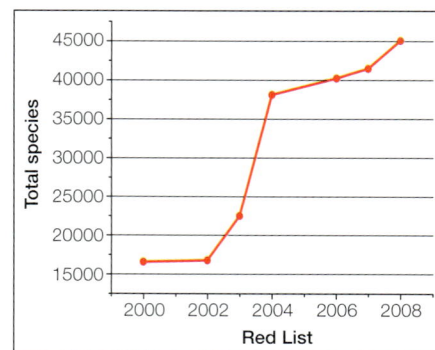

Figure 5. *Number of species appearing on each published IUCN Red List since 2000.*

and selected groups of plants and invertebrates.

Around 1.8 million species have been described, yet the estimates of the total number of species on earth range from 2 – 100 million. We are far from knowing the true status of the earth's biodiversity. Although, only a small proportion of the world's species has so far been assessed, this sample indicates how life on earth is faring, how little is known, and how urgent the need is to assess more species.

Despite the limited number of species assessed in relation to the total number of species known, and the significant number of Data Deficient species included in it,

Terraphosa leblondi, the world's largest spider, and Equadorian plants. Plant and invertebrate species are currently under-represented on the Red List but a dedicated effort is being made to increase their number.
© Jean-Christophe Vié

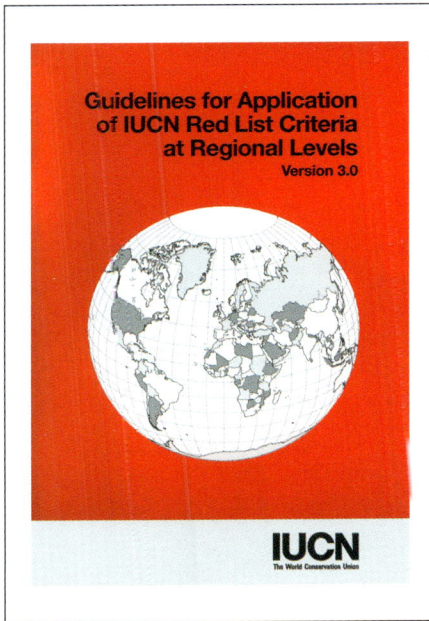

Guidelines for Application
of IUCN Red List Criteria
at Regional Levels
Version 3.0

IUCN
The World Conservation Union

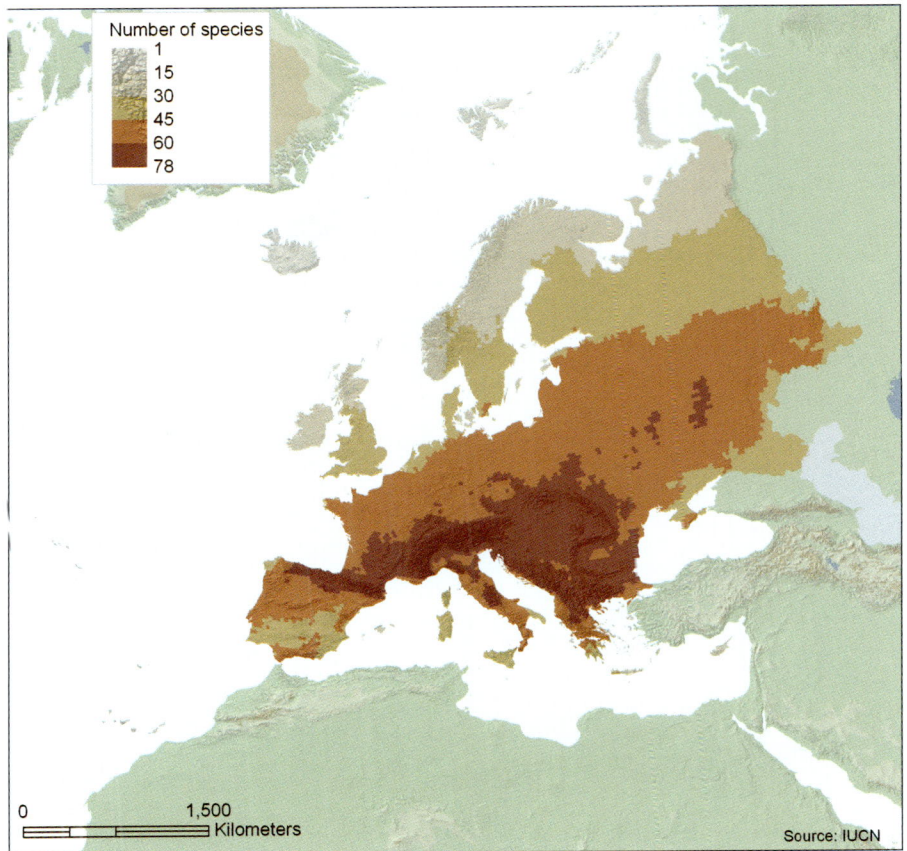

Number of species
1
15
30
45
60
78

0   1,500
Kilometers

Source: IUCN

Figure 6. An example of a regional biodiversity analysis: threatened terrestrial mammal species richness in Europe.

the Red List is still the largest dataset of current information on species. It allows us to measure how little the diversity of life on our planet is known and how urgent the need is to expand the assessment work if we want to be in a position to track progress towards reducing biodiversity loss.

Better links with regional and national Red Lists

The global IUCN Red List only includes information on species, subspecies or populations that have been globally assessed; regional and national level assessments are currently not included unless these are also global assessments (for example, a species that is only found in one country, (i.e., is endemic) and therefore has the same Red List status at both national and global levels).

For non-endemics, it is important to note that the status of a species at the global level may be different to that at a national level. In certain situations, a species may be listed as threatened on a national Red List even though it is considered Least Concern at the global level by IUCN and vice versa.

An increasing number of regional and national Red Lists are compiled following the *Guidelines for Application of IUCN Red List Criteria at Regional Levels* (Gärdenfors *et al.* 2001; IUCN 2003). IUCN is increasingly undertaking

regional Red List projects, for example in Europe and in the Mediterranean region (Temple and Terry 2007; Cuttelod *et al.* this volume) (Figure 6). IUCN is also collaborating with other national Red List projects to incorporate their data, especially on national endemics, into the global IUCN Red List.

Regional and national lists are usually country-led initiatives, and are not centralized in any way; they differ from each other widely in terms of scope and quality but are very useful to guide conservation work at sub-global levels. IUCN and its Red List Partners are currently discussing how to disseminate the data in the national and regional Red Lists more effectively, especially those that are conducted using the IUCN standards.

A multitude of uses

The IUCN Red List can help answer many important questions including:

- What is the overall status of biodiversity, and how is it changing over time?

Fungi represent a very diverse component of biodiversity which is too often overlooked.
© Jean-Christophe Vié

7

Black-browed Albatross Thalassarche melanophrys - *Endangered.* © Richard Thomas

- How does the status of biodiversity vary between regions, countries and sub-national areas?

- What is the rate at which biodiversity is being lost?

- Where is biodiversity being lost most rapidly?

- What are the main drivers of the decline and loss of biodiversity?

- What is the effectiveness and impact of conservation activities?

The IUCN Red List is used in many different applications, some of which are outlined below as examples.

An indicator of biodiversity trends: The IUCN Red List Index

Governments have agreed various targets to reduce biodiversity loss. A global target of reducing or stopping biodiversity loss by 2010 has been adopted respectively by the Parties to the Convention on Biological Diversity (CBD) and the European Union. In 2000, the United Nations adopted the Millennium Development Goals (MDG) with Goal 7 aiming at ensuring environmental sustainability by 2015; this goal underpins the others, in particular those related to health, poverty and hunger. Tools are needed to monitor our progress towards achieving these targets and to highlight where we need to focus our conservation efforts. Indicators are vital in tracking progress in achieving these targets. The IUCN Red List Index (RLI) provides such an indicator and reveals trends in the overall extinction risk of sets of species (Brooks and Kennedy 2004; Butchart *et al.* 2005ab, 2007).

The development of reliable indicators requires robust baseline data; species data are still scarce for most species groups and have been collected in a variety of formats. Collecting the baseline information is certainly what requires the largest effort in terms of time, expense and the number of people involved. To respond to this challenge, IUCN and its partners have been putting extensive efforts in biodiversity assessment initiatives at global and regional levels to develop The IUCN Red List in a manner that allows the Red List Index (including various cuts of it) to be calculated and measured over time.

The *IUCN Red List Index (RLI)* has been officially included in various sets of indicators to measure progress towards the 2010 CBD target. It has also been recently adopted as an indicator to measure progress towards the UN MDG 7 goal. It will play a vital role in tracking progress towards achieving these targets, and beyond.

The RLI shows trends in the overall extinction risk of sets of species. It is based on the number of species that move between Red List Categories as a result of genuine improvements in status (e.g., owing to successful conservation action) or genuine deteriorations in status (e.g., owing to declining population size). The RLI shows the net balance between these two factors. It excludes non-genuine changes in Red List status resulting, for example, from improved knowledge, taxonomic changes, or correction of earlier errors (Butchart *et al.* 2004; 2007).

The proportion of species threatened with extinction is a measure of human impacts on the world's biodiversity, as human activities and their consequences drive the vast majority of threats to biodiversity.

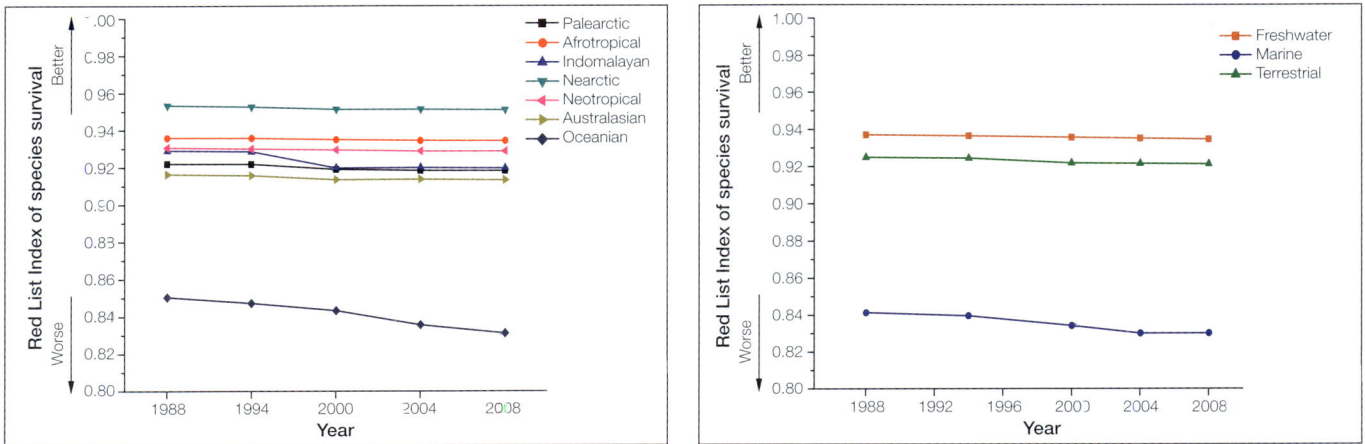

Figure 7. *The Red List Index for the world's birds shows that their overall status deteriorated steadily during 1988-2008. Declines have occurred worldwide but regions and biomes differ in the overall extinction risk of their bird fauna, and in the rate of declines (source BirdLife International). Similar graphs will be available shortly for mammals, amphibians, corals and cycads.*

Birds are the class of organisms for which all species (9,990) have been assessed the largest number of times (five times between 1988 and 2008). For this group, the percentage threatened increased from 11.1% in 1988 to 12.2% in 2008.

The RLI for the world's birds shows that their overall status (extinction risk) deteriorated steadily during 1988-2008. The RLI for birds in different regions shows that declines have occurred worldwide but regions differ in the overall extinction risk of their bird fauna, and in the rate of declines (Figure 7).

Birds are excellent, although not perfect, indicators for trends in other forms of biodiversity. Several other classes of organisms have been comprehensively assessed for The IUCN Red List and found to be even more threatened than birds.

This is the case for mammals (Schipper *et al.* 2008), amphibians (Stuart *et al.* 2004), reef-building corals (Carpenter *et al.* 2008), sharks and rays, freshwater crustaceans, cycads and conifers. A preliminary RLI has already been calculated for mammals, amphibians and corals (Hilton-Taylor *et al.* this volume).

For other groups (e.g., reptiles, fishes, molluscs, dragonflies, and selected groups

A preliminary assessment of all plant species have been called for by the Convention on Biological Diversity. © Jean-Christophe Vié

The IUCN Red List is a useful tool for infrastructure development and planning. © Jean-Christophe Vié

Informing Development and Conservation Planning

In regional and national resource management and development, The IUCN Red List can be used to guide management at scales ranging from local to national and sometimes regional levels. Examples include setting policies and developing legislation related to land-use planning, certification, transport, energy, river-basin management, and poverty reduction.

For site-development and planning, The IUCN Red List is a key input into the Environmental Impact Assessment process and can guide site level management and planning. There is growing interest by the corporate sector in using the Red List information to inform the selection and management of sites in which they operate.

of plants) assessment work is being undertaken with the aim of developing RLIs for each of these groups. For species groups that are composed of very large numbers of species (e.g., plants and invertebrates), a Red List Index will be calculated on the basis of a random sample of 1,500 species. This approach, pioneered by the Zoological Society of London, will allow trends in the status of a broader spectrum of biodiversity to be determined (Baillie *et al*. 2008; Collen *et al*. this volume).

Advising Policy and Legislation

The IUCN Red List data is used to inform the development of national, regional and sub-national legislation on threatened species protection, and also the development of national biodiversity strategies and action plans. It is also used to inform multi-lateral agreements such as the Convention on International Trade in Endangered Species of Wild Fauna and Flora (CITES), the Convention on Migratory Species (CMS), the Ramsar Convention on Wetlands, and the Convention on Biological Diversity (CBD). The Red List is recognized as a guiding tool to revise the annexes of some agreements such as the Convention on Migratory Species.

The IUCN Red List is also an important tool for implementing some elements of the Global Strategy for Plant Conservation adopted by the CBD in 2002, for example, Target 2 which calls for a preliminary assessment of all plant species and Target 7 aiming at conserving 60 per cent of the world's threatened species *in situ* (Callmander *et al*. 2005).

The wealth of information contained in The IUCN Red List on the distribution and ecological requirements of species can be used in large-scale analyses such as identifying gaps in threatened species coverage by the existing protected area network (Rodrigues *et al*. 2004). The data has long been used at various scales in conservation planning , especially for defining specific requirements of species at

Fergusson Island Striped Possum Dactylopsila tatei *– Endangered.* © Pavel German

A significant effort is being made to increase the number of marine species on the Red List. Scorpionfish Scorpaenopsis sp. on Pavona clavus coral in the Maldives. © Jerker Tamelander

site, landscape/seascape level, and global levels. For example, Red List data are used to support the identification of site-scale conservation priorities, such as Important Bird Areas, Key Biodiversity Areas, Important Plant Areas, Ramsar Sites, and Alliance for Zero Extinction sites (Eken et al. 2004; Hoffmann et al. 2008).

The Red List also helps to inform the conservation planning of wide-ranging species for which site-based approaches are not suitable strategies. Red List data have been used in the identification of global priorities (e.g., Endemic Bird Areas) and for setting geographical priorities for conservation funding, for example the Global Environment Facility (GEF) Resource Allocation Framework, which determine each country's GEF funding allocation.

Informing conservation action for individual species

Red List data (including information on habitat requirements, threats that need to be addressed, and conservation actions that are recommended) can be used to identify species that require specific conservation action, and to help develop the conservation programmes and recovery plans. The data have also been used in the identification of Evolutionary Distinct and Globally Endangered (EDGE) species, unique animals that are often not the focus of significant conservation support (http://www.edgeofexistence.org/).

Red for Danger... Red as a 'Wake up' Call?

Biological diversity goes beyond species and encompasses ecosystems and genes. However, species remain the well-identified building blocks of biodiversity, and they are easily understood by the public and policy makers alike. By enhancing knowledge on the state of biodiversity, explaining complex species-conservation issues, and highlighting species at risk, The IUCN Red List is attracting increasing attention to

the important role that species play if ecosystems are to function properly.

The Red List is increasingly informing academic work (e.g. school home-work assignments, undergraduate essays and dissertations) and many key websites rely on information from The IUCN Red List to help spread their messages and educate the world about conservation issues. Examples include ARKive, Encyclopedia of Life (EOL), Wikipedia, Alliance for Zero Extinction (AZE) and many more. IUCN strives to make The IUCN Red List an important companion to other sites, thus increasing their ability to have conservation impact. The Red List also provides a solid factual basis when drafting funding proposals which seek support for meaningful conservation work.

Guiding scientific research

A significant number of species are listed in the Data Deficient Category and could well

The Golden Mantella Mantella aurantiaca - Critically Endangered - has a very restricted distribution in east-central Madagascar. Amphibians are one of the most threatened groups of species worldwide. © Jean-Christophe Vié

be threatened. These species represent a priority for future research including species-specific survey work and research into threatening processes across multiple species. The Red List is therefore used to identify species-specific survey work and ecological studies that need to be done. Using data gaps identified in the assessment process helps guide research and funding opportunities.

The IUCN Red List data also highlight general overarching threatening processes, such as emerging threats like climate change. The use of these data could greatly improve the quality of models predicting the impacts of climate change on biodiversity (Foden *et al*. this volume).

Guidelines for data use
The IUCN Red List is not intended to be used alone as a system for setting conservation priorities. Red List assessments simply measure the relative extinction risk faced by species, subspecies, or subpopulations. The Red List Category is not on its own sufficient to determine priorities for conservation action. To set conservation priorities, additional information must be taken into account (Miller *et al*. 2006)

The IUCN Red List is freely available; however, it contains copyrighted material and/or other proprietary information that are protected by intellectual property agreements and copyright laws and regulations worldwide. In order to obtain the information, users are requested to comply with a User Agreement and in so-doing are granted a license to use, download and print the materials contained in the Red List solely for conservation or educational purposes, scientific analyses, and research.

References

Baillie, J.E.M., Collen, B., Amin, R., Akçakaya, H.R., Butchart, S.H.M., Brummit, N., Meagher, T.R., Ram, M., Hilton-Taylor, C. and Mace, G.M. 2008. Toward monitoring global biodiversity. *Conservation Letters* 1: 18-26.

Baillie, J.E.M., Hilton Taylor, C. and Stuart, S.N. (Editors). 2004. *2004 IUCN Red List of Threatened Species. A Global Species Assessment*. IUCN Gland, Switzerland and Cambridge, UK.

Brooks, T. and Kennedy, E. 2004. Biodiversity barometers. *Nature* 431: 1046-1047.

Butchart, S.H.M., Statterfield, A.J., Bennun, L.A., Shutes, S.M., Akçakaya, H.R., Baillie, J.E.M., Stuart, S.N., Hilton-Taylor, C. and Mace, G.M. 2004. Measuring Global Trends in the Status of Biodiversity: Red List Indices for Birds. *PloS Biology* 2(12): e383.

Butchart, S.H.M., Akçakaya, H.R., Kennedy, E. and Hilton-Taylor, C. 2005a. Biodiversity indicators based on trends in conservation status: strengths of the IUCN Red List Index. *Conservation Biology* 20(2): 579-581.

Butchart, S.H.M., Stattersfield, A.J., Baillie, J., Bennun, L.A., Stuart, S.N., Akçakaya, H.R., Hilton-Taylor, C. and Mace, G.M. 2005b. Using Red List indices to measure progress towards the 2010 target and beyond. *Philosophical Transactions of the Royal Society* 360: 255-268.

Butchart, S.H.M., Stattersfield, A.J. and Brooks, T.M. 2006. Going or gone: defining "Possibly Extinct" species to give a truer picture of recent extinctions. Bulletin of the British Ornithologists' Club 126A: 7-24.

Butchart, S.H.M., Akçakaya, H.R., Chanson, J., Baillie, J.E.M., Collen, B., Quader, S., Turner, W.R., Amin, R., Stuart, S.N. and Hilton-Taylor, C. 2007. Improvements to the Red List Index. *PLoS ONE* 2(1) :e140. doi:10.1371/journal.pone.0000140.

Callmander, M.W., Schatz, G.E. and Lowry, P.P. 2005. IUCN Red List assessment and the Global Strategy for Plant Conservation: taxonomists must act now. *Taxon* 54(4): 1047-1050.

Carpenter, K.E., Abrar, M., Aeby, G., Aronson, R.B., Banks, S., Bruckner, A., Chiriboga, A., Cortés, J., Delbeek, J.C., DeVantier, L., Edgar, G.J., Edwards, A.J., Fenner, D., Guzmán, H.M., Hoeksema, B.W. *et al*. 2008. One third of reef-building corals face elevated extinction risk from climate change and local impacts. *Science* 321: 560-563.

Colyvan, M., Burgman, M.A., Todd, C.R., Akçakaya, H.R. and Boek, C. 1999. The treatment of uncertainty and the structure of the IUCN threatened species categories. *Biological Conservation* 89: 245-249.

De Grammont, P.C. and Cuarón, A.D. 2006. An evaluation of threatened species categorization systems used on the American continent. *Conservation Biology* 20(1): 14-27.

Eken, G., Bennun, L., Brooks, T.M., Darwall, W., Fishpool, L.C.D., Foster, M., Knox, D., Langhammer, P., Matiku, P., Radford, E., Salaman, P., Sechrest, W., Smith, M.L., Spector, S. and Tordoff, A. 2004. Key

The Alpine Ibex Capra ibex *is endemic to Europe. It was driven very close to extinction in the early 19th century and is now listed as Least Concern.*
© Jean-Christophe Vié

biodiversity areas as site conservation targets. *BioScience* 54: 1110-1118.

Fitter, R. and Fitter, M. 1987. *The road to extinction: problems of categorizing the status of taxa threatened with extinction*. IUCN Gland, Switzerland and Cambridge, UK.

Gärdenfors, U., Hilton-Taylor, C., Mace, G.M. and Rodríguez, J.P. 2001. The application of IUCN Red List Criteria at regional levels. *Conservation Biology* 15(5): 1206-1212.

Hilton-Taylor, C. (Compiler). 2000. *2000 IUCN Red List of Threatened Species*. IUCN Gland, Switzerland and Cambridge, UK.

Hoffmann, M., Brooks, T.M., da Fonseca, G.A.B., Gascon, C., Hawkins, A.F.A., James, R.E., Langhammer, P., Mittermeier. R.A., Pilgrim, J.D., Rodrigues, A.S.L. and Silva, J.M.C. 2008. Conservation planning and the IUCN Red List. *Endangered Species Research* doi: 10.3354/esr00087.

IUCN. 1994. *IUCN Red List Criteria*. Prepared by the IUCN Species Survival Commission. IUCN, Gland, Switzerland.

IUCN. 2001. *IUCN Red List Categories and Criteria: Version 3.1*. IUCN Species Survival Commission. IUCN, Gland, Switzerland and Cambridge, UK.

IUCN. 2003. *Guidelines for Application of IUCN Red List Criteria at regional levels: Version 3.0*.

IUCN Species Survival Commission. IUCN, Gland Switzerland and Cambridge, UK.

Lamoreux, J., Akçakaya, H.R., Bennun, L., Collar, N.J., Boitani, L., Brackett, D., Bräutigam, A., Brooks, T.M., Fonseca, G.A.B. Mittermeier, R.A., Rylands, A.B., Gärdenfors, U., Hilton-Taylor, C., Mace, G., Stein, B.A. and Stuart, S. 2003. Value of the IUCN Red List. *Trends in Ecology & Evolution* 18: 214-215.

Mace, G.M., Collar, N.J., Gaston, K.J., Hilton-Taylor, C., Akçakaya, H.R., Leader-Williams, N., Milner-Gulland, E.J. and Stuart, S.N. 2008. Quantification of extinction risk: IUCN's system for classifying threatened species. *Conservation Biology* 22(6): 1424-1442.

Miller, R.M., Rodríguez, J.P., Aniskowicz-Fowler, T., Bambaradeniya, C., Boles, R., Eaton, M.A., Gärdenfors, U., Keller, V., Molur, S., Walker, S. and Pollock, C. 2006. Extinction risks and conservation priorities. *Science* 313: 441.

Rodrigues, A.S.L., Andelman, S.J., Bakarr, M.I., Boitani, L., Brooks, T.M., Cowling, R.M., Fishpool, L.D.C., Fonseca, G.A.B., Gaston, K.J., Hoffmann, M., Long, J.S., Marquet, P.A., Pilgrim, J.D., Pressey, R.L., Schipper, J. *et al.* 2004. Effectiveness of the global protected

area network in representing species diversity. *Nature* 428: 640-643.

Rodrigues, A.S.L., Pilgrim, J.D., Lamoreux, J.F., Hoffmann, M. and Brooks, T.M. 2006. The value of the IUCN Red List for conservation. *Trends in Ecology and Evolution* 21(2): 71-76.

Salafsky, N., Salzer, D., Statterfield, A.J., Hilton-Taylor, C., Neugarten, R., Butchard, S.H.M., Collen, B., Cox, N., Master, L.L., O'Connor, S. and Wilkie, D. 2008. A standard Lexicon for biodiversity conservation: unified classifications of threats and actions. *Conservation Biology* 22(4): 897-911.

Schipper, J., Chanson, J., Chiozza, F., Cox, N., Hoffmann, M., Katariya, V., Lamoreux, J., Rodrigues, A.S.L., Stuart, S.N., Temple, H.J., Baillie, J., Boitani, L., Lacher, T.E., Mittermeier, R.A., Smith, A.T. *et al.* 2008. The Status of the world's terrestrial and aquatic mammals. *Science* 322(5899): 225-230.

Stuart, S.N., Chanson, J.S., Cox, N.A., Young, B., Rodrigues, A.S.L., Fischman, D.L. and Waller, R.W. 2004. Status and trends of amphibian declines and extinctions worldwide. *Science* 306: 1783-1786.

Temple, H.J. and Terry, A. 2007. *The Status and Distribution of European Mammals*. Luxembourg: Office for Official Publications of the European Communities.

State of the world's species

Craig Hilton-Taylor, Caroline M. Pollock, Janice S. Chanson, Stuart H.M. Butchart, Thomasina E.E. Oldfield and Vineet Katariya

A species rich world

The magnitude and distribution of species that exist today is a product of more than 3.5 billion years of evolution, involving speciation, radiation, extinction and, more recently, the impacts of people. Estimates of the total number of eukaryotic species in existence on Earth today vary greatly ranging from 2 million to 100 million, but most commonly falling between 5 million and 30 million (May 1992, Mace et al. 2005), with a best working estimate of about 8 to 9 million species (Chapman 2006). But of these, just under 1.8 million are estimated to have been described (Groombridge and Jenkins 2002, Chapman 2006) although it has been argued that the number may be closer to 2 million (Peeters et al. 2003).

While scientists debate how many species exist, there are growing concerns about the status of biodiversity, particularly population declines (e.g., the Living Planet Index which monitors population trends in 1,686 animal species shows an overall decline of 30% for the period 1970 to 2005 (Loh et al. 2008)) and the increasing rates of extinction of both described and undescribed species as a direct and indirect result of human activities. Although only a very small proportion (2.7%; Table 1) of the world's described species have been assessed so far, The IUCN Red List provides a useful snapshot of what is happening to species around the world today and highlights the urgent need for conservation action.

The 2008 IUCN Red List

There have been some marked increases in the taxonomic coverage of The IUCN Red List in the last eight years (Vié et al. this volume). In 2000, The IUCN Red List included assessments for 16,507 species, 11,406 of which were listed as threatened (Hilton-Taylor 2000); in 2004 the list included 38,047 species, 15,589 of which were threatened (Baillie et al. 2004); and in 2008 the list includes 44,838 species, 16,928 of which are threatened (Box 1, Table 1). However, the conservation status for most of the world's species remains poorly known, and there is a strong bias in those that have been assessed so far towards terrestrial vertebrates and plants and in particular those species found in biologically well-studied parts of the world. Efforts are underway to rectify these biases (Darwall et al., Policoro et al. and Collen et al. this volume).

Comprehensive assessments (in which every species has been evaluated) are now available for an increased number of taxonomic groups, namely amphibians, birds, mammals, cycads and conifers, warm water reef-forming

The Endangered Western Prairie Fringed Orchid Platanthera praeclara is declining across much of its range in North America as a result of habitat loss and degradation owing to agricultural expansion and intensification. © Jim Fowler

Box 1. Summary of the 2008 IUCN Red List update

The 2008 update of The IUCN Red List (as released on 6th October 2008) includes conservation assessments for 44,838 species (see Table 1 for break-down):

- There are 869 recorded extinctions, with 804 species listed as Extinct and 65 listed as Extinct in the Wild;

- The number of extinctions increases to 1,159 if the 290 Critically Endangered species tagged as 'Possibly Extinct' are included;

- 16,928 species are threatened with extinction (3,246 are Critically Endangered, 4,770 are Endangered and 8,912 are Vulnerable);

- 3,796 species are listed as Near Threatened*;

- 5,570 species have insufficient information to determine their threat status and are listed as Data Deficient;

- 17,675 species are listed as Least Concern, a listing which generally indicates that these have a low probability of extinction, but the category is very broad and includes species which may be of conservation concern (e.g., they may have very restricted ranges but with no perceived threats or their populations may be declining but not fast enough to qualify for a threatened listing).

Note that The IUCN Red List is a biased sample of the world's species, and for the incompletely assessed groups, there is a a general tendency to assess species that are more likely to be threatened. It is therefore not possible to take the Red List as a whole (in which 38% of listed species are threatened), and say that this means that 38% of all species in the world are likely to be threatened.

* Includes species listed as Conservation Dependent (LR/cd), an old Red List Category which is now subsumed under the Near Threatened category.

---

corals, freshwater crabs, and groupers. In addition, taxonomic coverage is being broadened through a randomized sampled approach which provides representative samples (Collen et al. this volume). Closer examination of some of these taxonomic groups reveals that the proportions of threatened species differ markedly between groups, with the percentage threatened ranging from 12% for birds to 52% for cycads (Figure 1). Generally, it seems that the more mobile groups (birds and dragonflies) are less threatened, although once the

status of the Data Deficient dragonflies is resolved that group may have a much higher proportion of threatened species. Currently the two groups with the highest proportions of threatened species are the amphibians and cycads. Species in these groups generally are less mobile and have smaller ranges and are hence more easily impacted by threats e.g., a pathogenic disease (chytridiomycosis) caused by the chytrid fungus *Batrachochytrium dendrobatis* in the case of amphibians, and illegal collection in the case of the cycads.

For further discussion of the results for those taxonomic groups where a randomized sampling approach has been used (reptiles, freshwater fishes, dragonflies and freshwater crabs) see Collen et al. (this volume); the freshwater taxa are also covered in more detail by Darwall et al. (this volume); and the warm water reef-building corals are discussed along with other marine species by Polidoro et al. (this volume).

In addition to the species level assessments, the 2008 IUCN Red List also includes 1,804 assessments of infra-specific taxa (i.e., taxa below the level of a species) or discrete subpopulations, of which 1,197 (66%) are listed as threatened. These assessments are useful, particularly in the case of widespread Least Concern species, for helping to draw conservation attention to those parts of their geographic ranges where they are threatened.

The primary function of The IUCN Red List is not to document extinctions, but rather to draw attention to those species facing a high risk of extinction in the wild

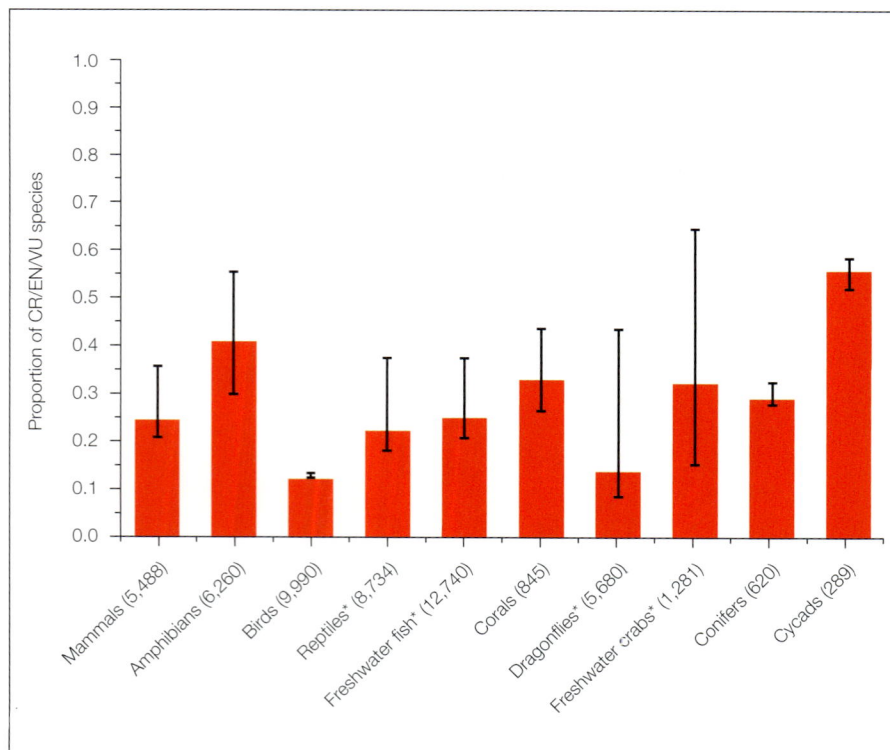

Figure 1. Proportion of species threatened with extinction in different taxonomic groups. Asterisks indicate those groups in which estimates are derived from a randomized sampling approach. The estimates assume that Data Deficient species are equally threatened as non-Data Deficient species; error bars show minimum and maximum estimates if all Data Deficient species are Least Concern or threatened, respectively. Numbers on the horizontal axis indicate the total number of described species in each group. Corals include only warm water reef-building species.

	Estimated Number of described species[7]	Number of species evaluated	Number of threatened species[8]	Number threatened, as % of species described[8]	Number threatened, as % of species evaluated[8,9]
Vertebrates					
Mammals[1]	5,488	5,488	1,141	21%	21%
Birds	9,990	9,990	1,222	12%	12%
Reptiles	8,734	1,385	423	5%	31%
Amphibians[2]	6,347	6,260	1,905	30%	30%
Fishes	30,700	3,481	1,275	4%	37%
Subtotal	**61,259**	**26,604**	**5,966**	**10%**	**22%**
Invertebrates					
Insects	950,000	1,259	626	0%	50%
Molluscs	81,000	2,212	978	1%	44%
Crustaceans	40,000	1,735	606	2%	35%
Corals	2,175	856	235	11%	27%
Arachnids	98,000	32	18	0%	56%
Velvet Worms	165	11	9	5%	82%
Horseshoe Crabs	4	4	0	0%	0%
Others	61,040	52	24	0%	46%
Subtotal	**1,232,384**	**6,161**	**2,496**	**0.20%**	**41%**
Plants[3]					
Mosses[4]	16,000	95	82	1%	86%
Ferns and allies[5]	12,838	211	139	1%	66%
Gymnosperms	980	910	323	33%	35%
Dicotyledons	199,350	9,624	7,122	4%	74%
Monocotyledons	59,300	1,155	782	1%	68%
Green Algae[6]	3,962	2	0	0%	0%
Red Algae[6]	6,076	58	9	0%	16%
Subtotal	**298,506**	**12,055**	**8,457**	**3%**	**70%**
Others					
Lichens	17,000	2	2	0%	100%
Mushrooms	30,000	1	-	0%	100%
Brown Algae[6]	3,040	15	6	0%	40%
Subtotal	**50,040**	**18**	**8**	**0.02%**	**50%**
TOTAL	**1,642,189**	**44,838**	**16,928**	**1%**	**38%**

Table 1. Numbers and proportions of species assessed and species assessed as threatened on the 2008 IUCN Red List by major taxonomic group.

so that appropriate conservation actions can be taken (Mace *et al*. 2008). Given this focus together with the uneven taxonomic coverage and the fact that it may take many years to prove that a species is truly Extinct and can be listed as such on The IUCN Red List (Baillie *et al*. 2004), the number of extinctions on the Red List is significantly under-recorded. In order to record probable extinctions a 'Possibly Extinct' tag has been introduced which is used only against Critically Endangered listings (Butchart *et*

Notes:

1. The number of described and evaluated mammals excludes domesticated species like sheep (*Ovis aries*), goats (*Capra hircus*), Dromedary (*Camelus dromedarius*), etc.

2. It should be noted that for certain amphibian species endemic to Brazil, it has not yet been possible to reach agreement on the Red List Categories between the Global Amphibian Assessment (GAA) Coordinating Team, and the experts on the species in Brazil. The numbers for Amphibians displayed here include those that were agreed at the GAA Brazil workshop in April 2003. However in the subsequent consistency check conducted by the GAA Coordinating Team, many of the assessments were found to be inconsistent with the approach adopted elsewhere in the world, and a "consistent Red List Category" was also assigned to these species. The "consistent Red List Categories" are yet to be accepted by the Brazilian experts; therefore the original workshop assessments are retained here. However, in order to ensure comparability between results for amphibians with those for other taxonomic groups, the data used in various analyses (e.g., Baillie *et al*. 2004; Stuart *et al*. 2008; the Global Amphibians analysis on the Red List web site) are based on the "consistent Red List Categories". Therefore, numbers for Amphibians in Table 1 above will not completely match numbers that appear in other analyses, including the analysis later in this chapter.

3. The plant numbers do not include species from the *1997 IUCN Red List of Threatened Plants* (Walter and Gillett 1998) as those were all assessed using the pre-1994 IUCN system of threat categorization. Hence the numbers of threatened plants are very much lower when compared to the 1997 results. The results from this Red List and the 1997 Plants Red List should be combined together when reporting on threatened plants.

4. Mosses include the true mosses (Bryopsida), the hornworts (Anthocerotopsida), and liverworts (Marchantiopsida).

5. Ferns and allies include the club mosses (Lycopodiopsida), spike mosses (Sellaginellopsida), quillworts (Isoetopsida), and true ferns (Polypodiopsida).

6. Seaweeds are included in the green algae (Chlorophyta), red algae (Rhodophyta), and brown algae (Ochrophyta).

7. The sources used for the numbers of described plant and animal species are listed in Appendix 3.

8. The numbers and percentages of species threatened in each group do not mean that the remainder are all not threatened (i.e., are Least Concern). There are a number of species in many of the groups that are listed as Near Threatened or Data Deficient (see Appendices 4-8). These numbers also need to be considered in relation to the number of species evaluated as shown in column two (see note 9).

9. Apart from the mammals, birds, amphibians and gymnosperms (i.e., those groups completely or almost completely evaluated), the numbers in the last column are gross over-estimates of the percentage threatened due to biases in the assessment process towards assessing species that are thought to be threatened, species for which data are readily available, and under-reporting of Least Concern species. The true value for the percentage threatened lies somewhere in the range indicated by the two right-hand columns. In most cases this represents a very broad range; the percentage of threatened insects for example, lies somewhere between 0.07% and 50%. Hence, although 38% of all species on The IUCN Red List are listed as threatened, this percentage needs to be treated with extreme caution given the biases described above.

The Radiated Tortoise Astrochelys radiata *is found only on Madagascar. In 2008 its Red List status changed from Vulnerable to Critically Endangered. Wild Radiated Tortoises are collected for the international pet trade, and also for local use (food and pets), which is of greater concern for the species. Habitat loss due to agricultural expansion and invasive plant species also threaten the remaining wild population. © Anders Rhodir*

The very rare Peacock Parachute Spider Poecilotheria metallica *is Critically Endangered. Habitat loss through logging in its only known location (Eastern Ghats of Andhra Pradesh, India) is the main threat to this species.* © Sanjay Molur

al. 2006a, IUCN Standards and Petitions Working Group 2008). If the species tagged as 'Possibly Extinct' are included, then the number of probable extinctions recorded on The IUCN Red List increases from 869 to 1,159 species.

Highlights of the 2008 IUCN Red List

Some of the highlights of the 2008 update of The IUCN Red List include the following:

- A complete reassessment of the world's mammals showed that nearly one-quarter (22%) of the world's mammal species are globally threatened or Extinct and 836 (15%) are Data Deficient (Schipper et al. 2008).

- The addition of 366 new amphibian species, many listed as threatened, and the confirmed extinction of two species,

which reaffirms the extinction crisis faced by amphibians; nearly one-third (31%) are threatened or Extinct and 25% are Data Deficient.

- A complete reassessment of the world's birds indicates that more than one in eight (13.6%) are considered threatened or Extinct; birds are one of the best-known groups with less than 1% being listed as Data Deficient (BirdLife International 2008a).

- For the first time 845 species of warm water reef-building corals have been included on the Red List with more than one-quarter (27%) listed as threatened and 17% as Data Deficient (Carpenter et al. 2008).

- All 161 species of groupers are now assessed; over 12% of these highly

sought after luxury live food fish species are threatened with extinction as a result of unsustainable fishing; a further 30% are Data Deficient.

- All 1,280 species of freshwater crabs have been assessed, 16% of which are listed as threatened with extinction, but a further 49% are Data Deficient (Cumberlidge et al. 2009).

- 359 freshwater fishes endemic to Europe, with 24% listed as threatened and only 4% listed as Data Deficient (Kottelat and Freyhof 2007).

The 2008 IUCN Red List also includes a number of notable new individual species assessments, for example 14 tarantula assessments from India, 12 threatened freshwater fishes from Lake Dianchi in China, orchids from the Americas, a striking *Rafflesia* species (a close relative of which has the largest single flower of any flowering plant in the world) from the Philippines, and a bumble bee which has undergone dramatic declines in North America and exemplifies what is happening to other key pollinators world-wide.

The status of amphibians, birds, mammals and plants

In previous analyses of the Red List, the general analysis has looked at facts, figures and trends across all the major taxonomic groups. However, a more thematic approach has been adopted in this review and hence because freshwater and marine groups are covered in other chapters, the main focus of the rest of this chapter is on the terrestrial groups. In particular an analysis is presented of the three comprehensively assessed vertebrate groups for which we have a relatively rich knowledge, namely the amphibians, birds and mammals. Plants are also included, but are not analyzed to the same extent as the vertebrates because much of the supporting documentation for such an analysis is not yet available. The only invertebrate groups for which there is reasonable assessment coverage are the corals, dragonflies and freshwater crabs, but as these are all covered in other chapters, they are not discussed any further here.

Amphibians

CURRENT STATUS

The first comprehensive assessment of the conservation status of all amphibians was completed in 2004, and the results were included in the 2004 IUCN Red List. The amphibian assessment is one of several initiatives led by IUCN and its partners with the aim of rapidly expanding the geographic and taxonomic coverage of The IUCN Red List. Since 2004 there have been two updates of the amphibian data, one in 2006, and the most recent in 2008.

Ninety-nine per cent of all known amphibian species (6,260 species; see Table 1) have been assessed, and of these, nearly one-third (32.4%) are globally threatened or Extinct, representing 2,030 species (Figure 2, Appendix 4). Thirty-eight are considered to be Extinct (EX), and one Extinct in the Wild (EW). Another 2,697 species are not considered to be

Rafflesia magnifica *is among the group of plants that produce the largest single flowers in the world. Endemic to the Philippines, only a few individuals of R. magnifica have been recorded, all of them male. The species is listed as* Critically Endangered. © H. Calalo

threatened at present, with 381 being listed as Near Threatened (NT) and 2,316 listed as Least Concern (LC), while sufficient information was not available to assess the status of an additional 1,533 species (Data Deficient (DD)). It is predicted that a significant proportion of these Data Deficient species are likely to be globally threatened.

Documenting population trends is key to assessing species status, and a special effort was made to determine which amphibians are declining, stable, or increasing. The assessment found declines to be widespread among amphibians, with 42.5% of species reported to be in decline. In contrast, 26.6% of species appear to be stable and just 0.5% are increasing. Because trend information is not available for 30.4% of species, the percentage of amphibians in decline may be considerably higher.

Extinctions are often difficult to confirm. Using the most conservative approach to documenting extinctions, just 38 amphibians are known to have become Extinct since the year 1500. Of greater concern, however,

are the many amphibians that can no longer be found. Until exhaustive surveys to confirm their disappearance can be carried out, these species cannot be classified as Extinct, but rather are flagged as 'Possibly Extinct' within the Critically Endangered category. Currently there are 120 such 'Possibly Extinct' amphibian species.

Unfortunately, there is strong evidence that the pace of extinctions is increasing.

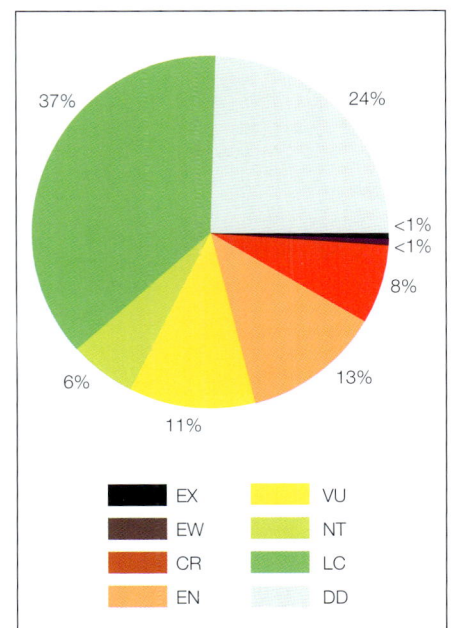

Figure 2. *IUCN Red List assessment for 6,260 amphibian species.*

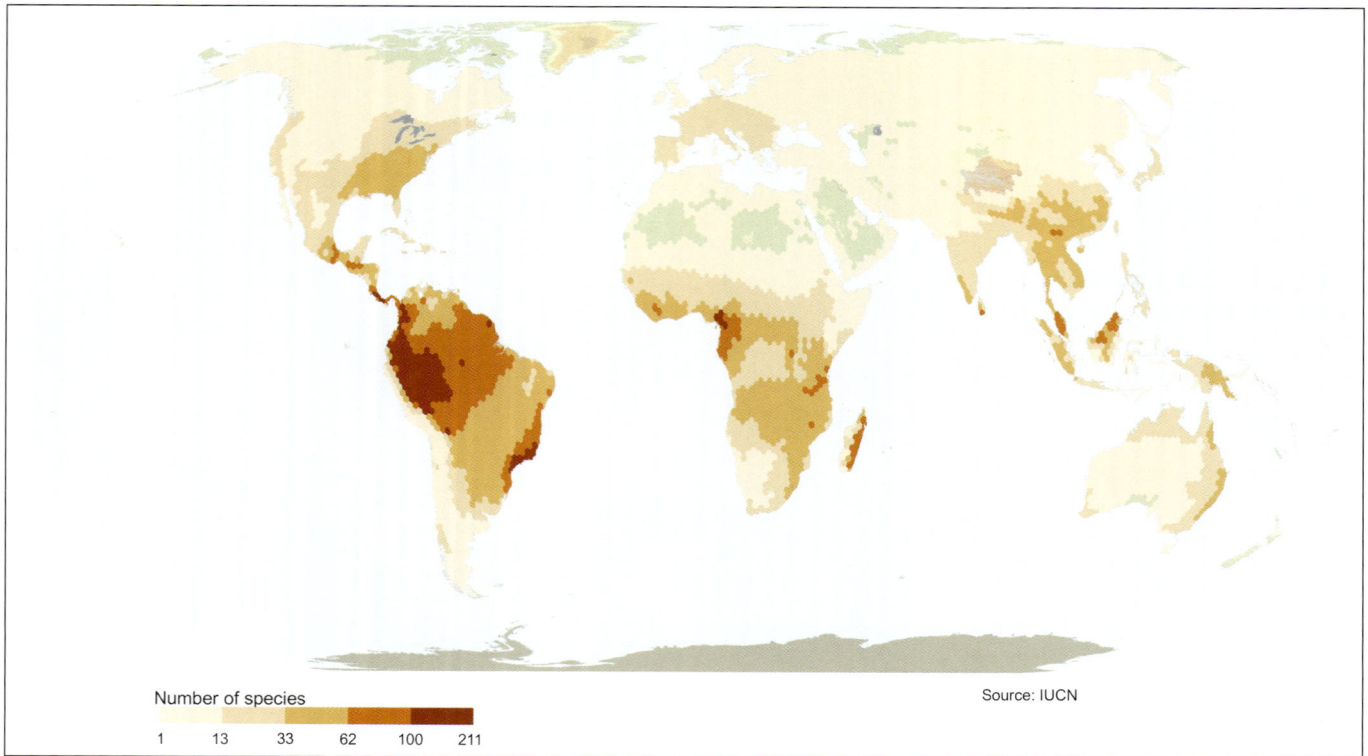

Number of species

| 1 | 13 | 33 | 62 | 100 | 211 |

Source: IUCN

Figure 3. *Global diversity of amphibian species.*

Of the 38 known extinctions, 11 have occurred since 1980, including such species as the Golden Toad *Incilius periglenes* of Monteverde, Costa Rica. Among those amphibians regarded as 'Possibly Extinct', most have disappeared and have not been seen since 1980. Fortunately, a few amphibians that previously were thought to be Extinct have been rediscovered. For example, *Atelopus cruciger* was not seen in its native Venezuela after 1986, until a tiny population was found in 2003.

GEOGRAPHIC PATTERNS

Amphibian diversity
Global patterns of amphibian diversity are shown in Figure 3. This map clearly indicates certain areas of high amphibian diversity, including tropical South America and tropical West Africa. In contrast to the usual pattern of high species diversity occurring in the tropics, the southeastern United States is a global centre for amphibian diversity, being particularly rich in salamanders. However, the problem of uneven survey efforts around the world complicates interpretation of this map. Regions such as Indonesia, New Guinea and the Congo Basin are especially likely

to be under represented on this map due to lack of adequate surveys.

Looking at amphibian diversity from a country perspective, Brazil, with at least 798 species, has the greatest number of amphibians of any country on Earth, followed by Colombia. Table 2 lists the 20 most diverse countries and reveals some interesting findings, although, these results must be considered in relation to the level of survey effort and the size of the countries. Both Colombia and Brazil have benefitted from extensive survey efforts in recent decades, and although both countries can be expected to add significantly more species to their totals, the levels of increase are likely to be less than in some of the other highly diverse countries. In South America, Peru in particular is relatively poorly sampled and is almost certain to rise very substantially in its species total, and can be predicted to pass the level of Ecuador before too long. The diversity in Ecuador is, however, remarkable for such a small country.

Table 2. *Top twenty countries* with the most amphibian species.*

Rank	Country	Number of amphibians
1	Brazil	798
2	Colombia	714
3	Ecuador	467
4	Peru	461
5	Mexico	364
6	Indonesia	363
7	China	333
8	Venezuela	311
9	United States	272
10	Papua New Guinea	266
11	India	252
12	Madagascar	242
13	Bolivia	230
14	Australia	223
15	Congo, The Democratic Republic of the	215
16	Malaysia	212
17	Cameroon	199
18	Panama	197
19	Costa Rica	186
20	Tanzania, United Republic of	178

* The country and territory names used in Tables 2–10 are based on the short country names specified by the International Organization for Standardization (ISO) Maintenance Agency for ISO 3166 country codes (see http://www.iso.org/iso/country_codes/ iso_3166_code_lists/english_country_names_and_ code_elements.htm).

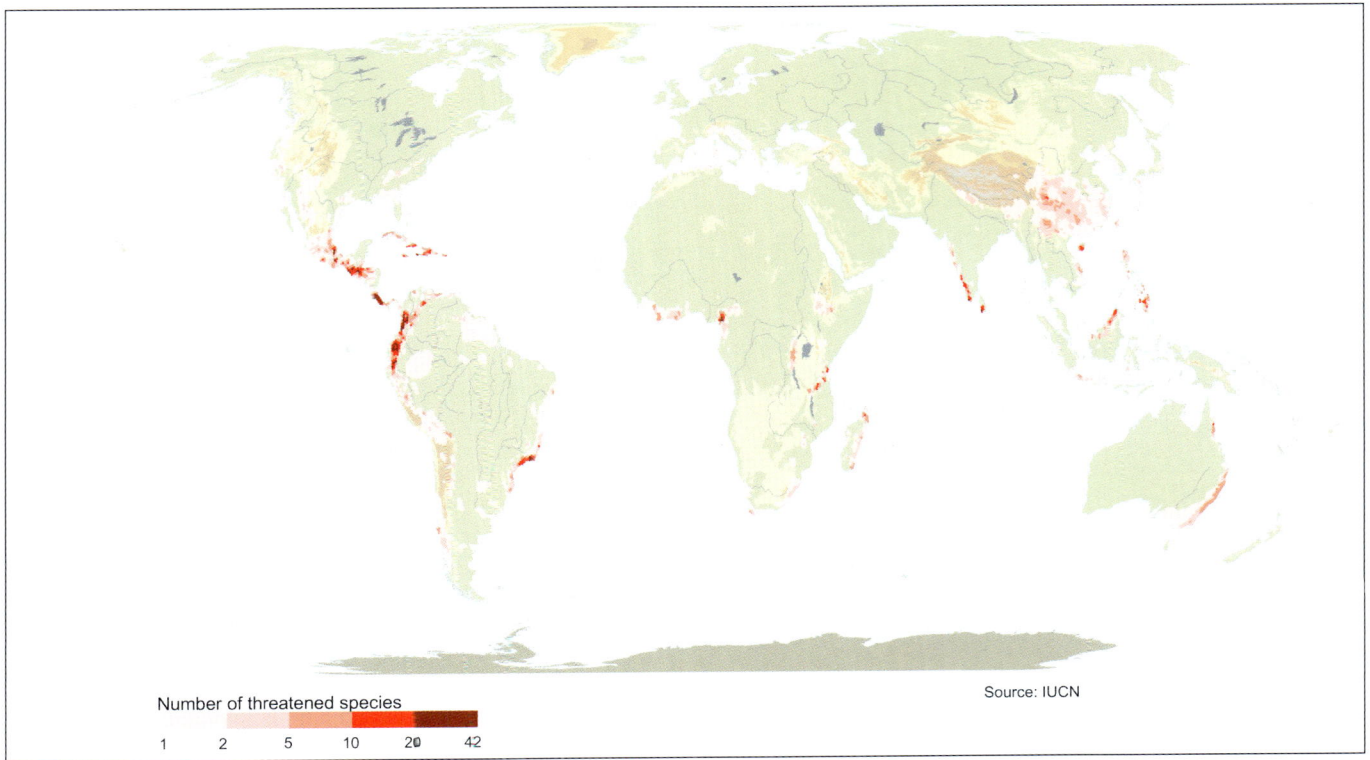

Number of threatened species

1   2   5   10   20   42

Source: IUCN

Figure 4. *Global distribution of threatened amphibians.*

Among the Old World countries, the level of survey effort is often much lower than in the Americas. Indonesia can be predicted to be the richest country outside the Americas, but it is doubtful if even half of its species are yet known. After more dedicated survey efforts, Indonesia may prove to have a level of amphibian diversity comparable with Brazil and Colombia. Very large increases in species totals can also be predicted for Papua New Guinea and The Democratic Republic of the Congo, the latter country having received almost no amphibian survey work in the last 40 years.

Countries that are likely to pass the 200 species mark include Cameroon, Panama, Costa Rica and the United Republic of Tanzania. The United States and Australia can be predicted to fall down the ranking over time, though the United States along with Mexico will remain the most important countries for salamander diversity.

Geography of threatened amphibian species

A map showing the global distribution of threatened amphibians (Figure 4) reveals very different patterns compared with the depictions of overall species diversity (Figure 3). The greatest concentration of such species, including well over half of the currently known threatened amphibians, is in a relatively limited area running from southern Mexico south to Ecuador and Venezuela, and in the Greater Antilles. This region is dominated by species with small ranges, often living in montane areas. Many of these species have been subjected to severe habitat loss and exposure to the fungal disease chytridiomycosis (Stuart *et al.* 2008).

Other important concentrations of threatened species are in the Atlantic Forests of southern Brazil, the Upper Guinea forests of western Africa, the forest of western Cameroon and eastern Nigeria, the Albertine Rift of central Africa, the Eastern Arc Mountains of the United Republic of Tanzania, Madagascar, the Western Ghats of India, Sri Lanka, central and southern China, the Bornean parts of Indonesia and Malaysia, the Philippines and eastern Australia.

Table 3 lists the 20 countries with the highest number of threatened amphibians. These countries are in many cases different to those listed in Table 2, suggesting that either amphibians in some countries are more susceptible to threats,

Rank	Country	Number of threatened amphibians
1	Colombia	214
2	Mexico	211
3	Ecuador	171
4	Brazil	116
5	Peru	96
6	China	92
7	Guatemala	80
8	Venezuela	72
9	India	65
10	Madagascar	64
11	Costa Rica	59
	Honduras	59
13	United States	56
14	Cameroon	53
	Sri Lanka	53
16	Tanzania, United Republic of	50
17	Panama	49
	Cuba	49
19	Australia	48
	Philippines	48

Table 3. *Countries with the largest number of threatened amphibian species.*

Bolitoglossa franklini *is an Endangered salamander from Mexico and Guatemala. Its range is becoming severely fragmented as forest habitats are lost to agricultural lands and human settlements.* © Gabriela Parra

that threats vary between countries, or that there are other factors influencing the distribution of threatened species.

The countries listed in Table 3 have a particularly great responsibility for protecting the world's threatened amphibians. Colombia, the second most diverse country, has the highest number of threatened species. The major threat to amphibians in Colombia is habitat loss although there have also been many declines due to chytridiomycosis. The dramatic topography of the Andes means that many of the amphibians found there have very restricted ranges making them more vulnerable to threatening processes. Brazil, the most diverse country, is ranked only fourth for number of species threatened, most of which are in the Atlantic Forest region, and has a significantly lower percentage of its amphibians threatened than the global average.

In Table 3, only the number of threatened species is given, and the number of Extinct species has been excluded. This is to highlight those countries that currently have the greatest responsibility towards protecting globally threatened species. If we also consider the Extinct species, Sri Lanka, with 21 Extinct amphibians,

would jump from being 14th on the list to 8th, ahead of several countries with much greater amphibian diversity. Sri Lanka is only the 28th most diverse country for amphibians.

Considering the percentage of a country's amphibian fauna that is threatened or Extinct provides a stark contrast to the previous table, which focuses on the number of threatened species. Table 4 lists the countries with the highest percentage of threatened and Extinct amphibians.

The top five countries are all in the Caribbean, and at least 70% of all the amphibians in these countries are threatened (no species are listed as Extinct for these five countries at present, but nine are tagged as 'Possibly Extinct'). Compared with other regions, the Caribbean stands out with by far the highest percentage of threatened species. This is mostly a result of extensive habitat loss as well as some incidence of chytridiomycosis, particularly in Puerto Rico.

In Mexico, ranked fifth for diversity, but second for the number of threatened species, more than 50% of amphibians are threatened (no species are considered

Rank	Country	% threatened & Extinct
1	Haiti	92.0
2	Dominican Republic	83.3
3	Jamaica	81.0
4	Cuba	80.3
5	Puerto Rico	73.7
6	Sri Lanka	70.5
7	Mexico	58.0
8	Guatemala	57.1
9	Seychelles	54.5
10	Honduras	48.8
11	Philippines	48.0
12	Ecuador	37.0
13	Chile	36.2
14	Japan	35.7
15	Turkey	34.5
16	Costa Rica	33.3
17	El Salvador	31.3
18	Colombia	30.0
19	Taiwan, Province of China	29.4
20	Tanzania, United Republic of	28.1

Note: only countries with 10 or more species are included in the analysis.

Table 4. *Countries with the highest percentage of threatened and Extinct amphibians.*

Madagascan Mantella Mantella madagascariensis *is a Vulnerable amphibian from Madagascar. This species lays its eggs on the ground.* © Jean-Christophe Vié

Extinct at present, but 26 are tagged as 'Possibly Extinct'). Severe habitat loss and the outbreak of chytridiomycosis in some regions are the main threats. Most of the other countries in Table 4 are in Mesoamerica or South America, with the main causes of threat here also being chytridiomycosis and habitat loss.

Sri Lanka is the highest ranked country outside of Mesoamerica or South America with over 70% of species in this country either threatened or Extinct, primarily as a result of habitat loss.

HABITAT PREFERENCES

A summary of the most important habitats for amphibians is shown in Figure 5.

The vast majority of amphibians, almost 5,000, depend on forests. Other terrestrial habitats are much less preferred by amphibians, in particular the drier habitats, such as savannas and deserts. These results are not surprising, as amphibians are well known for their preference for moist habitats.

Perhaps a more surprising result is that only 4,224 amphibians depend on freshwater during some stage of their life cycle. Amphibians are renowned for their dual lifestyle, starting off as youngsters in aquatic habitats then undergoing a metamorphosis to become terrestrial adults. However, although this is the most common life history strategy for amphibians, there are also many species that develop directly from eggs without a larval stage (and a few live-bearing species). Many of these species do not rely on freshwater habitats at any stage of their lives.

The freshwater habitats preferred by amphibians have been split depending on whether they are still or flowing, or swamp/marsh. Flowing freshwater habitats for amphibians are usually streams. Still freshwater habitats are often temporary rain pools or other small pools of freshwater. This distinction between freshwater habitats has a major influence on the likelihood that a species is threatened.

Species that are associated with flowing water are more threatened than those that use still water (indeed, stream-associated species are particularly susceptible to chytridiomycosis for reasons that are still not understood).

THREATS

A summary of the number of species affected by each threatening process is shown in Figure 6. Habitat loss and degradation are by far the greatest threats to amphibians at present, affecting nearly 61% of all known amphibians (nearly 4,000 species), including 87% of the threatened amphibian species. The vast majority of amphibians depend on tropical forest habitats, which are the same habitats that are subject to the highest rates of forest loss (Stuart *et al.* 2008).

The next most common threat to amphibians is pollution, which affects around one fifth (19%) of amphibian species overall and 29% of threatened species. These percentages are much higher than those recorded for birds or mammals (see Figures 10 and 15), but this is probably because most amphibians are semi-aquatic (Stuart *et al.* 2008).

Figure 5. Major habitat preferences of amphibians.

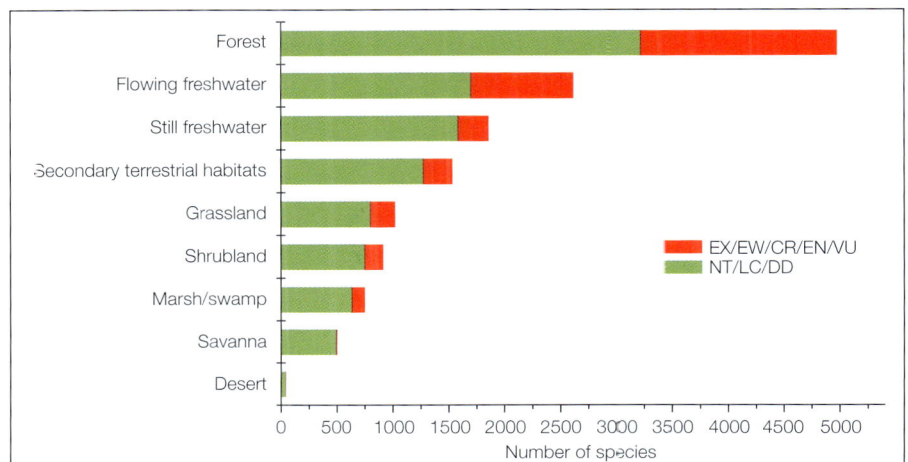

Plectrohyla dasypus *is a Critically Endangered amphibian from Honduras. The population is undergoing drastic declines as a result of chytridiomycosis.* © Silviu Petrovan

Although the disease chytridiomycosis appears to be a relatively less significant threat for amphibians, for those species affected, it can cause sudden and dramatic population declines resulting in very rapid extinction (Cunningham and Daszak 2008). In comparison, although habitat loss and degradation affect a much greater number of species, the rate at which a species declines due to these causes is usually much slower than is the case with disease. There are a number of strategies and mitigation measures that can be adopted, such as the creation of protected areas, to counter the threats of habitat loss and degradation. By contrast, there is no practical solution available as yet for dealing with chytridiomycosis in the wild; pathogens do not stop at the boundaries of protected areas.

Birds

CURRENT STATUS

Birds are probably the best known taxonomic group. Since 1988, the BirdLife International Partnership, working with a global network of experts and organizations, including the IUCN SSC bird Specialist Groups, has conducted five comprehensive assessments of birds, with the most recent assessment of all 9,990 known species being completed in 2008. Less than one per cent of bird species on the 2008 IUCN Red List have insufficient information available to be able to assess them beyond Data Deficient.

It is clear, however, that being well-studied does not provide immunity from decline and high extinction risk. More than one in seven bird species (13.6%) are globally threatened or Extinct, representing 1,360 species (Figure 7, Appendix 4). Of these, 134 species are Extinct, four species no longer occur in the wild, and a further 15 are Critically Endangered species flagged as

Figure 6. *Major threats to amphibians.*

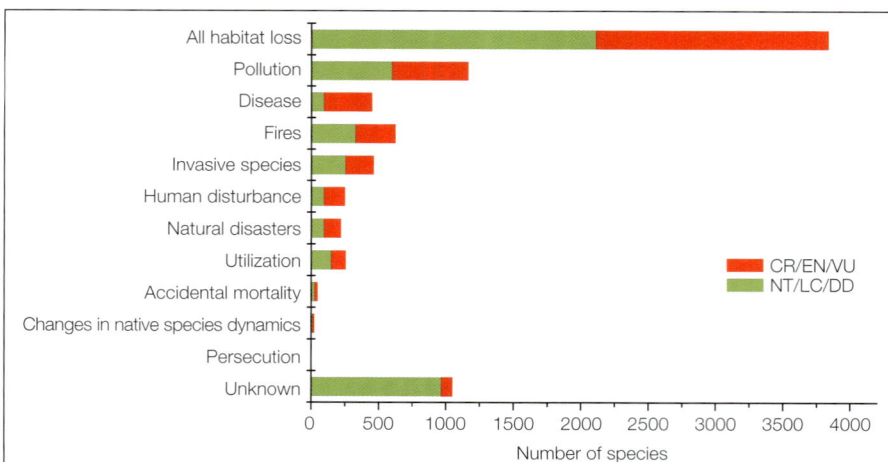

Box 2. Summary of results for amphibians

- Nearly one-third (32%) of the world's amphibian species are known to be threatened or Extinct, 43% are known not to be threatened, and 25% have insufficient data to determine their threat status.

- As many as 159 amphibian species may already be Extinct. At least 38 amphibian species are known to be Extinct, one is Extinct in the Wild, while at least another 120 species have not been found in recent years and are 'Possibly Extinct'.

- At least 42% of all species are declining in population, indicating that the number of threatened species can be expected to rise in the future. In contrast, less than one per cent of species show population increases.

- The largest numbers of threatened species occur in Latin American countries such as Colombia (214), Mexico (211), and Ecuador (171). The highest levels of threat, however, are in the Caribbean, where more than 80% of amphibians are threatened or extinct in the Dominican Republic, Cuba, and Jamaica, and a staggering 92% in Haiti.

- Although habitat loss clearly poses the greatest threat to amphibians, the fungal disease chytridiomycosis is seriously affecting an increasing number of species. Perhaps most disturbing, many species are declining for unknown reasons, complicating efforts to design and implement effective conservation strategies.

'Possibly Extinct', making a probable total of 153 bird extinctions since the year 1500.

Although 8,564 bird species (85.7%) currently are not considered threatened, 835 of these (8.4% of all known birds) are Near Threatened; the remaining 7,729 species are Least Concern.

Examining the current population trends for birds provides further confirmation that it is not just the threatened birds that are at risk as 40.3% of extant birds are recorded to be declining. A further 44.4% of bird species have stable populations and 6.2% are increasing. The population trend for 9.1% of birds is unknown or uncertain.

GEOGRAPHIC PATTERNS

Bird diversity
Birds occur in all regions of the world, from the tropics to the poles. They also occur in virtually every habitat, from the lowest deserts to the highest mountains (BirdLife International 2008a). Patterns of bird diversity are driven by fundamental biogeographic factors, with variety and extent of different habitats being particularly influential. For example, tropical forests are especially rich in species, hence the very high avian diversity found in the equatorial regions (BirdLife International 2008b). Not all bird species have been mapped hence a global map of bird diversity is not included here.

The highest numbers of bird species occur in the Neotropical realm. This concentration of high species richness can

The Philippine Eagle Pithecophaga jefferyi *has an extremely small population as a result of rapid declines caused by extensive deforestation. The species is listed as Critically Endangered.* © Nigel Voaden

Figure 7. *IUCN Red List assessment for 9,990 bird species.*

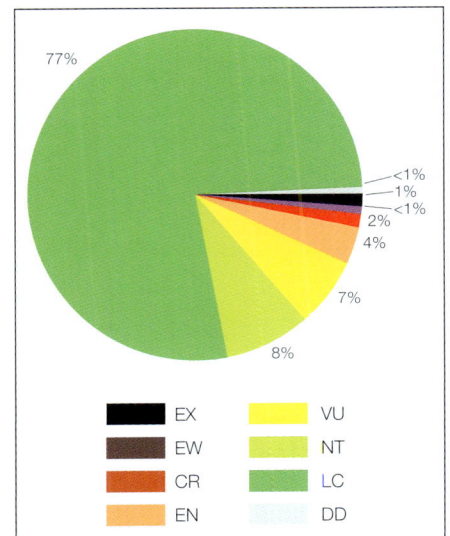

EX	VU
EW	NT
CR	LC
EN	DD

Rank	Country	Number of birds
1	Colombia	1,799
2	Peru	1,772
3	Brazil	1,704
4	Ecuador	1,578
5	Indonesia	1,561
6	Bolivia	1,416
7	Venezuela	1,347
8	China	1,237
9	India	1,178
10	Congo, The Democratic Republic of the	1,084
11	Mexico	1,077
12	Tanzania, United Republic of	1,050
13	Kenya	1,019
14	Myanmar	1,003
15	Argentina	993
16	Uganda	988
17	Sudan	919
18	Thailand	918
19	Panama	913
20	Angola	894

Table 5. *Top twenty countries with the largest number of bird species.*

Colombia supporting the highest bird diversity in the world. Eighteen per cent of the world's bird species occur in Colombia (1,799 species), closely followed by Peru (1,772 species), Brazil (1,704 species) and Ecuador (1,578 species). The other regions with high bird species diversity are Africa and Asia. Six of the top 20 countries in Table 5 are in Africa, with The Democratic Republic of the Congo, Kenya and the United Republic of Tanzania having more than 1,000 bird species each. Within Asia, Indonesia supports the highest bird diversity (1,561 species), followed by China (1,237 species) and India (1,178 species).

Geography of threatened bird species

The global distribution of threatened bird species is shown in Figure 8. Nearly all countries and territories of the world (97%) hold one or more globally threatened species, which are national priorities for conservation action (BirdLife International 2008b). Regions that stand out as having particularly high densities of threatened species include the tropical Andes, Atlantic Forests of Brazil, the eastern Himalayas,

eastern Madagascar, and the archipelagos of Southeast Asia (BirdLife International 2008b).

The majority of threatened birds (60%) are single-country endemics (i.e., they occur in only one country), and most of these species have small ranges and small population sizes (BirdLife International 2008d,e). Not all threatened birds have restricted ranges: 14 threatened species have ranges spanning more than 30 countries, including the Lesser Kestrel *Falco naumanni* with a native range that includes 96 countries in Europe, Asia and Africa (BirdLife International 2008b). Hence the political responsibility for conserving threatened species rests both nationally and, as a shared effort, internationally.

Table 6 lists the 20 countries with the highest numbers of globally threatened birds. Asia and South America emerge as the regions with the highest numbers of threatened bird species (nine of the top twenty countries are within Asia, and five are South American countries). The ten countries with the most threatened avifauna include seven of the most important in terms of absolute numbers of birds, with Brazil and Indonesia heading the list, holding 122 and

be seen by examining the 20 countries with the highest number of birds (Table 5). Six of the richest seven countries for birds are within South America, with

Figure 8. *Global distribution of threatened birds. The red shades indicate terrestrial species and the blue shades indicate marine species.*

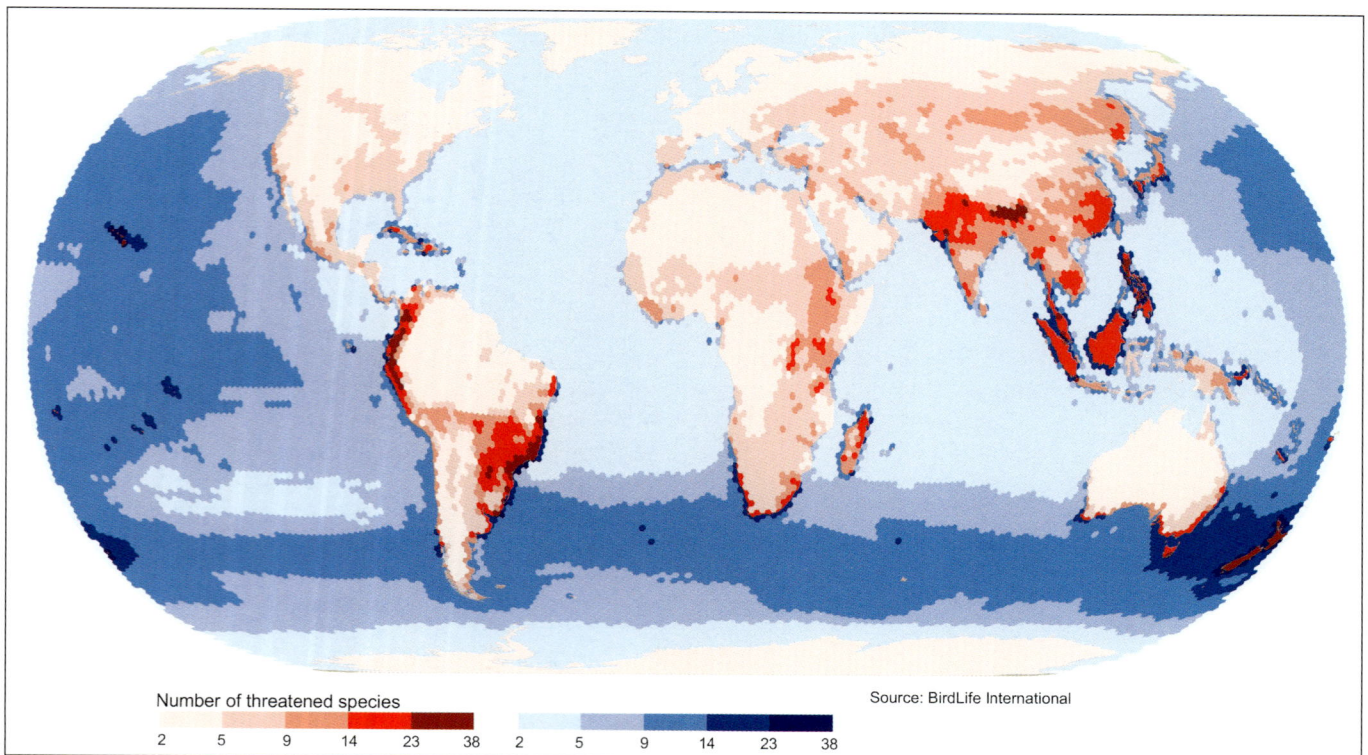

Number of threatened species

2 5 9 14 23 38   2 5 9 14 23 38

Source: BirdLife International

Rank	Country	Number of threatened birds
1	Brazil	122
2	Indonesia	115
3	Peru	93
4	Colombia	86
5	China	85
6	India	76
7	United States	74
8	New Zealand	69
	Ecuador	69
10	Philippines	67
11	Mexico	54
12	Russian Federation	51
13	Argentina	49
	Australia	49
15	Thailand	44
16	Malaysia	42
17	Myanmar	41
18	Tanzania, United Republic of	40
	Japan	40
20	Viet Nam	39

Table 6. *Countries with the largest number of threatened bird species.*

115 threatened species respectively. These two countries also support high numbers of threatened endemic birds: Brazil has 71 and Indonesia has 67 threatened endemics (see Appendix 12), which places a particular responsibility on these countries to protect these species.

In Table 6 only the number of threatened species is given, and the number of Extinct species has been excluded. This is to highlight those countries that currently have the greatest responsibility towards protecting globally threatened bird species.

Combining the numbers of threatened birds (Table 6) and the proportion of threatened and Extinct birds in each country (Table 7) highlights those countries that are most severely affected by declines and losses of bird species.

Countries with the highest proportions of threatened and Extinct birds include territories with low overall avian diversity. For example, French Polynesia, the Cook

Figure 9. *Major habitat preferences of birds. Occurrence in marginal habitats is included here, hence the number of species occurring in artificial terrestrial landscapes is over-emphasized.*

Islands, Saint Helena and Pitcairn all have fewer than 100 bird species, but very high percentages of their national avifauna are either globally threatened or already Extinct (more than 40% for each country). Another striking feature of Table 7 is the heavy dominance of oceanic islands with high percentages of threatened and Extinct species. The majority (88%) of known bird extinctions since the year 1500 have been on islands (Butchart *et al.* 2006), often as a result of introduced invasive species such as cats, rats and goats, which either preyed upon the native birds or degraded their habitat (BirdLife International 2008c). The extinction rate on islands now appears to be slowing thanks to ongoing efforts to eradicate established invasive alien species and to control species introductions on islands and conservation efforts to improve the status of native island species (BirdLife International 2008d).

HABITAT PREFERENCES

A summary of the most important habitats for birds is shown in Figure 9.

Birds occur in all major habitat types, and although some species may use a range of different habitats, many species are specialist to just one. Forests are the most important habitat, supporting 75% of all bird species, with tropical and subtropical forests being the richest bird habitats. Grasslands, savanna and inland wetlands are all important habitats for birds, each supporting about 20% of species, while shrublands support 39% of birds. Around 45% of birds are found in 'artificial' habitats (those that have been modified by humans, such as agricultural land), hence although birds appear to be more adaptable

Rank	Country	% threatened & Extinct
1	French Polynesia	47.8
2	Cook Islands	44.4
3	Saint Helena	42.2
4	Pitcairn	41.7
5	Norfolk Island	39.6
6	Mauritius	38.9
7	Heard Island and McDonald Islands	38.5
8	New Zealand	38.0
9	Niue	33.3
10	Réunion	29.1
11	French Southern Territories	27.5
12	United States Minor Outlying Islands	27.3
13	Wallis and Futuna	25.7
14	American Samoa	19.5
15	Samoa	15.6
16	Madagascar	14.8
17	Antarctica	14.7
18	Kiribati	13.5
19	Guam	13.1
20	New Caledonia	12.4

Note: only countries with 10 or more species are included.

Table 7. *Countries with the highest percentage of threatened and Extinct birds.*

or tolerant to such disturbance than amphibians or mammals, the importance of these habitats is low for a substantial proportion of these species.

Wetlands are very important habitats for the many waterbird species that tend to congregate in these areas in particular seasons for feeding and nesting, often regularly returning to the same site year after year. An example of one such habitat

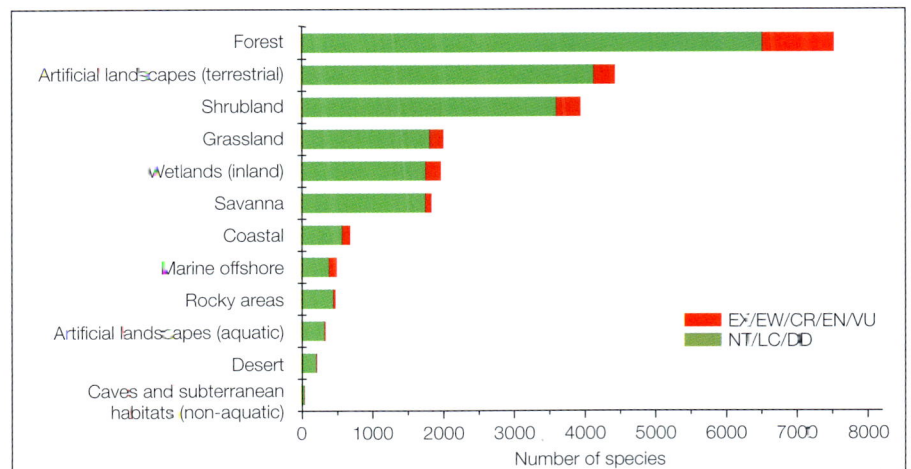

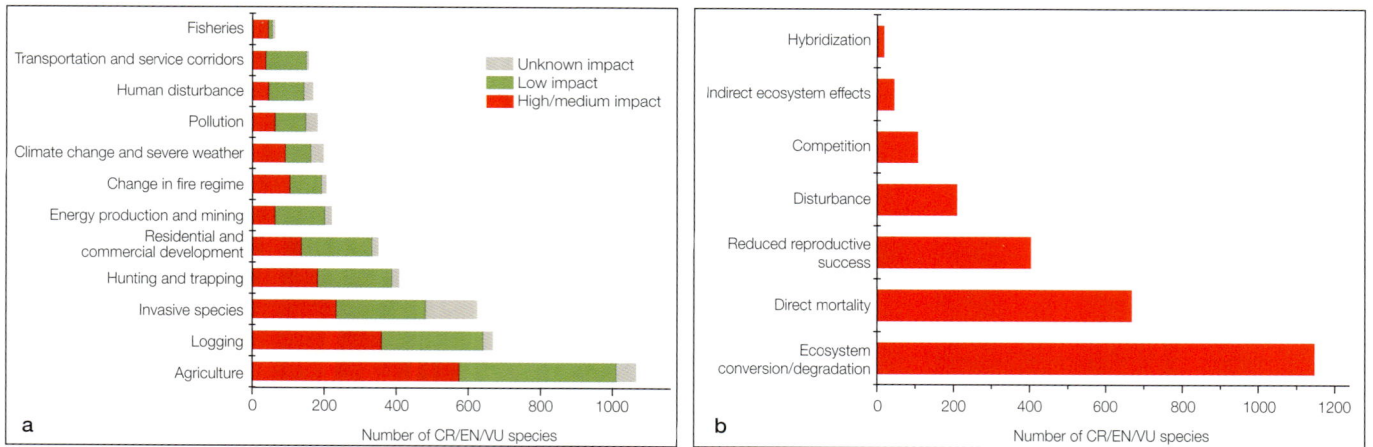

Note that the threats analysis presented here is not directly comparable to the analyses for amphibians and mammals presented in Figures 6 and 15 respectively. The threats to birds have been recorded using a new threats classification scheme that was recently adopted by IUCN (Salafsky *et al.* 2008). Under this new scheme the relative importance of threats is determined using an impact coding scheme, and the affects (or stresses) of the direct threats on the species are coded separately.

Figure 10. *Main threats* **(a)** *and stresses* **(b)** *affecting globally threatened bird species worldwide (modified from BirdLife International 2008a).*

is Lake Natron in the United Republic of Tanzania, where around 2.5 million individuals (75% of the global population) of Lesser Flamingo *Phoeniconaias minor* return each year to nest (Koenig 2006, BirdLife International 2008a).

THREATS

The threats leading to population declines in birds are many and varied (Figure 10a; BirdLife International 2008a): but agriculture, logging and invasive species are the most severe, respectively affecting 1,065 (87%), 668 (55%) and 625 (51%) globally threatened species. These threats affect bird populations in a range of ways (referred to in Figure 10b as the stresses), the commonest being habitat destruction and degradation, which affect 1,146 (93%) threatened species.

Humans are responsible for most of the threats to birds. Expanding and intensifying agriculture and forestry, the biggest problems, cause habitat destruction,

degradation and fragmentation. Fisheries degrade the marine environment and kill seabirds through incidental bycatch. The spread of invasive alien species, pollution and overexploitation of wild birds are also major threats. In the long term, human-induced climate change may be the most serious threat of all (BirdLife International 2008a).

Mammals

CURRENT STATUS

The mammal data on the 2008 IUCN Red List includes 5,488 species, 412 subspecies and 21 subpopulations. The primary focus of the current assessment, and hence this analysis, is at the species level. This is the second time that all mammals have been assessed, the first being in 1996 (Baillie and Groombridge 1996).

Nearly one-quarter of species (22%) are globally threatened or Extinct, representing

1,219 species (Figure 11, Appendix 4). Seventy-six of the 1,219 species are considered to be Extinct (EX), two are Extinct in the Wild (EW), and a further 29 are flagged as 'Possibly Extinct', making a total of 107 mammal extinctions since the year 1500.

Although 3,433 mammal species (63%) are not considered to be threatened at present, 323 of these (6% of all known mammals) are listed as Near Threatened (NT); the remaining 3,110 species are listed as Least Concern (LC).

Documenting population trends is a key part of assessing the status of species. Looking at current population trends in the extant mammal species, 30% are recorded to be decreasing. In contrast 25% of species are said to be stable and only 1.5% are increasing. Trend information is not available for 44% of species, hence the percentage of species in decline may be significantly higher.

Box 3. Summary of results for birds

- Birds are the best-known group of species, with less than 1% having insufficient data to determine their threat status. More than one in seven (14%) bird species are globally threatened or Extinct, 86% are not threatened.

- At least 134 birds have become Extinct since the year 1500, four species have become Extinct in the Wild, and a further 15 species are 'Possibly Extinct'.

- The highest numbers of bird species are found in South America, with Colombia supporting 18% of the world's birds (1,799 species). Africa and Asia are the next most diverse regions for bird species.

- 97% of the world's countries hold at least one globally threatened bird species. The highest numbers of threatened birds occur in Brazil (122 threatened species) and Indonesia (115 threatened species).

- Although they are much less diverse than tropical countries on the continents, oceanic island nations hold the highest proportions of threatened and extinct species. The majority (88%) of known extinctions since the year 1500 have been on islands.

- Agriculture, logging and invasive species are the most severe threats driving bird species towards extinction. The most common stress affecting bird populations is habitat loss and degradation.

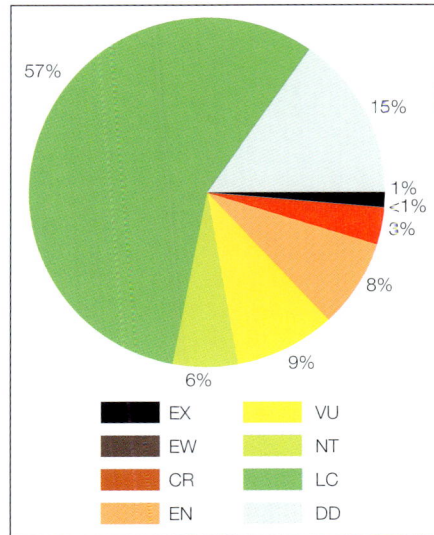

Figure 11. *IUCN Red List assessment for 5,488 mammal species.*

There was insufficient information available to assess the status of 836 species (15%) hence these are listed as Data Deficient (DD). While a number of these DD listings are due to taxonomic uncertaintities, in many cases they are due to inadequate information on population size, trends, distribution and/or threats. Most (80%) of the Data Deficient mammals occur in the tropics and 69% are bats and rodents

which are hard to catch because of their nocturnal habits and difficult to identify.

GEOGRAPHIC PATTERNS

Mammalian diversity

Mammal species are found all across the globe, with the exception of the land mass of Antarctica. The global pattern of land and marine mammal diversity is shown in Figure 12. Regions with high diversity are clearly visible as darker patches on the global map. For land species, these regions are found in Mesoamerica and tropical South America, sub-Saharan Africa and South and Southeast Asia. Marine mammals occur throughout the world's oceans but peaks in diversity are found along all continental coastlines, as well as Japan, New Zealand, the Caribbean Sea, and the southern Indian Ocean and the ocean west of Mesoamerica.

Looking at mammal diversity from a country perspective (Table 8), the country with the highest number of mammal species is Indonesia, with 670 species. Close behind is Brazil with 648 species; China (551) and Mexico (523) are the only other two countries with more than 500 native mammal species. Four of the top six countries, and seven of the top 20

countries are in Mesoamerica or tropical South America. Although a large part of sub-Saharan Africa is very rich in mammal diversity, only five African countries appear in Table 8, and only two of these are among the ten most diverse. However, many of the African countries in this mammal-rich region have a relatively small land area compared with other mammal rich countries on other continents (for example, Brazil, China and Mexico), so the diversity of Kenya with 376 mammals is impressive when its total area is taken into account. There are five Asian countries in Table 8, three of which are among the ten most diverse countries. Indonesia's place at the top of the list is unlikely to be challenged as there are undoubtedly many more species remaining to be described in this megadiverse country.

The United States is highly ranked in seventh place, but in years to come as survey efforts increase, currently less well-surveyed countries such as The Democratic Republic of the Congo may be expected to overtake the United States in the rankings.

Geography of threatened mammal species

The global distribution of threatened land and marine mammals is shown in Figure 13. Compared with the distribution of

Figure 12. *Global diversity of mammal species. Brown shades indicate terrestrial species and blue shades marine species.*

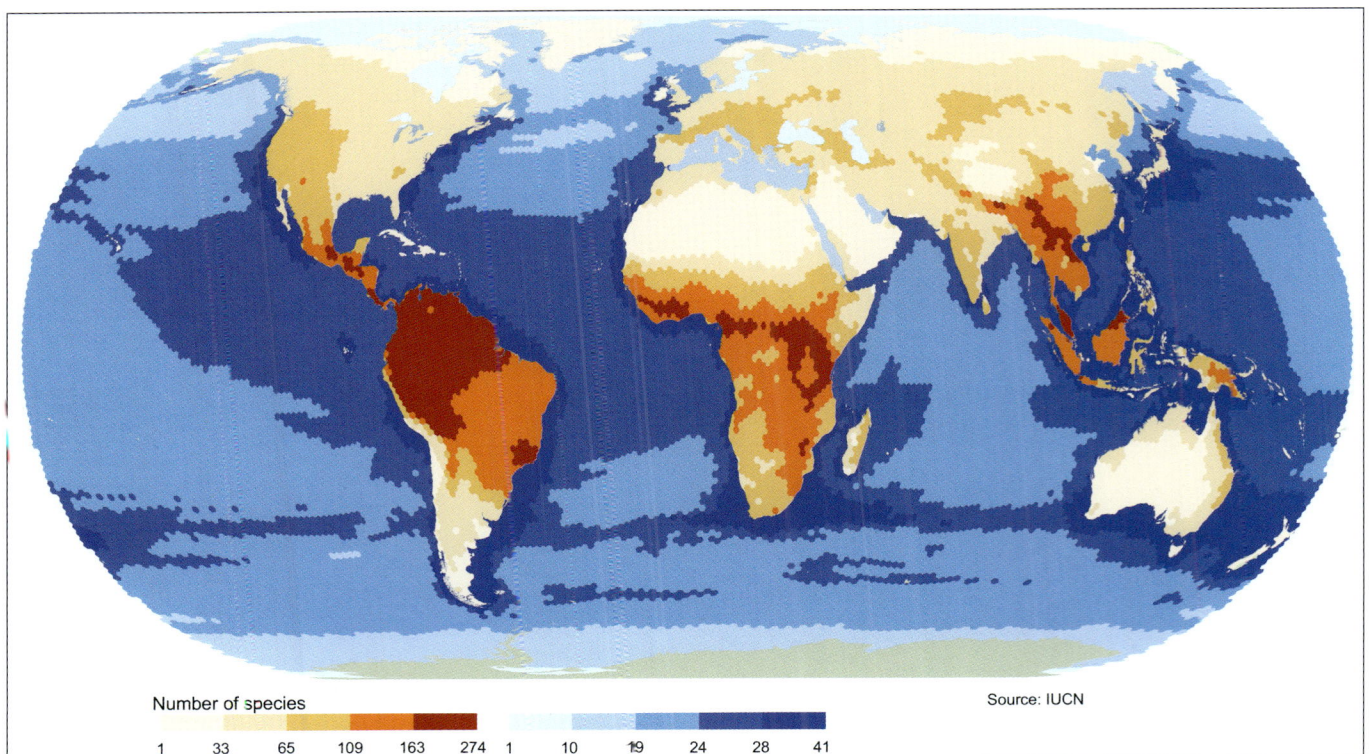

Rank	Country	Number of mammals
1	Indonesia	670
2	Brazil	648
3	China	551
4	Mexico	523
5	Peru	467
6	Colombia	442
7	United States	440
8	Congo, The Democratic Republic of the	430
9	India	412
10	Kenya	376
11	Argentina	374
12	Ecuador	372
13	Bolivia	363
	Venezuela	363
15	Tanzania	359
16	Australia	349
17	Malaysia	336
18	Cameroon	335
19	Uganda	319
20	Thailand	311

Table 8. Top twenty countries with the largest number of mammal species.

all mammal species, there are some similarities and some striking differences. Most noticeably, the density of threatened

mammals in Southeast Asia is much higher than anywhere else. Most of these species are threatened by over-utilization (e.g., hunting) and habitat loss.

Other regions that have a high number of threatened species include: the Western Ghats in southern India, Sri Lanka, the Cameroonian Highlands in West Africa, the Albertine Rift in central Africa, parts of Madagascar, and the tropical Andes. The major threat to species in these regions is habitat loss.

For marine mammals, concentrations of threatened species are found in Southeast Asia as well as the North Atlantic and North Pacific.

The 20 countries with the highest number of threatened mammal species are listed in Table 9. These countries have a particularly great responsibility for protecting the world's threatened mammals. Indonesia again is at the top of the list with 183 species, many more than the second country on the list, Mexico, with 100 species. Interestingly, there are now ten Asian countries in the top 20 list for threatened mammals, and these countries all rank higher for the number of threatened

species than they do for their species diversity, with the exception of China. In contrast, the African and South American countries have mostly dropped down the rankings when comparing the number of threatened species with overall species diversity. Madagascar has risen from being outside of the top twenty countries for diversity, to being number seven in the rankings for the number of threatened species.

In Table 9 only the number of threatened species is given, and the number of Extinct species has been excluded. This is to highlight those countries that currently have the greatest responsibility towards protecting globally threatened mammals.

In Table 10 the countries with the highest percentage of threatened and Extinct mammals are listed. This list of countries is very different to the list of the number of threatened species (Table 9), as well as the list of countries with the highest diversity of mammals (Table 8).

The top three countries for the highest percentage of threatened and Extinct mammals are all islands in the southwest Indian Ocean. Island nations dominate

Figure 13. Global distribution of threatened mammals. Red shades indicate terrestrial species and blue shades marine species.

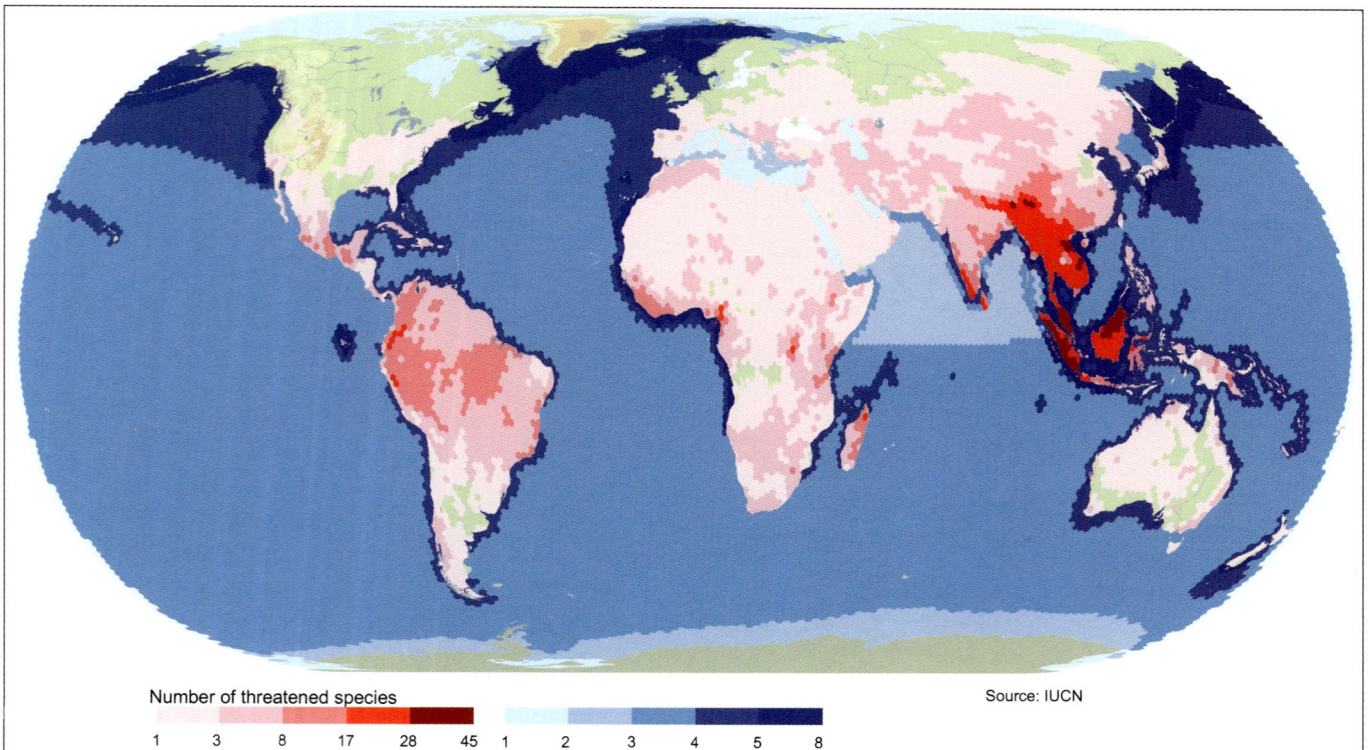

Number of threatened species
1 3 8 17 28 45   1 2 3 4 5 8

Source: IUCN

The Fishing Cat Prionailurus viverrinus is an Asian species found mainly in wetland habitats. In 2008, this species moved up from Vulnerable to Endangered because of the severe decline throughout much of its range over the last ten years. Over 45% of protected wetlands in Southeast Asia are now considered threatened. In addition, clearance of coastal mangroves over the past decade has been rapid. © Mathieu Ourioux

this list, and there are actually only three mainland countries listed in the top twenty. This is a stark reminder of the inherent vulnerability of small range island endemic species to threatening processes. For most of these species, habitat loss is the most important threat, but invasive species are also having a significant impact and have in some instances led to rapid extinctions. Not surprisingly, Indonesia, ranked first for diversity as well as the number of threatened species, is still within the top 20 for percentage of threatened species.

HABITAT PREFERENCES

A summary of the most important habitats for mammals is shown in Figure 14.

For terrestrial mammals by far the most common habitat is forest. Shrublands and grasslands are the next most favoured natural habitats. Notably rocky areas and caves are quite common habitat preferences too, especially for bats. The least favoured habitats are those that are arid or semi-arid. Interestingly, almost 1,500 species occur in disturbed or artificial (human created) habitats. This apparent tolerance of disturbance and adaptation to human-created habitats does

Table 9. Countries with the most threatened mammal species.

Rank	Country	Number of threatened mammals
1	Indonesia	183
2	Mexico	100
3	India	96
4	Brazil	82
5	China	74
6	Malaysia	70
7	Madagascar	62
8	Australia	57
	Thailand	57
10	Viet Nam	54
11	Peru	53
12	Colombia	52
13	Lao People's Democratic Republic	46
14	Myanmar	45
15	Ecuador	43
16	Papua New Guinea	41
	Cameroon	41
18	Philippines	39
19	Cambodia	37
	United States	37

Table 10. Countries with the highest percentage of threatened (including Extinct) mammals.

Rank	Country	% threatened & Extinct
1	Mauritius	63.6
2	Réunion	42.9
3	Seychelles	38.5
4	Vanuatu	33.3
5	Cuba	30.8
6	Madagascar	28.9
7	Dominican Republic	28.6
	Haiti	28.6
9	Bhutan	28.3
10	Solomon Islands	27.8
	Faroe Islands	27.8
12	Indonesia	27.5
13	New Caledonia	27.3
14	Sri Lanka	25.6
15	Brunei Darussalam	25.4
16	Micronesia, Federated States of	25.0
	Bahrain	25.0
18	Bangladesh	24.3
19	India	23.3
20	Montserrat	23.1

Note: only countries with 10 or more species are included.

a

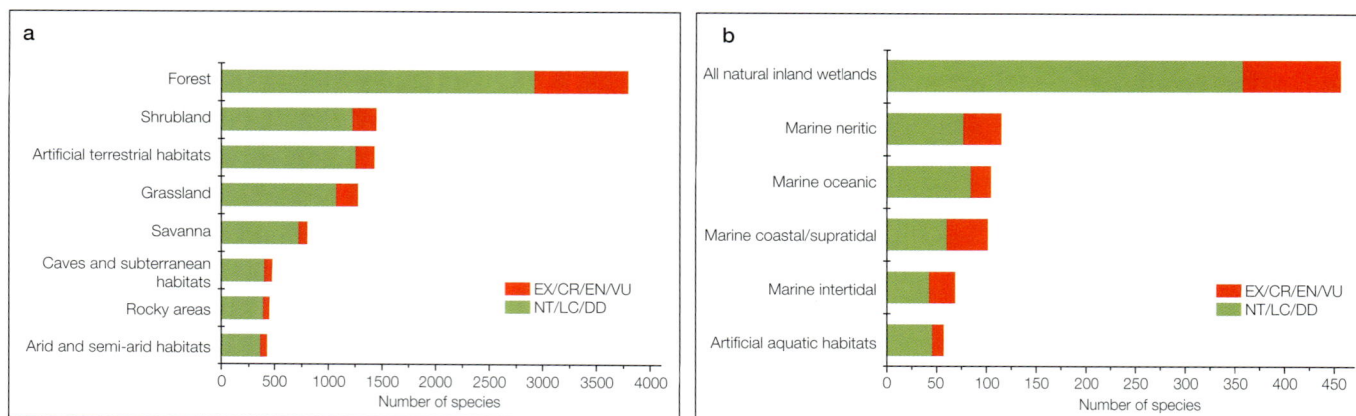

b

Figure 14. Habitat preferences of mammals: **(a)** terrestrial habitats, and **(b)** aquatic habitats.

not necessarily guarantee that a species will not be threatened; even though the impact of habitat loss may be lessened, some of these species are still being heavily impacted by utilization.

For aquatic species, the most common habitat is natural wetlands (mostly inland systems). Only 134 mammal species are recorded to occur in the marine environment and unsurprisingly occur in all the major marine habitats except for the deep benthic zone.

THREATS

A summary of the numbers of mammal species affected by each threatening process is shown in Figure 15. By far the most significant threat to mammals is habitat loss with over 2,000 species (45% of which are listed as threatened) being negatively impacted. The second most important threat is utilization (primarily for food or medicine), with almost 1,000

species (50% threatened) affected, especially in Asia. The impact of invasive species is probably a little underestimated as only threats to extant species are included here, and a significant proportion of species now considered Extinct were driven to extinction by invasive species.

Plants

The 2008 *IUCN Red List* includes assessments for 12,055 species of plants, 8,457 of which are listed as threatened. However, as only about 4% of the estimated 298,506 described plant species have been assessed, it is not possible to say that based on The IUCN Red List that 3% of the world's flora is threatened.

Since the plant and animal Red Lists were combined in the *2000 IUCN Red List of Threatened Species*™ the number of plant assessments on the Red List has increased very slowly compared to other

taxonomic groups. Of the 12,055 plants evaluated, 70% are listed as threatened (Table 1). This partially reflects a bias amongst the botanical community to focus primarily on the threatened species, but there is also a tendency to not report on the species that have been assessed as Least Concern. The focus on threatened

Box 4. Summary of results for mammals

- Nearly one-quarter (22%) of the world's mammal species are known to be globally threatened or Extinct, 63% are known to not be threatened, and 15% have insufficient data to determine their threat status.

- There are 76 mammals which have gone Extinct since 1500, two are Extinct in the Wild and 29 are 'Possibly Extinct'.

- The most diverse country for mammals is Indonesia (670), followed closely by Brazil (648). China (551) and Mexico (523) are the only other two other countries with more than 500 species.

- The country with by far the most threatened species is Indonesia (184). Mexico is the only other country in triple figures with 100 threatened species. Half of the top 20 countries for numbers of threatened species are in Asia; for example, India (96), China (74) and Malaysia (70). However, the highest levels of threat are found in island nations, and in particular the top three are islands or island groups in the Indian Ocean: Mauritius (64 %), Réunion (43 %) and the Seychelles (39%).

- Habitat loss, affecting over 2,000 mammal species, is the greatest threat globally. The second greatest threat is utilization which is affecting almost 1,000 mammal species, especially those in Asia.

Figure 15. Major threats to mammals.

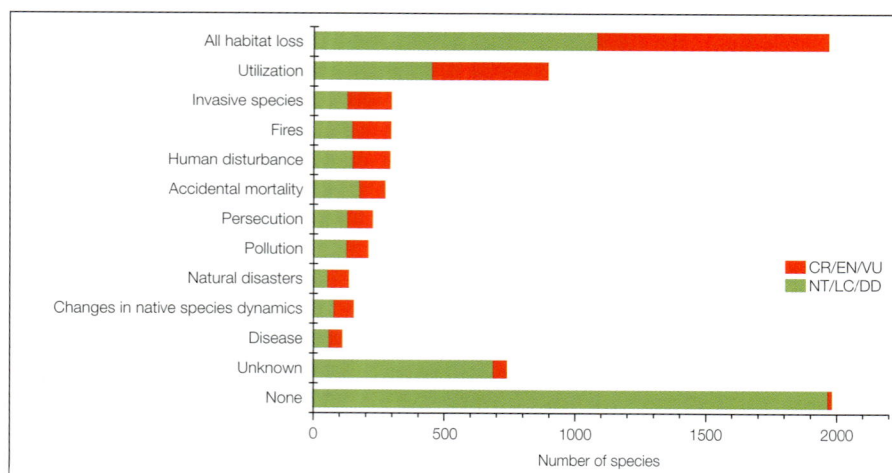

Lions Panthera leo *in South Africa. The population of this Vulnerable species is declining, mainly because of retaliatory or pre-emptive killing by humans to protect life and livestock from this top predator.* © Troy Inman

species is clearly illustrated by the assessments of bryophytes (mosses, liverworts and hornworts), where the subset of 95 species was specifically chosen in order to "provide the public with general information as to which bryophytes are threatened with extinction" (Tan *et al.* 2000). The same is partly true of the assessments for ferns and fern allies (includes club mosses, spike mosses, quillworts and true ferns); in this case, the 211 species assessed (although only 1% of the species) represent a widely distributed geographic sample and so might be more representative of the threats faced by this plant group, but it would be misleading to extrapolate from these results to the whole group.

A strong bias in the plant assessments in the 2000 *IUCN Red List* was towards threatened tree species because of the inclusion of the 7,383 species (includes species in all categories from Data Deficient to Extinct) listed in *The World*

List of Threatened Trees* (Oldfield *et al.* 1998). That bias has been reduced slightly through the inclusion of non-tree assessments. However, the trees still form 66% of the plants on the 2008 *IUCN Red List* (7,977 species), 5,643 of which are listed as threatened.

Many of the recent plant assessments are now introducing a geographic bias

The Endangered Premnanthes amibilis *is endemic to the island of Soqotra (Yemen). It has a very small range, being restricted to a very specific part of the island where precipitation and mists are caught from monsoons. The trend for lower rainfall in the region is a particular threat to this plant.* © Anthony Miller

as they are single country or sub-country endemics (e.g., Cameroon, China, Ecuador, Madagascar, Mauritius, Namibia, Saint Helena, South Africa, Yemen (Soqotra), and the United States (Hawaii)).

The seemingly very large figure of 8,457 threatened plant species is proportionally very small relative to the total number of described plant species worldwide (Table 1). The proportion threatened is even smaller if the higher estimate for the number of described plants is used (422,127 as opposed to 298,506 species; see Appendix 3). It is therefore premature at this stage to attempt any detailed analysis of the plants as the low numbers assessed and the strong biases towards trees and certain geographic areas misrepresents the overall picture for plants. For further details on the numbers of plants in each category, see Table 1 and the detailed order and family results in Appendices 5 and 7.

STATUS OF CONIFERS AND CYCADS

Despite the low numbers of plant assessments and the biases in these, some trends are evident. Two classes of plants have been fully assessed, namely the cycads and the conifers. Whether these gymnosperm groups are representative of what is happening to plants generally is doubtful. However, both are relatively ancient lineages and clearly illustrate very different threats and trends (Figures 1, 16a and 16b). Although there is not a major difference between the numbers of threatened conifers and cycads (172 and 150 respectively), the

proportion of cycads that are threatened is considerably higher. For the conifers, 28% are listed as threatened (21 Critically Endangered, 54 Endangered and 97 Vulnerable). For cycads, 52% are listed as threatened (45 Critically Endangered, 40 Endangered and 65 Vulnerable). In addition, a further 23% of cycad species are considered Near Threatened.

At present the cycads are the most threatened plant group known, and are the most threatened taxonomic group on the Red List. The cycads in particular are a unique lineage of plants that survived the last major extinction event and many species are now facing imminent extinction in the wild as a direct result of human activities.

GEOGRAPHIC PATTERNS IN CONIFERS AND CYCADS

The conifers although, in many respects, a relictual group, are widely distributed across the globe (Farjon and Page 1999). They form the dominant elements in almost all of the world's temperate rainforests, but are most notable in the high latitude boreal regions of Europe, Asia and North America where conifer forests cover vast areas. The major gaps in distribution are large areas in Africa and South America, the arid parts of Asia and Australia, and the Arctic and Antarctica.

The geographic patterns described here are the result of a preliminary analysis. The cycad distribution maps are not finalized and mapping of the conifer distribution ranges has not yet started, hence global

maps of diversity and distribution of threatened species for these two groups are not included here.

Examination of the countries with highest diversity of conifers and highest numbers of threatened species reveals several conifer 'hotspots' (Table 11). North America has 98 conifer species, with a particular concentration of threatened species in California (United States). Mesoamerica has a rich diversity of conifers (83 species), with most of the diversity and threatened species occurring in Mexico (80 species, 16 threatened). South America on the other hand is relatively species-poor with only 36 species - Guatemala has the highest diversity (18 species, 5 threatened) and Argentina the highest number of threatened conifers (11 species, 7 threatened). Oceania has the second highest number of conifers of any region (142 species) with Australia (39 species, 10 threatened) and New Caledonia (45 species, 17 threatened) being the main centres of richness. The main areas for conifer richness are in Asia, particularly the mountainous regions of western China and

Figure 16. Red List assessment for **(a)** all known conifers (620 species), and **(b)** all known cycads (289 species).

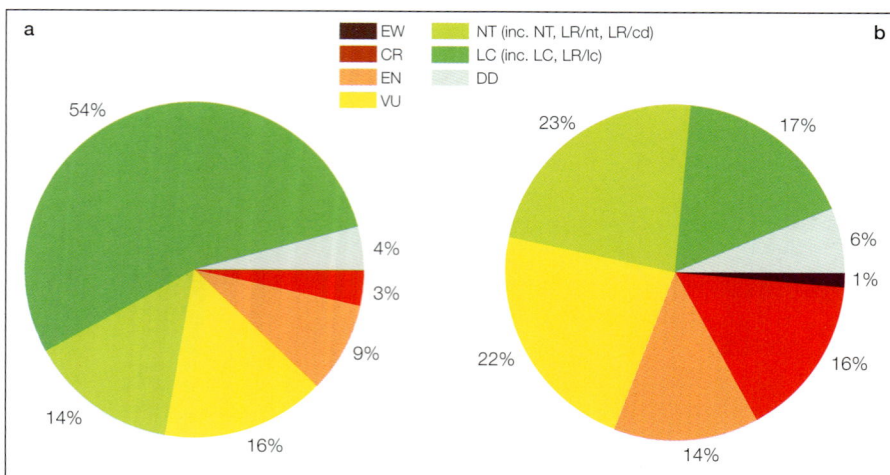

Table 11. Top twenty countries with the most conifer species and number of threatened conifers in each.

Rank	Country	Number of conifers	Number of threatened conifers
1	China	130	34
2	United States	98	14
3	Mexico	80	16
4	Indonesia	54	6
5	New Caledonia	45	17
6	Malaysia	44	15
7	Australia	39	10
	Japan	39	5
9	Canada	34	1
10	Papua New Guinea	33	0
11	India	29	3
	Russian Federation	29	0
13	Viet Nam	27	13
14	Taiwan, Province of China	26	10
15	Philippines	21	5
16	Greece	20	0
17	New Zealand	19	1
	Turkey	19	0
19	Guatemala	18	5
	Myanmar	18	4

The native range of the Vulnerable Monkey Puzzle tree Araucaria araucana is from the Coastal Cordillera of Chile to the Andes in Argentina. The wood from this conifer is very resistant, making it an attractive material for building construction and furniture. Populations have declined and become severely fragmented. © Peter Hollingsworth

Rank	Country	Number of cycads	Number of threatened cycads
1	Australia	69	18
2	Mexico	44	38
3	South Africa	38	24
4	Viet Nam	25	16
5	China	20	12
6	Colombia	18	9
7	Mozambique	14	10
8	Panama	12	3
9	Swaziland	10	8
10	Guatemala	9	7
	Peru	9	6
	Thailand	9	5
13	Congo, The Democratic Republic of the	7	2
	Cuba	7	1
	Indonesia	7	0
16	Papua New Guinea	6	0
	Tanzania, United Republic of	6	2
18	India	5	2
	Kenya	5	1
	Philippines	5	1

Table 12. Top twenty countries with the most cycad species and number of threatened cycads in each.

the neighbouring areas in Myanmar and India; China alone has 130 species 34 of which are threatened. Other countries in Asia with high conifer diversity include Indonesia (54 species, 6 threatened), Malaysia (44 species, 15 threatened) and Japan (39 species, 5 threatened).

The distribution of cycads is much more restricted and patchy than the conifers with all species being confined to the tropical and sub-tropical parts of the world (Donaldson 2003). A few countries stand out as critical centres of cycad diversity (Table 12), notably Australia (69 species, 18 threatened), Mexico (44 species, 38 threatened), South Africa (38 species, 24 threatened), Viet Nam (25 species, 16 threatened) and China (20 species, 12 threatened). Together these five countries account for 68% of the world's cycads.

THREATS TO CONIFERS AND CYCADS

The conifer and cycad assessments are not yet fully documented, hence a detailed analysis of their threats is not possible. But it is possible to draw some generalities about the threats.

Conifers are mostly constituents or dominants of forests, hence factors that negatively influence forests also threaten conifer species. These factors include:

direct exploitation through logging operations, particularly in the coastal forests of the Pacific Rim; uncontrolled forest fires and subsequent grazing of seedlings and saplings by domestic animals or introduced wild animals; conversion of forest ecosystems to pasture, arable land and human habitation; exploitation of conifers for non-timber resources e.g., resin, edible seeds and medicines; and destruction or disturbance of forests by large scale mining or hydroelectric projects (Farjon and Page 1999).

The threatened cycad species generally have small and declining populations and/or small ranges and are frequently targeted by collectors and/or impacted by habitat loss and degradation. One exception is Cycas micronesica, a relatively widespread species on islands in the Pacific, which is declining rapidly as a result of the spread of an invasive species of scale (Aulacaspis sp.); infections are fatal.

Are species becoming more or less threatened with extinction?

In those taxonomic groups about which we know most, species are sliding ever faster towards extinction. IUCN Red List Indices (RLIs; see description in Vié et al.

this volume) show that trends in extinction risk are negative for birds, mammals, amphibians and reef-building corals (Figure 17). Although successful conservation interventions have improved the status of some species (Box 5), many more are moving closer towards extinction, as measured by their categories of extinction risk on The IUCN Red List.

The groups vary in their overall level of threat; for example, amphibians have a higher proportion of species threatened (i.e., lower RLI values) than mammals or birds. Groups also vary in their rate of deterioration, with the rapid declines in reef-building corals since 1996 being driven primarily by the worldwide coral-bleaching events in 1998 (Polidoro et al. this volume, Carpenter et al. 2008). Whereas the RLI for birds shows that there has been a steady and continuing deterioration in the status of the world's birds between 1988 and 2008. Over these 20 years, 225 bird species have been uplisted to a higher category of threat because of genuine changes in status, compared to just 32 species downlisted.

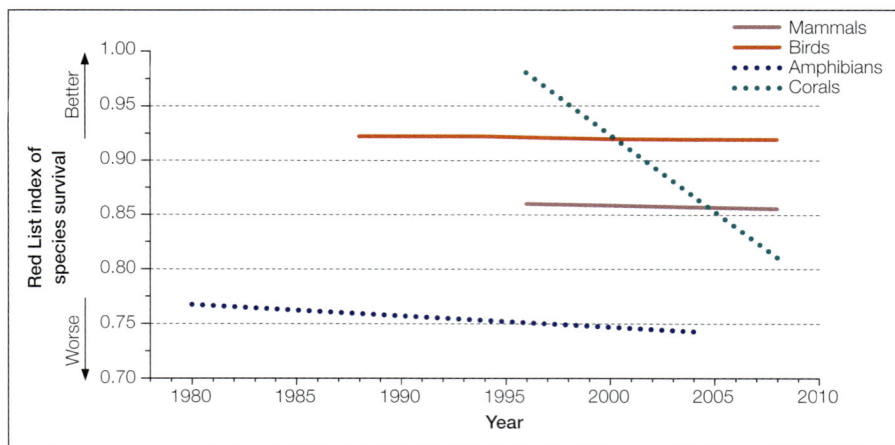

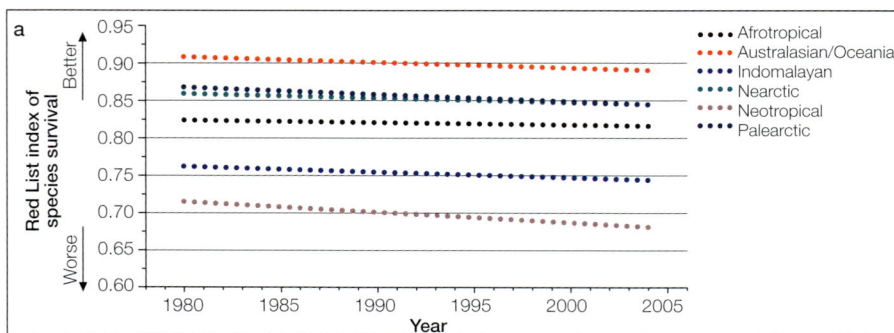

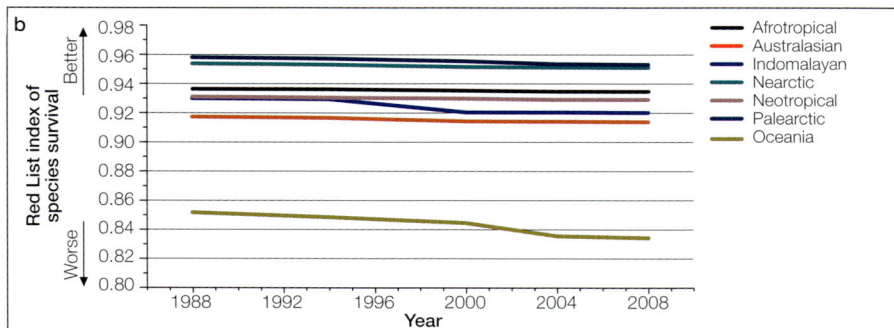

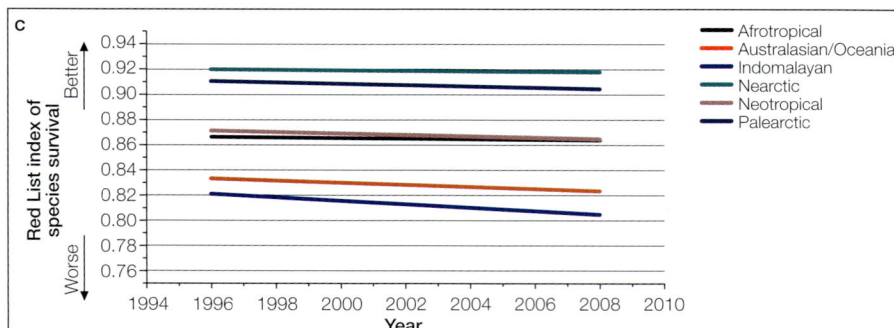

Figure 17. *Red List Index of species survival for corals, birds, mammals and amphibians, showing the proportion of species expected to remain extant in the near future without additional conservation action. An RLI value of 1.0 equates to all species being categorized as Least Concern, and hence that none are expected to go Extinct in the near future. An RLI value of zero indicates that all species have gone Extinct. (Number of non-Data Deficient species = 9,785 birds, 4,555 mammals, 4,416 amphibians and 704 corals (warm water reef-building species only); data are preliminary for amphibians and corals).*

species (as reflected in population trend-based indicators), it has global scope and coverage, and hence is not biased geographically in the way that global population trend-based indicators may be.

What are the geographic patterns in declines?

Species are deteriorating in status worldwide, but some regions have undergone steeper declines and have more threatened faunas (Figure 18). The Indomalayan realm showed rapid declines in birds and mammals, driven by the rapid increases in the rate of deforestation during the 1990s, particularly in the Sundaic lowlands, combined for mammals with high rates of hunting, particularly among medium- to large-bodied species. Amphibians are also highly threatened in the Indomalayan realm. Birds in the Oceania realm are substantially more threatened (with lower RLI values) than in other realms, largely owing to the impacts of invasive alien species. Amphibians are most threatened in the Neotropical realm, in particular owing to chytridiomycosis.

A downwards trend in the graph line (i.e., decreasing RLI values) means that the expected rate of species extinctions is increasing i.e. that the rate of biodiversity loss is increasing. A horizontal graph line (i.e., unchanging RLI values) means that the expected rate of species extinctions is

unchanged. An upward trend in the graph line (i.e., increasing RLI values) means that there is a decrease in expected future rate of species extinctions (i.e., a reduction in the rate of biodiversity loss).

While the RLI is not very sensitive to small-scale changes in the status of

Species loss and human health

The 2008 IUCN Red List clearly shows that many species are under threat of extinction mainly as a direct or indirect result of human activities. But why should humans be concerned about this and why should we invest time and money on saving species?

Figure 18. *Red List Index of species survival for **(a)** amphibians, **(b)** birds, and **(c)** mammals in different biogeographic realms, showing the proportion of species expected to remain extant in the near future without additional conservation action. An RLI value of 1.0 equates to all species being categorized as Least Concern, and hence that none are expected to go Extinct in the near future. An RLI value of zero indicates that all species have gone Extinct. (Number of non-Data Deficient amphibian/bird/mammal species = 395/1,706/776 Palearctic; 746/2,210/1,045 Afrotropical; 692/2,144/823 Indomalayan; 307/991/471 Nearctic; 2,187/3,972/1,335 Neotropical; 1,765 Australasian birds; 316 Oceania birds; totals for Australasia and Oceania are combined for amphibians/mammals = 384/692 species; data for amphibians are preliminary).*

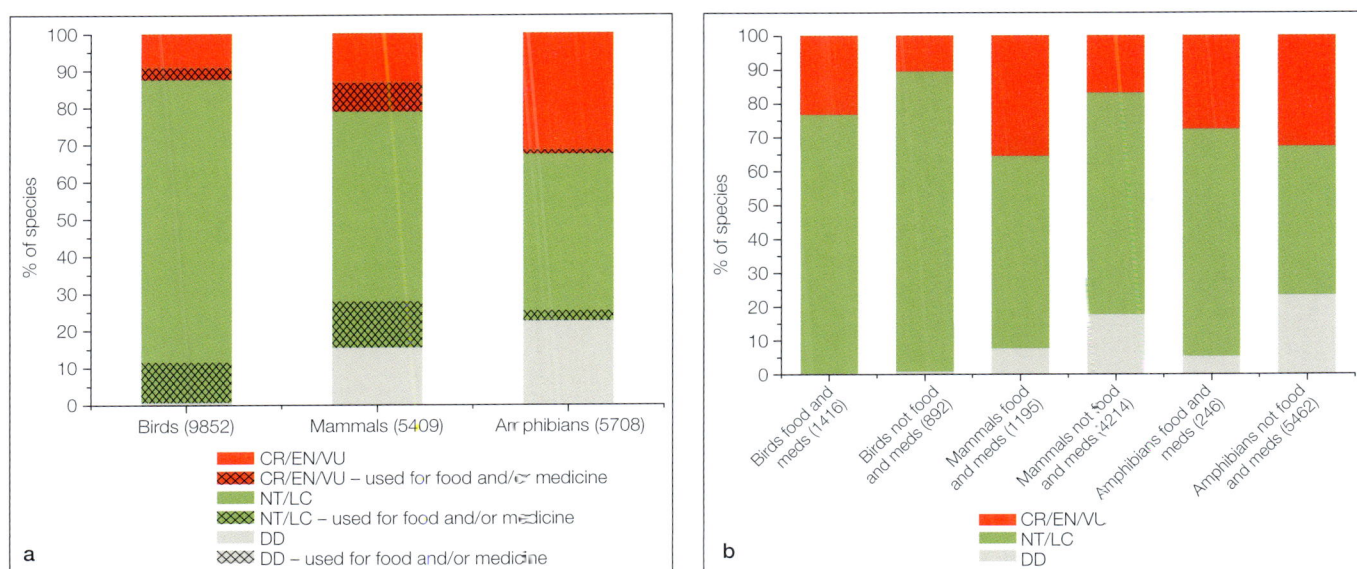

Note that for amphibians the data used is from 2004.

Figure 19. Proportion of all known birds, mammals and amphibians by threat status (i.e., threatened, not threatened and Data Deficient) and used for food and medicine (a); and a comparison of threat status of species used for food and medicine against threat status for those species not used in this way (b).

For as long as humans have existed we have used the species around us for our own survival and development. Even today, with vast numbers of people living in towns and cities, seemingly far removed from nature, we still need plants and animals for our food, materials, and medicines, as well as for recreation and inspiration for everything from the sciences to the arts. In the developing countries, where wild animal and plant species can make a significant contribution to human diets and healthcare, maintaining a healthy biodiversity is of particular importance.

Biodiversity for food and medicine

It is estimated that 50,000 to 70,000 plant species are used for traditional and modern medicine (Schippmann *et al.* 2006). These species are vital to traditional healthcare systems in less-developed countries. For example, in some Asian and African countries, up to 80% of the population depends on traditional medicine for primary health care (World Health Organization 2008). Medicinal plants are also increasingly recognized as effective alternative treatments in developed countries. Herbal treatments, for instance, are highly lucrative in the international marketplace. Annual

revenues for herbal treatments in Western Europe reached US$ 5 billion in 2003–2004; in China, sales totalled US$ 14 billion in 2005; and herbal medicine revenue in Brazil was US$ 160 million in 2007 (World Health Organization 2008).

Figure 19 shows the proportions of birds, mammals and amphibians used for food and medicine and compares threatened and non threatened species that are utilized in this way against species that are not utilized.

Figure 19a indicates that 14% of the world's birds are used for food and/or medicine although this is probably an underestimate. It

is difficult to know how many individual birds are used, but it is estimated that between half a billion and one billion songbirds are hunted each year in Europe alone, for sport and food (BirdLife International 2008f). Forty-five bird species are known to be used for medicinal purposes. More than a fifth (22%) of mammals, and 4% of amphibian species are used for food and/or medicine (Figure 19a). Although the proportion of amphibians known to be utilized in this way is small, this represents 218 species used for food (ranging from local and national use to the extensive international trade in frog legs) and 75 species used for medicine. At least 212 amphibians are used for subsistence food,

Baskets of frog skins from a market in Thailand. At least 218 amphibians are currently used for human consumption, although in some cases such use does not always represent the major threat to the species.
© Peter Paul van Dijk/Conservation International

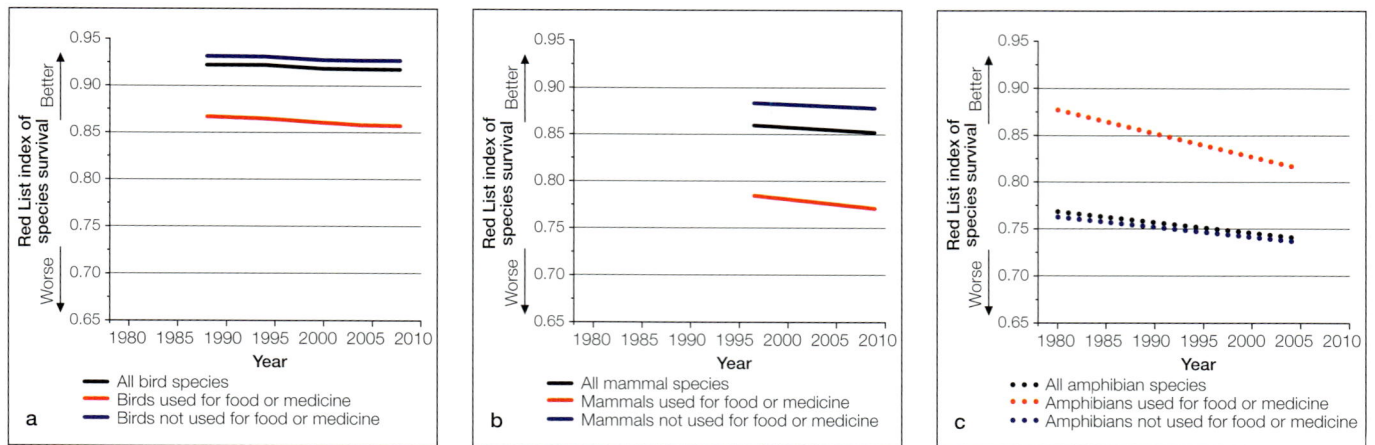

Figure 20. *Red List Indices showing the proportion of species expected to remain extant in the near future without additional conservation action for all species, for species used for food and/or medicine and for species not used for these purposes; broken down by **(a)** birds, **(b)** mammals, and **(c)** amphibians.*

but the diversity of species consumed is probably under-recorded and further studies will almost certainly reveal additional species not previously identified as being used in this way (Stuart *et al*. 2008). Amphibians have long been recognized for their value in traditional medicines and they are still collected for this purpose today. The potential value of amphibians to modern medicine is coming under increasing scientific study, with their diverse skin secretions being of particular interest (Stuart *et al*. 2008).

Threat status of species used for food and medicine

Figure 19b shows the proportions of threatened species within those bird, mammal and amphibian species that are used for food and medicine. Although 12% of all bird species are globally threatened with extinction (Table 1), a larger proportion (23%) of those species used for food and medicine are threatened. Mammals show a similar pattern: 21% of all species are known to be threatened (Table 1), but 36% of the species used for food and medicine are threatened. For amphibians, there is little difference between the proportion of threatened species within all known species (30%) (Table 1) and the proportion of species used for food and medicine that are threatened (28%). Many of the wild species used for food and medicine are threatened, some due to over-exploitation, some to different pressures such as habitat loss, and for others a combination of factors. Regardless of the causes, the diminishing availability of these resources threatens the health and well-being of the people who depend on them directly for food and medicine, and on wild collection as a source of income.

Trends in status of biodiversity for food and medicine

The RLI for birds used for food and medicine (Figure 20a) indicates that these species are more threatened than those that are not utilized in this way and that the conservation status of these species is also deteriorating at a slightly greater rate. The RLI for mammals shows a similar pattern (Figure 20b). In contrast to the birds and mammals, amphibians used for food and medicine appear overall to be less threatened than amphibians not used for these purposes (Figure 20c). However, the conservation status of these species is declining more rapidly than that of amphibian species not used for food and medicine.

At present, insufficient data are available to produce a meaningful Red List Index (RLI) for medicinal plants; only 109 species (0.7%) of medicinal plants have Red List assessments available for the years 1997 and 2008. Hence an analysis of these species is not included here.

The 2008 IUCN Red List – Good News or Bad?

The overwhelming message from the results presented in this chapter and in other chapters in this volume is that the world is losing species and that the rate of loss appears to be accelerating in many taxonomic groups (Box 6). The number of threatened species grows with each update of the Red List. Although this growth is to a large degree the result of increased taxonomic coverage, the downward Red List Index trends calculated for those groups that have been completely assessed clearly indicate that

the rate of biodiversity loss is increasing. Even a simple examination of the 223 species which changed status in 2008 for genuine reasons (i.e., become less threatened due to conservation efforts or become more threatened due to ongoing or increased threats), shows that while only 40 of these were species that became less threatened, 183 were listed in a higher category of threat (Appendix 12).

The 40 species that showed improved conservation status in 2008 do provide a glimmer of hope. Conservation actions are being taken for many species around the world. These range from species-specific actions to broad changes in national, regional or global policies. Measuring the efficacy of these actions in relation to individual threatened species is just beginning. But there are many case studies which show that well-focussed and concerted species-centred actions can succeed in reducing threats and improving the conservation status of species and their habitats (Box 5).

Thirty-seven of the recorded improvements in status in 2008 were for mammals, most if not all of these the result of direct conservation interventions. It is estimated that 16 bird species would have gone Extinct between 1994 and 2004 were it not for conservation programmes that tackled their threats, reduced rates of population decline and/or increased population sizes (Butchart *et al*. 2006b). In addition, during this 10-year period, 49 Critically Endangered bird species (28%) benefited from conservation action such that they declined less severely or even improved in status.

Box 5. Success stories show that conservation can work, but more is needed

Over the last 20 years, a number of species have been downlisted to lower categories of threat on the IUCN Red List as a consequence of successful conservation action that has mitigated threats, halted or reversed declines and hence increased the population and/or range size. Examples include:

In Mauritius, the Pink Pigeon and Mauritius Kestrel, Fody and Parakeet have all improved in status sufficiently to have been downlisted to lower categories of threat on the IUCN Red List during recent years. Control of alien invasive species, habitat restoration and captive breeding and release have been important actions behind these successes, leading to reduced threats, and reducing, halting and reversing population declines. In some cases, these interventions were only just in time: the Mauritius Kestrel was brought back from the brink of extinction when the population fell to just four individuals in 1974.

In Brazil, Lear's Macaw was until recently classified as Critically Endangered owing to its population having been reduced to little more than 200 birds by 2001 through unsustainable exploitation for the cage bird trade, and habitat loss. Successful conservation actions including control of trade, nest protection and habitat management have now increased numbers to almost 1,000 individuals, allowing it to be downlisted to Endangered.

African Elephant has moved from Vulnerable to Near Threatened, although its status varies considerably across its range. Poaching for ivory and meat has traditionally been the major threat to the species. Across the continent, the total population is believed to have suffered a decline of approximately 25% between 1979 and 2007, which falls short of the 30% threshold required for a Vulnerable listing. It is believed that the change in status reflects recent and ongoing population increases in major populations in southern and East Africa, largely due to the implementation of highly successful conservation efforts, and which have been of sufficient magnitude to outweigh the decreases that are taking place elsewhere across their vast range, especially in West and Central Africa.

Among amphibians, conservation success stories are few, but they do exist. The Mallorcan Midwife Toad occurs on the Balearic Island of Mallorca (Spain), where it is confined to the Serra de Tramuntana. The major threat to this species has been identified as predation and competition from introduced species such as green frogs and, more significantly, the Viperine Snake, a semi-aquatic serpent that preys upon both tadpoles and adult toads. In 1985, at the invitation of the Mallorcan government and the Durrell Wildlife Conservation Trust (DWCT) initiated a species recovery programme for the toad. This recovery programme has proven extremely successful in reversing the decline of the toad, and the species is now listed as Vulnerable.

There are also many examples of constructive policy responses, some of which are beginning to address the underlying causes of threat (see Vié *et al*. this volume). The analyses of threats in this chapter, however, show that there is a need to monitor threats very carefully, especially new emerging threats like diseases and climate change which can very rapidly have a marked impact. The new emerging threats are often hard to detect and address because of several factors operating in synergy.

It is clear that conservation actions do work; but to mitigate the extinction crisis much more needs to be done, and quickly. Conservation efforts need to be focussed and should make full and better use of the constantly improving information provided by *The IUCN Red List of Threatened Species™*.

References

Baillie, J. and Groombridge, B. (compilers and editors). 1996. *1996 IUCN Red List of Threatened Animals*. IUCN, Gland, Switzerland and Cambridge, UK.

Baillie, J.E.M., Stuart, S.N. and Hilton-Taylor, C. (eds). 2004. *2004 IUCN Red List of Threatened Species. A Global Species Assessment*. IUCN, Gland, Switzerland and Cambridge, UK.

BirdLife International. 2008a. *State of the World's Birds: indicators for our changing world*. BirdLife International, Cambridge, UK.

BirdLife International. 2008b. Threatened birds occur in nearly all countries and territories. Presented as part of the BirdLife State of the World's Birds website. Available from http://www.biodiversityinfo.org/sowb/casestudy.php?r=state&id=92. Accessed: 1st April 2009.

BirdLife International. 2008c. *Threatened Birds of the World 2008*. CD-ROM. BirdLife International, Cambridge.

BirdLife International. 2008d. Most threatened birds have small ranges. Presented as part of the BirdLife State of the World's Birds website. Available from http://www.biodiversityinfo.org/sowb/casestudy.php?r=state&id=91. Accessed 1st April 2009.

BirdLife International. 2008e. Most threatened birds have small populations. Presented as part of the BirdLife State of the World's Birds website. Available from http://www.biodiversityinfo.org/sowb/casestudy.php?r=state&id=89. Accessed: 1st April 2009.

BirdLife International. 2008f. Nearly half of all bird species are used directly by people. Presented as part of the BirdLife State of the World's Birds website. Available from http://www.biodiversityinfo.org/sowb/casestudy.php?r=state&id=80. Accessed: 1st April 2009.

Butchart, S.H.M., Stattersfield, A.J. and Brooks, T.M. 2006a. Going or gone: defining 'Possibly Extinct' species to give a truer picture of recent extinctions. *Bulletin of the British Ornithology Club*. 126A: 7–24.

Butchart, S.H.M. Statterfield, A.J. and Collar, N.J 2006b. How many bird extinctions have we prevented? *Oryx* 40(3): 266-278.

Carpenter, K.E., Abrar, M., Aeby, G., Aronson, R.B. Banks, S., Bruckner, A., Chiriboga, A., Cortés, J., Delbeek, J.C., DeVantier, L., Edgar, G.J., Edwards, A.J., Fenner, D., Guzmán, H.M., Hoeksema, B.W. *et al*. 2008. One-third of reef-building corals face elevated extinction risk from climate change and local impacts. *Science* 321: 560–563.

Chapman, A. 2006 (updated April 2007). Numbers of Living Species in Australia and the World. Report for the Department of the Environment and Heritage, Canberra, Australia. Available from: http://www.environment.gov.au/biodiversity/abrs/publications/other/species-numbers/. Accessed: 15 May 2009.

Cumberlidge, N., Ng, P.K.L., Yeo, D.C.J., Magalhães, C., Campos, M.R., Alvarez, F., Naruse, T., Daniels, S.R., Esser, L.J., Attipoe, F.Y.K., Clotilde-Ba, F.-L., Darwall, W., McIvor, A., Baillie, J.E.M., Collen, B. and Ram, M. 2009. Freshwater crabs and the biodiversity crisis: importance, threats, status, and conservation challenges. *Biological Conservation* 142(8): 1665-1673.

Cunningham, A.A. and Daszak, P. 2008. Essay 4.5. Chytridiomycosis: driver of amphibian declines and extinctions. In: S.N. Stuart, M. Hoffmann, J.S. Chanson, N.A. Cox, R.J. Berridge, P. Ramani and B.E. Young (eds), *Threatened Amphibians of the World*, pp 49–50. Lynx Edicions, Barcelona, Spain; IUCN, Gland, Switzerland; and Conservation International, Arlington, Virginia, USA.

Donaldson, J. (ed.) 2003. *Cycads. Status Survey and Conservation Action Plan*. IUCN/SSC Cycad Specialist Group. IUCN, Gland, Switzerland and Cambridge, UK.

Farjon, A. and Page, C.N. (compilers). 1999. *Conifers. Status Survey and Conservation Action Plan*. IUCN/SSC Conifer Specialist

In Brazil, Lear's Macaw Anodorhynchus leari *has moved from Critically Endangered to Endangered. Named after the English artist and poet Edward Lear, this spectacular blue parrot has increased four-fold in numbers as a result of a joint effort of many national and international non-governmental organizations, the Brazilian government and local landowners. © Andy and Gill Swash; www.worldwildlifeimages.com*

Group. IUCN, Gland, Switzerland and Cambridge, UK.

Groombridge, B. and Jenkins, M.D. 2002. *World Atlas of Biodiversity*. Prepared by the UNEP World Conservation Monitoring Centre. University of California Press, Berkeley, USA.

Hilton-Taylor, C. (compiler). 2000. *2000 IUCN Red List of Threatened Species*. IUCN, Gland, Switzerland and Cambridge, UK.

IUCN Standards and Petitions Working Group. 2008. *Guidelines for using the IUCN Red List Categories and Criteria. Version 7.0.* Prepared by the Standards and Petitions Working Group of the IUCN SSC Biodiversity Assessment Sub-Committee in August 2008. Downloadable from http://intranet.iucn.org/webfiles/doc/SSC/RedList/RedListGuidelines.pdf.

Koenig, R. 2006. The pink death: die-offs of the Lesser Flamingo raise concern. *Science* 313: 1724–1725.

Kottelat, M. and Freyhof, J. 2007. *Handbook of European Freshwater Fishes*. Kottelat, Cyrnol, Switzerland and Freyhof, Berlin, Germany.

Loh, J., Collen, B., McRae, L., Carranza, T.T., Pamplin, F.A., Amin, R. and Baillie, J.E.M. 2008. Living Planet Index. In: C. Hails (ed.) *Living Planet Report 2008*, pp. 6–20. WWF International, Gland, Switzerland. Downloadable from http://assets.panda.org/downloads/living_planet_report_2008.pdf.

Mace, G.M., Masundire, H., Baillie, J. *et al*. 2005. Biodiversity. In: R. Hassan, R. Scholes and N. Ash (eds), *Ecosystems and Human Well-being: Current State and Trends, Volume 1*, pp. 77–122. Findings of the Conditions and Trends Working Group of the Millennium Ecosystem Assessment. Island Press, Washington, DC.

Mace, G.M., Collar, N.J., Gaston, K.J., Hilton-Taylor, C., Akçakaya, H.R., Leader-Williams, N., Milner-Gulland, E.J. and Stuart, S.N. 2008. Quantification of extinction risk: IUCN's system for classifying threatened species. *Conservation Biology* 22(6): 1424–1442.

May, R.M. 1992. How many species inhabit the earth? *Scientific American* 267(4): 42–48.

Oldfield, S., Lusty, C. and MacKinven, A. (compilers). 1998. *The World List of*

Threatened Trees. World Conservation Press, Cambridge.

Peeters, M., Franklin, A. and Van Goethem, J.L. 2003. *Biodiversity in Belgium*. Royal Belgian Institute of Natural Resources, Brussels, Belgium.

Salafsky, N., Salzer, D., Stattersfield, A.J., Hilton-Taylor, C., Neugarten, R., Butchart, S.H.M., Collen, B., Cox, N., Master, L.L., O'Connor, S. and Wilkie, D. 2008. A standard lexicon for biodiversity conservation: unified classifications of threats and actions. *Conservation Biology* 22(4): 987–911.

Schipper, J., Chanson, J.S., Chiozza, F., Cox, N.A., Hoffmann, M., Katariya, V., Lamoreux, J., Rodrigues, A.S.L., Stuart, S.N., Temple, H.J., Baillie, J.E.M., Boitani, L., Lacher, T.E., Jr., Mittermeier, R.A., Smith, A.T. *et al*. 2008. The status of the world's land and marine mammals: diversity, threat and knowledge. *Science* 322(5899): 225–230.

Schippmann, U., Leaman, D. and Cunningham, A.B. 2006. Cultivation and wild collection of medicinal and aromatic plants under

Box 6. Key messages

- Not all species groups are equally threatened, but the proportion of species threatened is substantial in all groups that have been comprehensively assessed so far;

- Habitat loss (resulting in particular from agriculture, logging and residential and commercial development) remains the primary threat to most species, with over-exploitation and the impact of invasive alien species being additional significant threats;

- Assessing the conservation status of the most species-rich and less well-known groups remains a significant challenge, but new approaches are improving our understanding of the status, trends and threats to biodiversity;

- The Red List Index (RLI) shows that all species groups assessed to date are deteriorating in status: more species are slipping towards extinction than are improving in status as a result of successful conservation action;

- The fastest rate of decline of the groups measured so far is seen in the reef-building corals;

- For those groups with longer term data, the declines started to be documented over 20-30 years ago;

- The RLI shows that at a global scale the 2010 Target has not been met for the species groups we know most about: the risk of biodiversity loss is increasing rather than decreasing;

- The RLI shows that species are deteriorating in status in all biogeographic realms and ecosystems across the world;

- For birds, declines have been particularly steep in the Indomalayan and Oceania realms, and in the marine ecosystem;

- Among mammals, declines have also been most steep in the Indomalayan realm, as a result of the combined effects of hunting and habitat loss;

- Amphibians are most threatened, and have deteriorated fastest, in the Neotropical realm, in particular owing to chytridiomycosis; terrestrial amphibians are more threatened than freshwater species;

- Maintaining biodiversity is important to maintain a healthy human population as many thousands of species are used by societies all around the world for food and medicine;

- The bird, mammal and amphibian species used by humans for food and medicine are all showing declining trends in their conservation status similar to or higher than for species that are not used. The loss of these and other food and medicinal species could have a significant impact on human health in some parts of the world;

- Human use of plants and animals is not always the main threat to the species used; habitat loss and degradation or combinations of factors are often the drivers pushing these species towards extinction.

sustainability aspects. In: R.J. Bogers, L.E. Craker and D. Lange (eds), *Medicinal and Aromatic Plants: Agricultural, Commercial, Ecological, Legal, Pharmacological and Social Aspects*, pp. 75–95. Wageningen, The Netherlands, Springer, Dordrecht.

Stuart, S.N., Hoffmann, M., Chanson, J.S., Cox, N.A., Berridge, R.J. Ramani, P. and Young, B.E. (eds). 2008. *Threatened Amphibians of the World*. Lynx Edicions, Barcelona, Spain; IUCN, Gland, Switzerland; and Conservation International, Arlington, Virginia, USA.

Tan, B., Geissler, P., Hallingbäck, T. and Söderström, L. 2000 The 2000 IUCN World Red List of Bryophytes. In: T. Hallingbäck and N. Hodgetts (compilers), *Mosses, Liverworts and Hornworts. Status Survey and Conservation Action Plan for Bryophytes*, pp. 77–90. IUCN/SSC Bryophyte Specialist Group. IUCN, Gland, Switzerland and Cambridge, UK.

Walter, K.S. and Gillett, H. (eds) 1998. *1997 IUCN Red List of Threatened Plants*. Compiled by the World Conservation Monitoring Centre. IUCN, Gland, Switzerland and Cambridge, UK.

World Health Organization. 2008. Traditional medicine. Fact sheet No. 134. Revised December 2008. Available from: http://www. who.int/mediacentre/factsheets/fs134/en/ print.html. Accessed: 3rd April 2009.

The Black-and-White Ruffed Lemur Varecia variegata *from Madagascar is Critically Endangered because of habitat destruction and over-hunting; it is one of the more expensive and desired meats. © Jean-Christophe Vié*

Freshwater biodiversity: a hidden resource under threat

William R.T. Darwall, Kevin G. Smith, David Allen, Mary B. Seddon, Gordon McGregor Reid, Viola Clausnitzer and Vincent J. Kalkman

Biodiversity in hot water

While freshwater habitats cover less than 1% of the world's surface (Gleick 1996), they provide a home for 7% (126,000 species) of the estimated 1.8 million described species (Balian et al. 2008), including a quarter of the estimated 60,000 vertebrates.

Freshwater ecosystems not only provide habitat for the survival of their component species but also enable the storage and provision of clean water for human use. They also provide many important goods and services ranging from food and building materials, to water filtration, flood and erosion control, and are a critical resource for the livelihoods of many of the world's poorest communities (Millennium Ecosystem Assessment 2005). For example, tropical rivers and inland fisheries have been valued globally at US $5.58 billion per year (Neiland and Béné 2008). The goods and services provided by the world's wetlands are valued at $70 billion per year (Schuyt and Brander 2004) - a figure equivalent to the GDP of some countries ranked within the top third of the world's economies (World Bank 2008).

The high value and importance of freshwater ecosystems is often overlooked such that wetlands are frequently

Harvesting of gastropod snails, Cambodia. © Kong Kim Sreng

Box 1: Pro-poor conservation in wetlands

IUCN has produced a *toolkit* (Springate-Baginski *et al.* 2009) that will assist in wetland conservation and development decision-making. It provides an assessment approach that ensures the links between biodiversity, economics and livelihoods are captured, with a particular focus on strengthening pro-poor approaches to wetland management.

The *toolkit* was developed through integrated assessments in Cambodia's Stung Treng Ramsar Site and on the Rufiji floodplain in Tanzania. These wetlands are vital for the food security and nutrition of local communities. In the case of Stung Treng, previous biodiversity assessments had proposed total exclusion zones within the protected area, where fishing and other activities of local communities would be banned. The integrated assessment found that local communities, including migrant settlers, the landless and those depending on fish to provide daily nutrition, relied heavily on the natural resources from

within the proposed exclusion zones. The project's results are already helping to shape the management plan for the Stung Treng Ramsar Site, supporting pro-poor wetland conservation and sustainable use of the site's resources to the benefit of both local livelihoods and biodiversity. In the Rufiji, the assessment has provided a village community with vital information on the full value of their wetland resources, informing the development of their Village Environmental Management Plan.

The *toolkit* is targeted at providing policy-relevant information on individual wetland sites. Integrated assessments present the strongest case for conserving wetlands and allow local people to defend their livelihoods from developers. They can also act as an early warning system, highlighting areas of potential conflict between conservation and livelihoods.

considered as 'wastelands' ripe for conversion to alternative uses. As a result, many wetlands have been drained and converted for ostensibly more 'profitable' uses; 60% of Europe's wetlands have already been lost (UNEP/DEWA 2004) through conversion to alternative use or simply through lack of conservation over the last 50 to 100 years.

Globally, rapidly increasing human populations are putting ever-greater pressure on the goods and services supplied by freshwater ecosystems.

The long-term survival of many wetland-dependant species is therefore becoming more precarious as wetlands are increasingly exploited for human use. With the number of people living in water-scarce or water-stressed conditions projected to rise from 745 million in 2005 to 3.2 billion by 2025 (Population Action International 2006), it is therefore no surprise that global development objectives are firmly focused on the world's freshwater supply crisis. For example, the Millennium Development Goals (MDGs) include targets for halving the number of people without access to clean drinking water and sanitation by 2015. However, if we are not careful, the stage could be set for large-scale impacts to freshwater biodiversity. In order to avoid and mitigate major impacts to freshwater species and ecosystems, information on the status, distribution and value of freshwater biodiversity is urgently needed to inform the development planning process.

Data on freshwater species often exist, especially for the more developed catchment areas, but they are frequently widely dispersed in unpublished literature, and are hence effectively inaccessible, particularly in places where the greatest increase in development is taking place. Such data need to be easily and freely accessible, with species distributions available in a digital format, to enable a full understanding of the impact of developments on freshwater systems.

Girl selling fish at Stung Treng market, Cambodia.
© William Darwall

Aponogeton distachyos. *An edible aquatic plant native to the Western Cape in South Africa where it is used to prepare a local dish. It is listed as Least Concern.*
© Craig Hilton-Taylor

The information also needs to be more comprehensive (i.e., cover more taxonomic groups), reliable, robust and regularly updated. Without access to this information, development projects will not be able to mitigate or avoid actions that may have major negative impacts upon wetland biodiversity and the predominantly poor communities dependant on wetland resources.

Filling the information gap

IUCN is working with a number of partner organisations to fill the information gap on freshwater species by providing relevant data in a format suitable for use within development and conservation planning processes. This is being accomplished through conducting assessments of all known species within the following priority groups; freshwater fishes, freshwater molluscs, dragonflies and damselflies, crabs and selected aquatic plant families.

These groups were chosen because they represent a wide range of trophic levels and are amongst the better-known species within freshwater ecosystems. The biodiversity assessments collate and make available information on each species' taxonomy, ecology, distribution, conservation status (according to The IUCN Red List Categories and Criteria), use, and value to peoples' livelihoods. Given the wide range of ecological roles encompassed within these five taxonomic

groups, the information collated provides a useful indication of the overall status of the associated wetland ecosystems. Data on other species groups already assessed through this process, for example freshwater-dependent mammals, amphibians and birds, are also used to provide an increasingly informative picture on the status of freshwater species.

A regional approach (e.g., focussing on eastern Africa or Europe) has been

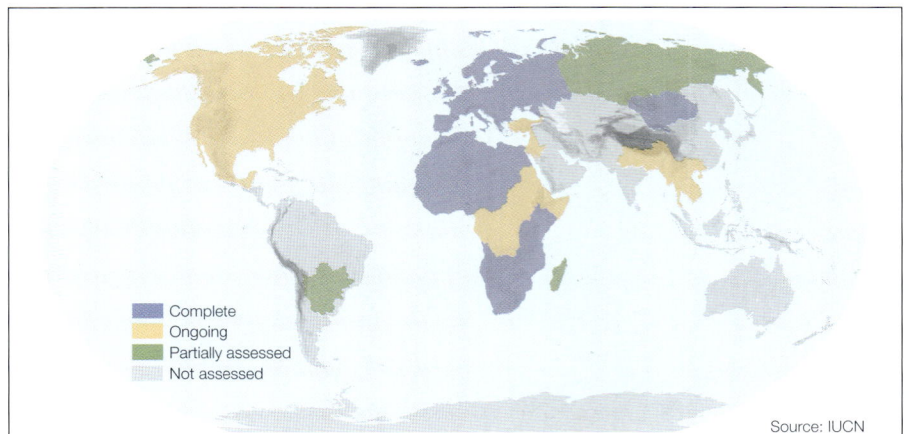

Figure 1. *The state of progress in completion of comprehensive regional assessments of the worlds freshwater fish species.*

Complete
Ongoing
Partially assessed
Not assessed

Source: IUCN

45

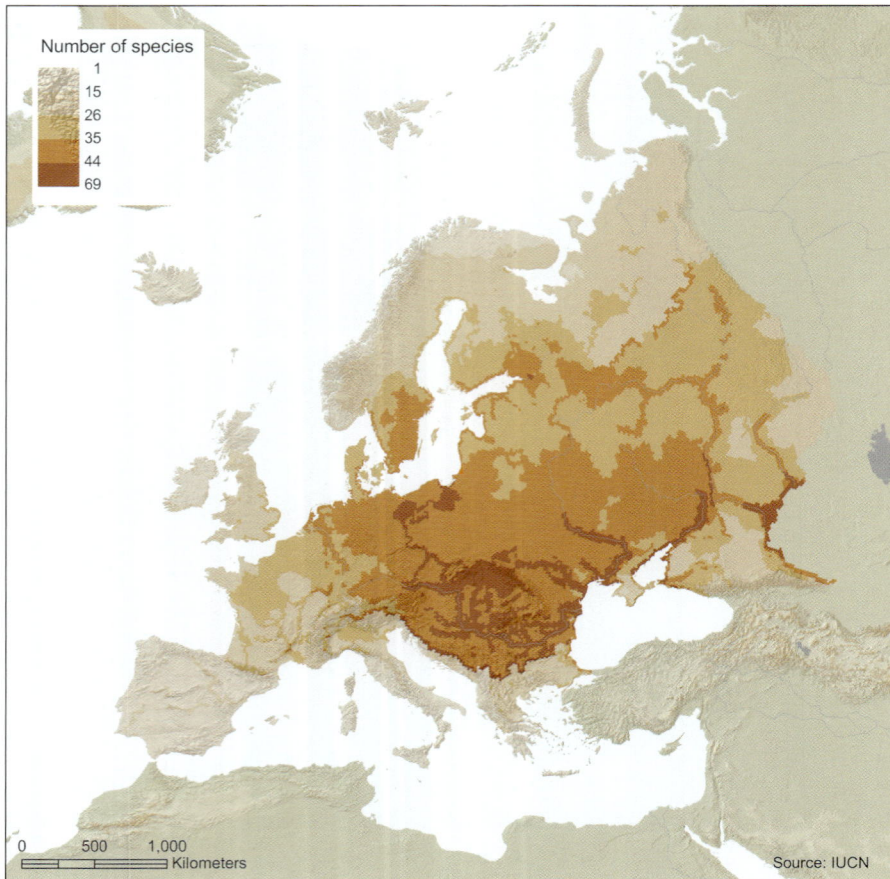

Figure 2. *Species richness of freshwater fishes in Europe.*

An estimated 2,000 of these species are currently in the assessment process and should be included in the 2009 Red List. The species still to be assessed are not evenly distributed worldwide, with major gaps including the two regions with the greatest freshwater fish species diversity, namely Asia and South America (Lévêque *et al.* 2008). These regions include some of the major river systems of the world, many of which are subject to substantial modifications (e.g. dam construction and canalization) both in place and planned. Species from many of the world's most extensive and species-rich wetland systems such as the Pantanal, the Mesopotamian marshes, the floodplain marshes of the Brahamaputra, and the Mekong Delta are yet to be assessed.

Results

The information collated through comprehensive regional assessments, where every described species from a taxonomic group within a region is assessed, has enabled identification of those river or lake basins (the logical management units for freshwater systems) containing the highest levels of species richness, threatened species, restricted range species, migratory species and/ or species important to the livelihoods of local communities. This information can be used to help prioritize conservation efforts and to inform the development planning process such that impacts of development might be minimized or

adopted. This approach provides a comprehensive picture of the status of freshwater biodiversity in the region concerned and enables IUCN to meet the information needs of regional bodies in the near term. At the same time the work continues towards completion of the longer term goal of globally comprehensive assessments for each species group. IUCN has so far completed freshwater regional assessments for eastern Africa (Darwall *et al.* 2005) and southern Africa (Darwall *et al.* 2009), and ongoing assessments for the rest of Africa are to be completed in 2009.

Global assessments for each taxonomic group are ongoing and have been completed for the amphibians (6,267 species; http://www.iucnredlist.org/amphibians) and freshwater crabs (all 1,281 species; Collen *et al.* this volume). Figure 1 shows the progress towards achieving a global assessment of all freshwater fishes; in addition to the regional assessments conducted for Africa, all known species in Europe (Kottelat and Freyhof 2007; Figure 2), Mongolia (Ocock *et al.* 2006), the endemic species of Madagascar

(IUCN 2004), and the Mediterranean basin (Smith and Darwall 2006) have now been assessed. Assessments for freshwater species in North America, Mexico, Indo-Burma and South Asia are now underway. Progress is also being made on the global dragonfly assessment with over 40% of the 5,680 dragonflies now assessed, and projects to assess all species of Europe and parts of Asia are underway. A particular strong point in the progress of the dragonfly assessment is the ongoing development of a number of large species distribution databases storing species point locality data in particular for Africa, Europe, Australia and large parts of Asia.

IUCN and Conservation International have recently joined forces to assess the estimated 27,394 freshwater species included in the five species groups mentioned above (Chambers *et al.* 2008; Bogan 2008; Strong *et al.* 2008; Yeo *et al.* 2008; Kalkman *et al.* 2008; Lévêque *et al.* 2008). Of these, only 6,000 species have so far been assessed on a global scale and included in The 2008 IUCN Red List of Threatened Species™ (Figure 3), leaving over 21,000 species still to be assessed.

Figure 3. *The cumulative total of freshwater species (fishes, odonates, molluscs, crabs, plants) on The IUCN Red List over the period 2000 - 2008.*

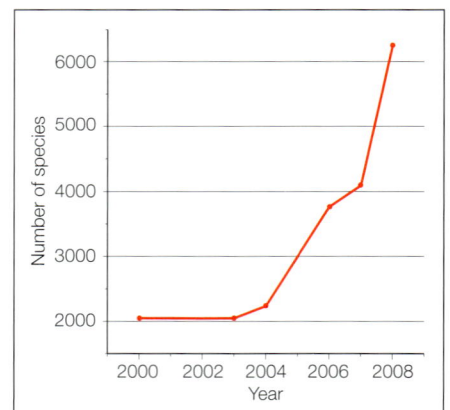

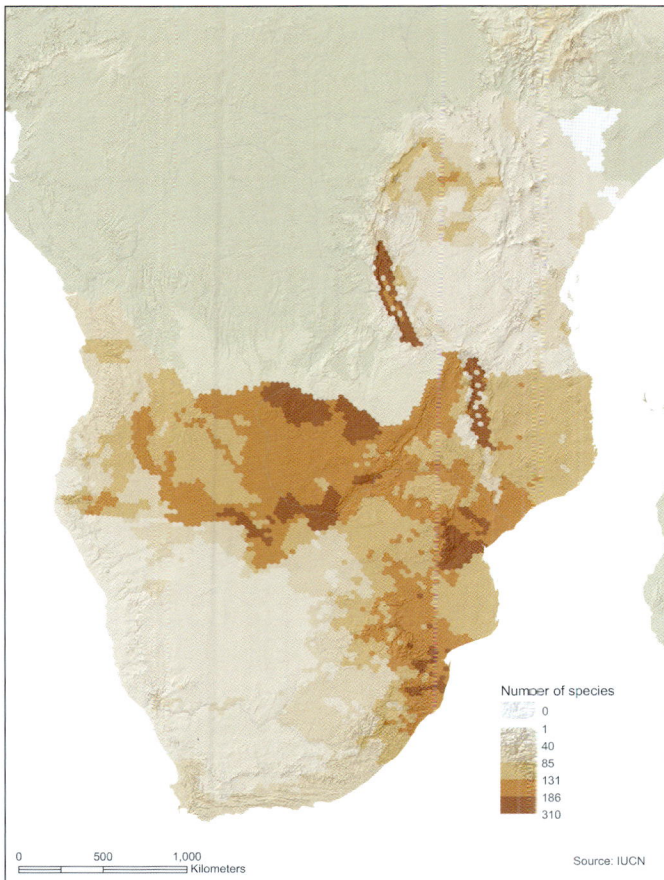

Figure 4. Distribution patterns of species richness for freshwater fishes, molluscs, odonates (dragonflies and damselflies) and crabs across eastern and southern Africa.

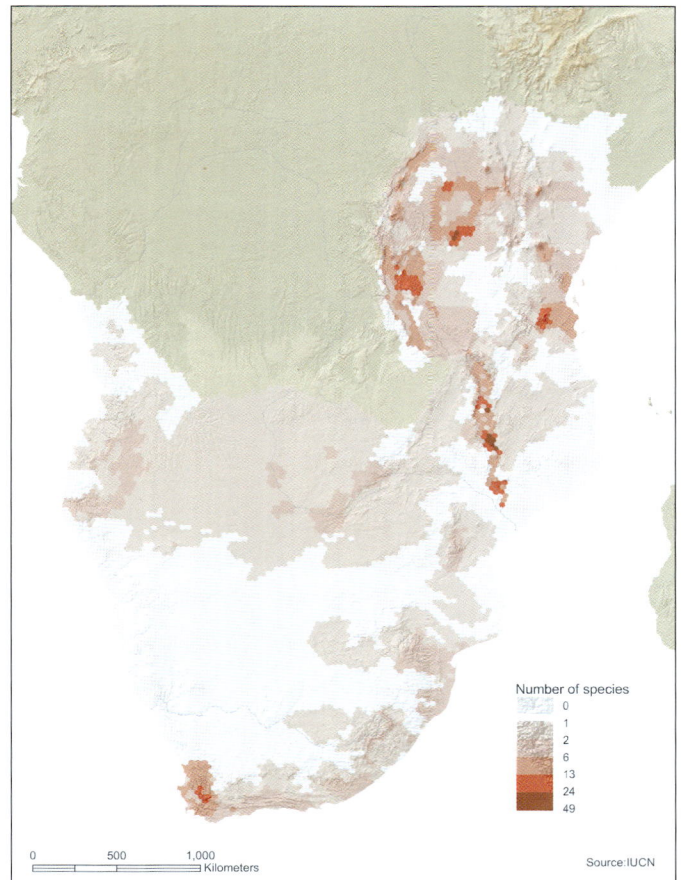

Figure 5. Distribution patterns of regionally threatened species for freshwater fishes, molluscs, odonates (dragonflies and damselflies) and crabs across eastern and southern Africa.

mitigated, and development of critical sites for biodiversity may be avoided. The results from two of the regional assessments are presented to, in brief, demonstrate the outputs and potential value of this approach.

Eastern and southern Africa

The assessments completed for eastern and southern Africa have identified Lakes Malawi and Tanganyika, and the headwaters of the Zambezi river, as containing exceptionally high numbers of freshwater species (Figure 4), whereas Lakes Malawi and Victoria, the lower Malagarasi drainage, Kilombero valley and the Western Cape in South Africa, contain some of the highest numbers of threatened freshwater species (Figure 5).

Where all species of fish, molluscs, dragonflies, damselflies and crabs have been assessed across a region, the overall level of threat to freshwater biodiversity can be better determined. Figures 6 and 7 show the Red List

status for these taxa in eastern and southern Africa respectively. In terms of the numbers of species threatened, freshwater biodiversity is more than twice as threatened in eastern Africa as

in southern Africa, with 21% of species Critically Endangered, Endangered or Vulnerable in comparison to 8%, respectively. Within taxonomic groups there are also regional variations with, for

Figure 6. Proportion of freshwater fishes, molluscs, odonates (dragonflies and damselflies) and crabs within southern Africa in each Red List Category. A total of 8% of species are assessed as regionally threatened.

Figure 7. Proportion of freshwater fishes, molluscs, odonates (dragonflies and damselflies) and crabs within eastern Africa in each Red List Category. A total of 21% of species are assessed as regionally threatened.

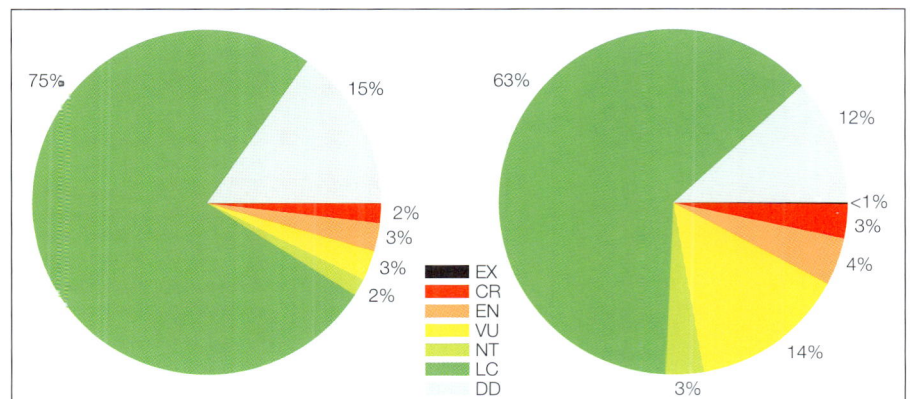

EX = Extinct, EW = Extinct in the Wild, Threatened = all Critically Endangered, Endangered and Vulnerable species; NT/LC = Near Threatened and/or Least Concern; DD = Data Deficient.

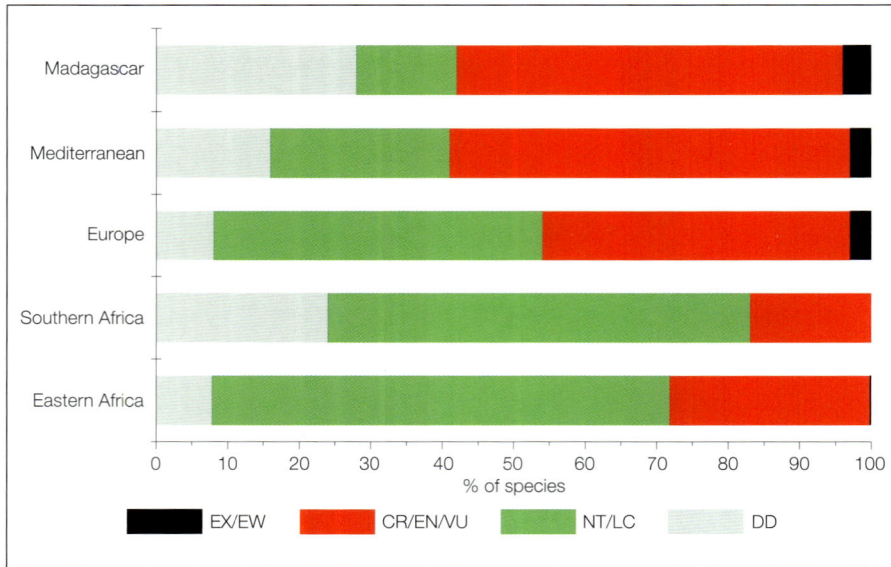

Figure 8. *Proportion of freshwater fish species by threat category in each of the regions assessed comprehensively. Only species assessed since 2000 which are endemic to each region are included in the analysis.*

Regional variation in threats to freshwater fishes
Madagascar

In Madagascar, the two most significant threats to freshwater fishes are sedimentation, which impacts over 60% of threatened species; and invasive alien species, which impact just under 45% of threatened species. Sedimentation results from the high degree of deforestation and regular burning of grasses on the 'pseudo-steppe' (Benstead *et al.* 2003). Alien invasive fish species are prevalent throughout Madagascar with at least 24 non-native freshwater fish species introduced, mostly tilapiine cichlids, as part of an ill-conceived plan to replace depleted native species fisheries which had declined largely as a result of overfishing (Benstead *et al.* 2003).

Eastern Africa

In eastern Africa, overharvesting of fishes for food is the primary threat, impacting 60% of threatened fish species; with water pollution, mainly in the form of increased sedimentation, affecting just over 40% of threatened species. These figures reflect the reported overexploitation of fisheries in a number of areas within the region (West 2001; UNEP-DEWA 2006). Increasing sedimentation of river and lake systems is largely a result of deforestation to make

example, 23% of all molluscs threatened and 28% Data Deficient in eastern Africa and only 8% threatened and 31% Data Deficient in southern Africa.

Identifying threats to freshwater biodiversity across the globe

Freshwater biodiversity is being threatened by a number of key impacts including overexploitation, water pollution, flow modification including water abstraction, destruction or degradation of habitat, and invasive alien species (Dudgeon *et al.* 2006; Millennium Ecosystem Assessment 2005). Compounding these threats are the predicted global impacts of climate change leading to temperature changes and shifts in precipitation and runoff patterns (Dudgeon *et al.* 2006).

Knowledge of current and predicted threats to species and areas they depend upon is vital to informing conservation action, policy development and the development planning process. The biodiversity assessment process allows for the major threats to species within regions to be identified and mapped.

Using freshwater fishes as an example, being one of the most widely assessed of the freshwater species groups, the level, nature, and distribution of major

threats can be identified. Of the regions assessed so far the Mediterranean and Malagasy endemic freshwater fishes are shown to have the highest proportions of globally threatened species with more than 50% of species threatened in each case, and southern Africa to have the lowest proportion with 17% of species threatened (Figure 8).

The types of threat acting upon species can also be analysed (Figure 9) and used to inform conservation and development planners. In the example of freshwater fish, the threats indentified in each region largely reflect the nature and scale of past and present development activities, as summarized below for each region.

Figure 9. *A regional breakdown of the major threats to freshwater fishes, which have led to species being assessed as threatened according to the IUCN Red List Criteria.*

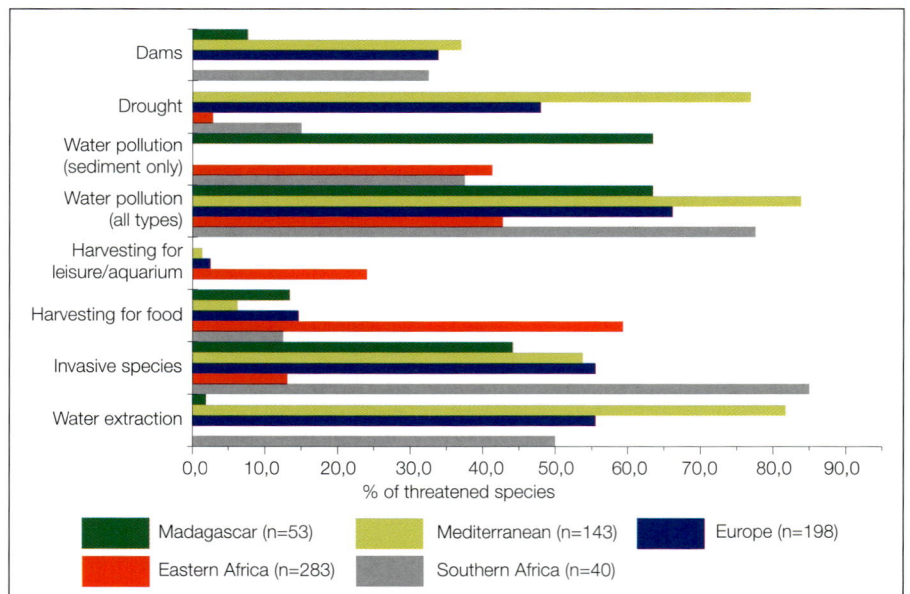

way for expanding agriculture and as a supply of fuel wood or charcoal. As an example, in Lake Tanganyika increased sedimentation has led to loss of rocky substrates along the lakeshore that provide important habitat for many of the endemic cichlid fishes (Gilbert 2003). Invasive alien species are also a major threat throughout the region in particular to many of the cichlid species endemic to Lake Victoria where a number of alien species, and in particular Nile Perch *Lates niloticus*, have been introduced to support fisheries. These species introductions have significantly changed the native species composition of the lake with many species reported as possibly extinct (Achieng 2006). The popularity within aquarium trade of many of the cichlid species in Lakes Malawi, Tanganyika and Victoria has also resulted in possible overharvesting and inadvertent impacts from fish translocations within the lakes, which again threaten many species. With more than 20% of threatened fish species in eastern Africa reported to be impacted by the aquarium trade, either now or predicted, this issue is significantly more important here than in any other region assessed to date.

Southern Africa

In southern Africa, invasive alien species are the greatest current threat to freshwater fishes, impacting nearly 85% of threatened species. Many native species in the Western Cape, Mpumalanga and the Upper Zambesi have been affected by introduced alien fishes including bass *Micropterus spp.*, Sharptooth Catfish *Clarias gariepinus*, Bluegill Sunfish *Lepomis macrochirus* and tilapia species *Tilapia* and *Oreochromis spp.* (Tweddle *et al*. 2009). Water pollution, primarily from agricultural sources, is increasing in southern Africa (UNEP-DEWA 2006) and has become a major threat to more than 60% of threatened species.

Mediterranean

Increasing human population combined with significant levels of tourism and agricultural intensification has led to high levels of water extraction and pollution throughout the region, which is impacting over 80% of the threatened freshwater fishes. Drought, already a major threat impacting more than 75% of threatened fishes (Smith and Darwall 2006) is expected to increase in severity, with many Mediterranean countries predicted to be in 'water scarcity' conditions by 2025 (UNECA 1999).

Europe

As Europe and the Mediterranean partially overlap it is no surprise that threats to freshwater fishes in Europe are similar to those in the Mediterranean. The greatest threats in Europe are water pollution, invasive species and water extraction which impact 66%, 55% and 55% of threatened freshwater fish species, respectively. There are currently 28 established alien freshwater fish species in Europe (Kottelat and Freyhof 2007). The good news, however, is that better waste water treatment, and changes in agricultural and industrial practices are leading to improvements in water quality in some parts of the region – hopefully leading to improvements in the status of associated freshwater species.

Threats to freshwater molluscs

The freshwater molluscs include a high proportion of range-restricted species, such as river rapid specialists and

Dense algal mats impacting freshwater biodiversity, fishing and transport in the Stung Treng Ramsar site in the Mekong, Cambodia – possibly a result of upstream run-off of agro-chemicals. © Alvin Lopez

Deepcheek Bream Sargochromis greenwoodi. This species is widespread and fairly common in the Okavango Delta, though rare elsewhere in the Upper Zambezi system. It is listed as Least Concern.
© Roger Bills

spring snails. The river rapid specialists require highly oxygenated clean water, and in Africa and South America a number of species are already listed as Extinct, Extinct in the Wild or Critically Endangered. The recent review of their status in western Africa and central Africa (work in progress) has shown that conditions have not improved and their habitats are typically targeted for dam construction, with water pollution from mining and increased sedimentation being secondary threats.

The spring snail group (family Hydrobiidae) is highly diverse with over 1,200 species in the family, representing around 25% of all freshwater molluscs. Currently the Red List has 283 of these species listed (182 threatened), and

in regions such as Australia, North America and Europe this group is disproportionately highly threatened. The springs where these species are found are typically exploited as water sources, with an emphasis on cleaning the point of water abstraction through actions such as concreting over habitats. Other threats include over-abstraction of the groundwaters that feed these artesian spring sources.

Climate change and desertification are increasingly recognized as important future threats to water sources of range restricted species on the edge of drylands and deserts. For example, the western African regional assessment shows that mollusc species restricted to Lake Chad are all threatened due to the rapid decline in lake

size and associated loss of mollusc habitat over the last 40 years.

Regional pattern of threats for dragonflies and damselflies

As part of an effort to expand the taxonomic coverage of The IUCN Red List, 1,500 species of dragonflies and damselflies were assessed through a sampled approach (Collen *et al.* this volume). This is about one quarter of the known dragonflies and damselflies and so provides a good insight into the status and trends of this freshwater group. About one in ten species was found to be threatened, a relatively low number compared to some other groups. The centres of species richness are the Neotropical and Indo-Malayan regions which support around two thirds of all known species. The main areas of threatened species are in the Indo-Malayan and Australian regions. The high proportion of threatened species in the Indo-Malayan area is mainly accounted for by the high number of restricted-range species in the Indonesian and Philippine archipelagos which are threatened by large-scale logging of lowland forest. In Australia the main threat is climate change, which is already resulting in the loss or degradation of freshwater ecosystems.

Box 2: Freshwater biodiversity Specialist Groups of the IUCN Species Survival Commission (SSC)

The following three Specialist Groups have been heavily involved in the biodiversity assessments reported on in this Chapter.

The Dragonfly Specialist Group (DSG)
The DSG is an active network of experts from over the world who all bring together their own regional networks. The group focuses on collating information on the 5,700 known damselflies and dragonflies. The DSG is currently active in building distribution databases in order to facilitate biodiversity assessments and conservation planning. Species distribution databases have been developed for Africa, Europe, Australia and large parts of Asia with projects for America underway. Other priorities of the group include production of field guides and training fieldworkers in particular in the tropics.

The Freshwater Fish Specialist Group (FFSG)
The FFSG was re-established in 2004. It includes a number of Regional Co-ordinators and several Special Advisors. Regional Co-ordinators each work with their own expert networks to provide the information

used in building the species assessments for The IUCN Red List. The FFSG's biggest challenge is the development of a practical global strategy for freshwater fish conservation in the face of increasing species extinction and rapidly declining fisheries worldwide. Since May 2008, the FFSG has been producing bi-monthly newsletters and is in the process of developing a dedicated new website to increase communications and the capacity building potential of the group.

The Mollusc Specialist Group (MSG)
The current focus of MSG at present is on assessment of species status and making information on the success of conservation breeding programmes accessible to others. Over the past 10 years the Group has completed 3,000 species assessments from the freshwater, terrestrial and marine biomes. The MSG newsletter *Tentacle* is published annually.

Global threat to freshwater crabs

All species of freshwater crabs have been globally assessed as part of the effort to increase the species diversity of The IUCN Red List; 16% of the species are threatened (Colen *et al*. this volume).

Key Messages

- *Freshwater biodiversity is extremely threatened*. Findings from the comprehensive assessments undertaken to date show freshwater biodiversity to be highly threatened, possibly more so than species in other systems. This is largely a result of: i) the high degree of connectivity within freshwater systems such that threats like pollution and invasive alien species spread more rapidly and easily than in terrestrial ecosystems, and ii) the rapidly increasing use and development of water resources with little regard to the requirements of the freshwater dependant species sharing the resource.

- *Public awareness of the threat to freshwater species needs to be raised*. The level of threat to freshwater biodiversity is extremely high, yet public awareness of this situation remains woefully low. Freshwater species are largely unseen by the general public, are not often considered as charismatic, and their values to people not well recognized. Conservation of freshwater species needs to be treated on a par with other more visible and charismatic species groups, such as birds and large mammals. Freshwater species need to be treated as being worthy of conservation in their own right, not simply as exploitable resources for human consumption. For example, in Europe fishes are primarily managed as agricultural resources and in many parts of the world molluscs are managed as fisheries resources, not as species of conservation significance – this is in stark contrast to the treatment of birds and mammals.

- *Freshwater species provide important ecosystem services*. Awareness of the ecosystem services provided by freshwater biodiversity needs to be raised. For example, the production of clean drinking water depends on the

Etheria eliptica. *An Endangered bivalve mollusc restricted to river rapids which is regionally threatened in southern Africa due to dam construction.* © Daniel Graf and Kevin Cummings

functions provided by many freshwater species yet this is rarely recognized. A single freshwater bivalve may filter more than seven litres of water a day – without keystone species such as these, the quality of water in river systems would most likely decline.

- *Freshwater species are important to peoples' livelihoods*. The value of freshwater species to peoples' livelihoods, which is extremely high in many countries, is not fully appreciated and is not often considered when decisions are taken on the potential development of wetland resources for alternative uses. We need to collate and make available the relevant information to demonstrate these values as a key part of future biodiversity assessments.

- *Management of water resources must take account of the requirements of freshwater biodiversity*. If we are to conserve and continue to benefit from the services provided by freshwater species we need to manage water resources as a resource for both people and freshwater biodiversity. This approach is encapsulated within the Environmental Flows concept, which aims to ensure that there is enough water to maintain environmental, economic and social benefits.

- *Protected areas must be better designed to protect freshwater species*. Existing protected areas are rarely designed to protect freshwater species. Even where species are identified by species driven legislation, without catchment based planning that extends the designated control areas to the edge of the river catchment, impacts such as from water pollution and invasive alien species will inevitably lead to species decline. Protected areas for freshwater species must be designed to employ the principles of catchment protection.

- *Support for in-situ conservation actions*. The collection of new data is essential for our increased understanding and monitoring of processes taking place within ecosystems, however

Afrithelphusa monodosa. *An Endangered freshwater crab which is restricted to a very small area of Guinea where it is threatened by habitat loss and degradation.* © Poitr Nastrecki

Box 3: Protected areas for freshwater biodiversity

The majority of protected areas, with the exception of Ramsar Sites, are designed for protection of terrestrial fauna often viewing rivers and lakes as useful park boundaries rather than as targets for inclusion and protection in their own right. For example, an analysis of eastern African freshwater fishes showed that only 1.7% of threatened species were completely included within a protected area, more than half were not included within any protected area, and for those that were partially included, on average only 13.7% of each species distribution range was included (Darwall *et al*. 2005).

Freshwater systems are unique in their high levels of connectivity and ability to transport threats rapidly and widely. This means that a freshwater species within a protected area could very easily be threatened by impacts that take place some distance away from and outside the protected area boundary. A protected area boundary will not stop the spread of threats such as pollution, invasive species, sedimentation, and altered flow regimes. If protected areas are to provide effective protection to freshwater species they need to be designed specifically to protect upper catchments and to include entire wetland systems within their boundaries.

This major gap in protection of freshwater biodiversity combined with the need to manage and protect freshwater systems at the basin scale led IUCN to develop a method for identifying important sites of freshwater biodiversity (Darwall and Vié 2005) that falls under the umbrella Key Biodiversity Area (KBA) approach. The criteria used are largely based on the framework of *vulnerability* and *irreplaceability* widely used in systematic conservation planning (Langhammer *et al*. 2007). Of course for effective conservation of biodiversity within these sites, management will need to be at the broader catchment scale to take account of the connectivity

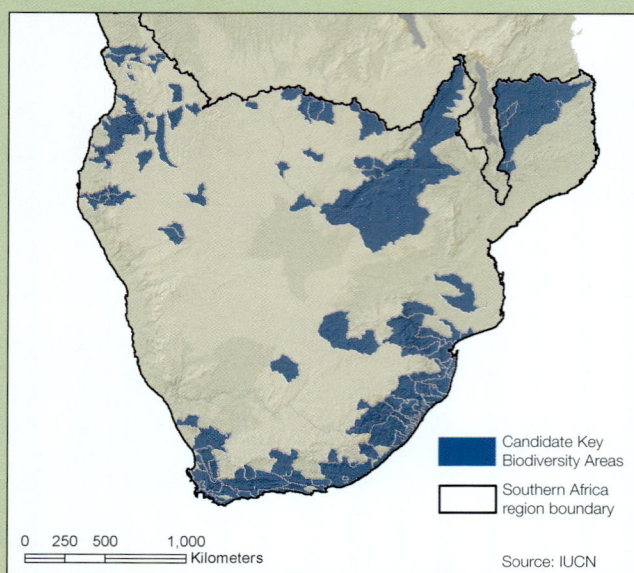

Figure 10. *Candidate freshwater Key Biodiversity Areas for southern Africa (Darwall et al. in prep.). Analysis is based on selection of level 6 river basins as derived by the Hydro1 K dataset (USGS EROS).*

and potential for rapid spread of external threats as mentioned above. Using this approach IUCN has identified candidate KBAs for southern Africa's freshwater biodiversity (Figure 10, Darwall *et al*. 2009).

---

assessments alone do not conserve species. Increased support of *in-situ* conservation initiatives capable of addressing immediate known problems is needed. Furthermore, support should be given to *in-situ* conservation education programmes which increase awareness of the problems among the local community, highlight potential ramifications for the future, build support and identify and develop practical solutions.

- *Environmental Impact Assessments (EIAs) need to take better account of impacts to freshwater species.* EIA guidelines and legislation should aim to highlight potential impacts to freshwater species. EIA specialists should be encouraged to consult the information being collated through the biodiversity assessments conducted by IUCN, its partners and others.

- *The lack of existing information for many freshwater species needs to be rectified.* A significant proportion of freshwater species remain Data Deficient, in particular

due to lack of taxonomic expertise to formally describe new species, and lack of spatial information on species distributions. This situation appears to be getting worse as the number of qualified taxonomists decreases and as opportunities for field survey become less frequent. For example, the provision of new location records for dragonflies has declined dramatically over the last 20 years. With an estimated 35% of the world's dragonflies assessed being classified as Data Deficient, there is currently little opportunity for of obtaining better information on these species. An increase in field survey combined with taxonomic training for local experts, and the publication of field guides are recommended.

References

Achieng, A.P. 2006. The impact of the introduction of Nile perch, *Lates niloticus* (L.) on the fisheries of Lake Victoria. *Journal of Fish Biology* 37: 17-23.

Balian, E.V., Segers, H., Lévêque, C. and Martens, K. 2008. The freshwater animal diversity assessment: an overview of the results. *Hydrobiologia* 595: 627-637.

Benstead, J.P., De Rham, P.H., Gattoliat, J.-L., Gibon, F.-M., Loiselle, P.V., Sartori, M., Sparks, J.S. and Stiassny, M.L.J. 2003. Conserving Madagascar's freshwater biodiversity. *Bioscience* 53(11): 1101-1111.

Bogan, A.E. 2008. Global diversity of freshwater mussels (Mollusca, Bivalvia) in freshwater. *Hydrobiologia* 595: 139-147.

Chambers, P.A., Lacoul, P., Murphy, K.J. and Thomaz, S.M. 2008. Global diversity of aquatic macrophytes in freshwater. *Hydrobiologia* 595: 9-26.

Darwall, W., Smith, K. and Vié, J.-C. 2005. *The Status and Distribution of Freshwater Biodiversity in Eastern Africa*. IUCN, Gland, Switzerland and Cambridge, UK.

Darwall, W. and Vié, J.-C. 2005. Identifying important sites for conservation of freshwater biodiversity: extending the species-based approach. *Fisheries Management and Ecology* 12: 287–293.

Darwall, W.R.T., Smith, K.G. and Tweddle, D. (eds.). 2009. *The Status and Distribution of Freshwater Biodiversity in Southern Africa*. IUCN, Gland, Switzerland.

Dudgeon, D., Arthington, A.H., Gessner, M.O., Kawabata, Z.I., Knowler, D.J., Lévêque, C., Naiman, R.J., Prieur-Richard, A.-H., Soto, D., Stiassny, M.L.J. and Sullivan, C.A. 2006. Freshwater biodiversity: importance, threats, status and conservation challenges. *Biological Review* 81: 163-182.

Gilbert, D. 2003. Impact of sediment pollution on the littoral zone of Lake Tanganyika: A case study of two cichlid fish, *Petrochromis polyodon* and *Tropheus brichardi*. The Nyanza Project, 2003 Annual Report. University of Arizona, pp. 97-104.

Gleick, P.H. 1996. Water resources. In: S.H. Schneider (ed.) *Encyclopaedia of Climate and Weather*, pp. 817-823. Oxford University Press, New York, USA.

IUCN. 2004. Red List assessment of Madagascar's freshwater fishes. IUCN, Gland, Switzerland.

Kalkman, V.J., Clausnitzer, V., Dijkstra, K.-D.B., Orr, A.G., Faulson, D.R. and van Tol, J. 2008. Global diversity of dragonflies (Odonata) in freshwater. *Hydrobiologia* 595: 351-363.

Kottelat, M. and Freyhof, J. 2007. *Handbook of European Freshwater Fishes*. Kottelat, Cornol, Switzerland and Freyhof, Berlin, Germany.

Langhammer, P.F., Bakarr, M.I., Bennun, L.A., Brooks, T.M., Clay, R.P., Darwall, W., Silva, N. de., Edgar, G.J., Eken, G., Fishpool, L.D.C., Fonseca G.A.B. da, Foster, M.N., Knox, D.H., Matiku, P., Radford, E.A.. Salaman, P., Sechrest, W. and Tordoff, A.W. 2007. *Identification and Gap Analysis of Key Biodiversity Areas: Targets for Comprehensive Protected Area Systems*. IUCN, Gland, Switzerland.

Lévêque, C., Oberdorff, T., Paugy, D., Stiassny, M.L.J. and Tedesco, P.A. 2008. Global diversity of fish (Pisces) in freshwater. *Hydrobiologia* 595: 545-567.

Millennium Ecosystem Assessment. 2005. *Ecosystems and Human Well-being: Wetlands and Water Synthesis*. World Resources Institute, Washington, DC.

Neiland, A.E. and C. Béné (eds.). 2008. Tropical river fisheries valuation: background papers to a global synthesis. The WorldFish Center

Studies and Reviews 1836, 290 pp. The WorldFish Center, Penang, Malaysia.

Ocock, J., Baasanjav, G., Baillie, J.E.M., Chimedtsen, O., Erbenebat, M., Kottelat, M., Mendsaikhan, B. and Smith, K. 2006. *Mongolian Red List of Fishes*. Regional Red List Series Vol. 3. Zoological Society of London, London.

Population Action International. 2006. People in the Balance - Population and Natural Resources at the Turn of the Millennium. Update 2006. Available at: http://www.populationaction.org/Publications/Reports/People_in_the_Balance/Summary.shtml.

Schuyt, K. and Brander, L. 2004. Living Waters Conserving the source of life. The economic value of the world's wetlands. WWF, Gland, Switzerland.

Smith, K.G. and Darwall, W.R.T. (compilers). 2006 *The Status and Distribution of Freshwater Fish Endemic to the Mediterranean Basin*. IUCN Red List of Threatened Species – Mediterranean Regional Assessment No.1. IUCN, Gland, Switzerland and Cambridge, UK.

Springate-Baginski, O., Allen, D. and Darwall, W.R.T. (eds.). 2009. *An Integrated Wetland Assessment Toolkit: A guide to good practice*. IUCN, Cambridge, UK and Gland, Switzerland.

Strong, E.E., Gargominy, O., Ponder, W.F. and Bouchet, P. 2008. Global diversity of gastropods (Gastropoda, Mollusca) in freshwater. *Hydrobiologia* 595: 149-166.

Tweddle D., Bills, R., Swartz, E., Coetzer, W., Da Costa, L., Engelbrecht, J., Cambray, J., Marshall, B., Impson, D., Skelton, P., Darwall, W.R.T. and Smith, K.G. 2009. Chapter 3. The status and distribution of freshwater fishes. In: W.R.T. Darwall, K.G. Smith and D. Tweddle (eds.). *The Status and Distribution*

of *Freshwater Biodiversity in Southern Africa*. IUCN, Gland, Switzerland.

UNECA. 1999. Freshwater stress and scarcity in Africa by 2025. United Nations Economic Commission for Africa, Addis Ababa; Global Environment Outlook (GEO) 2000, UNEO, Earthscan, London, 1999, Population Action International.

UNEP-DEWA. 2004. Freshwater in Europe. Facts figures and maps. Division of Early Warning and Assessment (DEWA), United Nations Environment Programme, Geneva.

UNEP-DEWA. 2006. *Africa Environment Outlook 2*. Division of Early Warning and Assessment (DEWA), United Nations Environment Programme. Available at: http://www.unep.org/dewa/Africa/publications/AEO-2/content/001.htm.

U.S. Geological Survey's Center for Earth Resources Observation and Science (USGS EROS) HYDRO1k Elevation Derivative Database. Available at: http://edc.usgs.gov/products/elevation/gtopo30/hydro.

West, K. 2001. Lake Tanganyika: results and experiences of the UNDP/GEF Conservation Initiative (RAF/92/G32) in Burundi, D.R. Congo, Tanzania, and Zambia. Lake Tanganyika Biodiversity Project. UNDP, UNOPS, GEF.

World Bank. 2008. Gross domestic product 2007. World Development Indicators database. World Bank. Available at: http://siteresources.worldbank.org/DATASTATISTICS/Resources/GDP.pdf.

Yeo, D.C.J., Ng, P.K.L., Cumberlidge, N., Magalhaes, C., Daniels, S.R. and Campos, M.R. 2008. Global diversity of crabs (Crustacea: Decapoda: Brachyura) in freshwater. *Hydrobiologia* 595: 275-286.

Common Hippos Hippopotamus amphibius *are found in many countries throughout sub-Saharan Africa. Recent estimates suggested that over the past 10 years there has been a 7–20% decline of their numbers due to illegal or unregulated hunting for meat and ivory, mainly in areas of civil unrest. This species is listed as Vulnerable.*
© Jean-Christophe Vié

Status of the world's marine species

Beth A. Polidoro, Suzanne R. Livingstone, Kent E. Carpenter, Brian Hutchinson, Roderic B. Mast, Nicolas J. Pilcher, Yvonne Sadovy de Mitcheson and Sarah V. Valenti

Introduction

The oceans are home to a large percentage of Earth's biodiversity, occupying 70 percent of its surface and, when volume is considered, an even larger percentage of habitable space. The oceans drive weather, shape planetary chemistry, generate 70 percent of atmospheric oxygen, absorb most of the planet's carbon dioxide, and are the ultimate reservoir for replenishment of fresh water to land through cloud formation. Trouble for the oceans means trouble for humankind.

In recent years, there has been growing concern in the scientific community that a broad range of marine species could be under threat of extinction and that marine biodiversity is experiencing potentially irreversible loss due to over-fishing, climate change, invasive species and coastal development (Dulvy *et al.* 2003; Roberts and Hawkins 1999). Governmental and public interest in marine conservation is increasing, but the information needed to guide marine conservation planning and policy is seriously deficient. The IUCN Red List of Threatened Species™ is the most commonly used global dataset for identifying the types of threat, and the levels of extinction risk to marine species (Hoffmann *et al.* 2008; Rodrigues *et al.* 2006). It forms the foundation for determining and validating marine conservation priorities, for example through the planning and management of protected area systems designed to reduce extinction risk in the sea (Edgar *et al.* 2008). However, as of 2007, the number of marine species assessed for their probability of extinction lagged far behind that of the terrestrial realm; out of more than 41,500 plants and animals currently assessed under the IUCN Red List Criteria, only approximately 1,500 were marine species. In many regions around the world, biodiversity conservation in the seas is currently taking place without the essential species-specific data needed to inform robust and comprehensive conservation actions.

Protection of our rapidly declining ocean ecosystems and species is one of the greatest challenges we face as stewards of our planet. In 2006, IUCN, Conservation International and Old Dominion University joined forces to address this gap and initiated an ambitious project (the Global Marine Species Assessment) to complete IUCN Red List assessments for a greatly expanded number of marine species. It is planned to complete Red List assessments for over 20,000 marine species by 2012. A great deal of progress has already been made, and approximately 1,500 marine species have been added to the 2008 Red List, including all of the world's known species of sharks and rays, groupers, and reef-building corals. These groups were completed in collaboration with a number of Red List Partners including the IUCN SSC Shark Specialist Group, the IUCN SSC Grouper and Wrasse Specialist Group, the IUCN SSC Marine Turtle Specialist Group.

Results

For the first time, every species in selected taxonomic groups is being assessed against the IUCN Red List Categories and Criteria. As of 2008, six major groups of marine species have been completed, and include all the world's known species of sharks and rays, groupers, reef-building corals, seabirds, marine mammals, and marine turtles (Figure 1).

Sharks and their relatives

Of the 1,045 species of sharks and their relatives (class Chondrichthyes), a high proportion (47%) are listed as Data

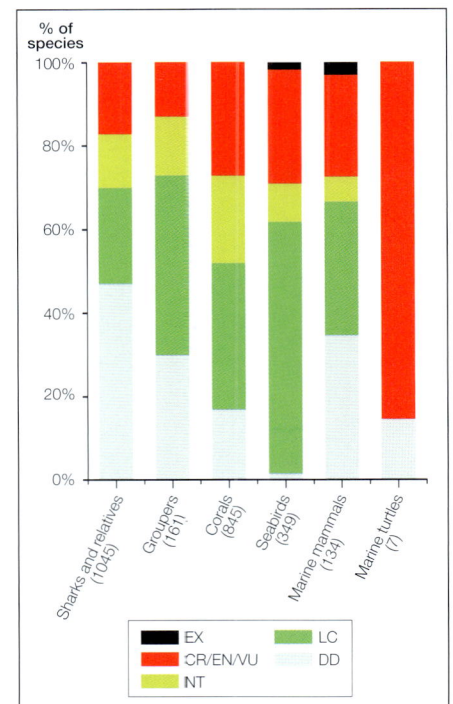

Figure 1. *Summary of 2008 Red List Categories for completed clades of marine species. Number of species assessed in each group in parentheses.*

Millions of sharks are caught each year for their fins, which are used to make the Asian delicacy shark fin soup.
© John Nightingale

Deficient compared to the five other marine groups shown. As many sharks and rays are deep-water pelagic species, they are harder to study in the wild, and less is known about their ecology and population status, including the impact of known and potentially unknown major threats. Approximately 17% of shark and ray species are in threatened categories (Critically Endangered, Endangered, and Vulnerable), and 13% are considered Near Threatened and may reach the thresholds for a threatened category in the near future if current threats are not reduced.

Much of what is currently known about sharks and rays comes from their capture in nets from both targeted and accidental catch, which is the primary threat to this species group. Sharks grow slowly, mature late, produce few young and have low rates of population increase, making them highly vulnerable to depletion with a low capacity for recovery from over-exploitation. Shark fisheries have proliferated around the world during recent decades, in response to increasing demand for shark products and as traditional fisheries come under stronger management. Millions of sharks are caught each year for their fins which are used to make the Asian delicacy shark fin soup. Sharks are being increasingly targeted for this purpose. Mortality from accidental catch (or 'bycatch') in fisheries targeting other species is just as much of a threat, if not more so, for many species. Populations of intrinsically threatened sharks can be driven to collapse un-noticed, whilst fisheries continue to be supported economically by more productive and plentiful target

species, such as bony fishes, crustaceans and squid. The life history characteristics of these species demand a precautionary approach to their exploitation; however, the lack of adequate shark fisheries management remains an over-whelming problem, exacerbated by largely unreported catches. Historically fisheries managers have given sharks low priority, but they are now receiving increasing international attention with growing concerns over the sustainability of shark fisheries.

Some species are affected by a combination of all these factors. For example, all seven species of sawfish (family Pristidae) are listed as Critically Endangered. These large unusual rays (sometimes in excess of 7 m long) are slow-growing, and populations are often isolated, with little migration between areas. They have long flattened snouts (or saws) edged with tooth-like serrations. This saw makes them extremely susceptible to bycatch in almost any fishing gear and they are also targeted for their very high value saws and fins. The 22 species of angel sharks (*Squatina spp.*) face similar threats and are among the most threatened families of sharks. Of the species of angel sharks with sufficient data for assessment, 80% are threatened and 20% are Critically Endangered.

Endemic sharks and rays with restricted habitats and geographic distributions also feature prominently among those most threatened. The endemic Brazilian Guitarfish *Rhinobatos horkelii*, Maltese Skate *Leucoraja melitensis* and Harrison's Dogfish *Centrophorus harrissoni* are all listed as Critically Endangered. All have undergone significant population declines as a result of bycatch and target fisheries. Although unsustainable exploitation appears to be the greatest threat to most sharks globally, endemic species with specific habitat preferences are also threatened by localized habitat degradation and destruction. For example, several inshore stingrays endemic to areas of Southeast Asia are being impacted by large-scale degradation and removal of mangroves, acting in combination with threats from fisheries activities.

Many wide-ranging oceanic species are also threatened. Both the Short-fin Mako *Isurus oxyrinchus* and the Long-fin Mako *Isurus paucus*, as well as the 3 species of thresher sharks (*Alopias spp.*) and the Porbeagle Shark *Lamna nasus* are all classified as Vulnerable, with some subpopulations of these species at even greater risk. Oceanic sharks are taken in large numbers in international waters. It is clear that wide-ranging, highly migratory sharks need international precautionary collaborative management, but very few countries have set catch limits for sharks and there are none in place on the High Seas. The Food and Agricultural Organization of the United Nations has urged countries and regional fishing bodies to develop and adopt Shark Management Plans, but only few have done so to date. The adoption of finning bans by fishing states, regional bodies and fisheries organizations is accelerating, which should increasingly prevent the fishing of sharks for their fins alone, but further coordinated measures are needed. A major obstacle to the formulation and implementation of management measures is the lack of data on a large proportion of species. Catches are largely unreported in many areas and improved monitoring systems are needed.

The IUCN/SSC Shark Specialist Group will continue to raise awareness about the plight of sharks and promote their effective management at national, regional and international levels. This will be done firstly, through the wide dissemination of the

The wide-ranging Short-fin Mako Isurus oxyrinchus *is listed as Vulnerable.* © Jeremy Stafford-Deitsch

results from this first complete assessment for the IUCN Red List, which can be used to inform decision makers; and secondly by continuing to advise on the development and implementation of Shark Plans and the application of conservation instruments such as the Convention on Migratory Species (CMS) and the Convention on Trade in Endangered Species (CITES).

Groupers

Groupers (family Serranidae) are found in rocky and coral reefs of the tropics and sub-tropics around the world, and are also subject to threats from over-exploitation from fishing, especially for the live fish trade, given their high commercial value. According to the Food and Agriculture Organization, about 250 thousand tonnes of groupers are

The Square-tailed Coral Grouper Plectropomus areolatus *is listed as Vulnerable. It is heavily targeted for the live reef fish food trade.* © J.E. Randall

Southern Elephant Seal pup Mirounga leonina. *The species is listed as Least Concern, as many populations are considered to be stable. However, the effects of climate change and the development of new fisheries could have a significant impact on their populations in the future. © Jean-Christophe Vié*

harvested annually, with 80% from Asia. In 1996, when the first Red List assessments were conducted on commercially important marine fishes, the groupers emerged as a particularly vulnerable group of fishes.

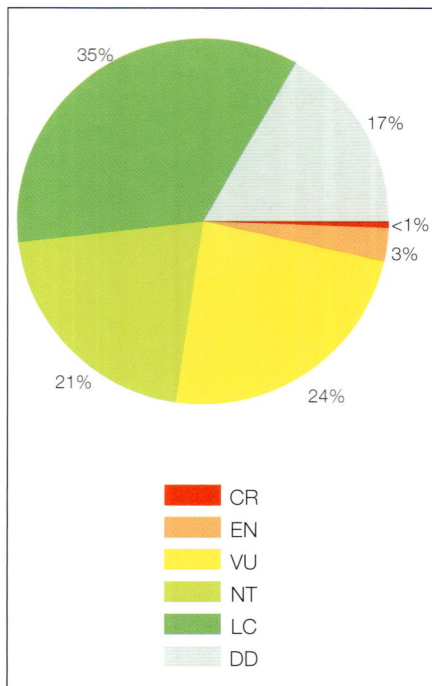

Figure 2. *Summary of 2008 Red List Categories for all 845 species of reef-building corals. Including Data Deficient species, approximately 27% of species are in threatened Categories, primarily due to climate change and anthropogenic impacts.*

Completion and updating of all 161 grouper Red List assessments has been ongoing since then, and was finalized in February 2007 at a workshop organized and held in Hong Kong by the IUCN SSC Grouper and Wrasse Specialist Group.

At least 12.4% of the world's 161 grouper species are now listed in threatened categories (Critically Endangered, Endangered, or Vulnerable), another 14% are Near Threatened, and 30% are considered to be Data Deficient. Given their long life span, with some species living up to 40 years, and late sexual maturation combined with specializations such as adult sex change ('protogyny') and aggregation-spawning, groupers are very vulnerable to fishing pressure and over-exploitation. In particular, major threats from over-fishing include targeting of spawning aggregations and uncontrolled fishing throughout the entire range of the

species on multiple life history phases, from small juveniles to adults. For example, in Southeast Asia juveniles are sometimes the major fishery target, as they are taken at sub-market size and grown-out in captivity (a practice often referred to as mariculture) until they reach a larger market size. As for other marine fishes, the most susceptible groupers to these threats are generally the longest-lived and largest species. In some cases, little is known about the species biology or impact of fishing on its population, including several species of considerable economic importance that are traded for the live seafood restaurant trade in Southeast Asia and are widely sourced in the Indian and Pacific Oceans. As a consequence, many populations are likely to be biologically over-fished, some of them very seriously, and several of these species are also considered to be threatened with extinction unless action is taken.

The Square-tailed Coral Grouper *Plectropomus areolatus*, along with such relatives as the Camouflage Grouper *Epinephelus polyphekadion*, are examples of Indo-Pacific groupers that form part of the live reef food fish trade (LRFFT) and are taken in massive numbers from their spawning aggregations and maintained alive during shipment to Hong Kong, the global trade centre for live marine fish. The demand for live fish for the luxury restaurant trade in China is massive and expected to grow in tandem with increasing wealth in the region. As much as 20% of groupers landed globally are destined for the LRFFT. However, the populations of many preferred groupers are limited and already beginning to show the strain in some areas, with several species in the trade now listed in threatened categories or as Near Threatened.

In the tropical western Atlantic, the Nassau Grouper *Epinephelus striatus*, once the most important of all groupers in the landings of Caribbean islands, is now considered Endangered. Living for several decades and taking about five years to become sexually mature and spawning in aggregations, this species has proven biologically unable to withstand decades of heavy and uncontrolled fishing and is severely reduced throughout most of its range. Regional discussions are now being

Figure 3. *The Indo-Malay-Philippine Archipelago or the "Coral Triangle" region has the highest coral species richness* **(a)** *and proportion of species in threatened categories* **(b)**.

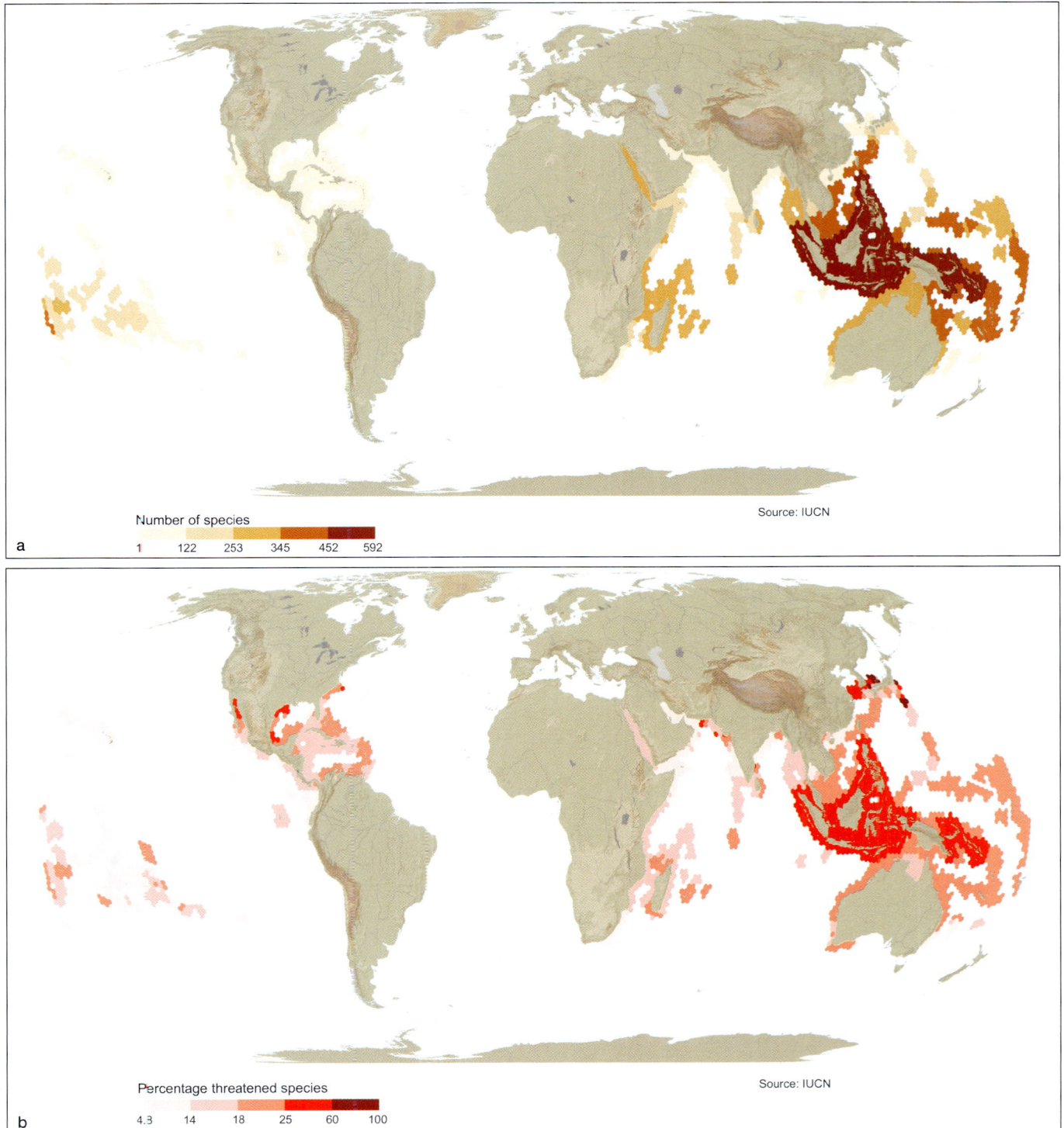

a

Number of species

1   122   253   345   452   592

Source: IUCN

b

Percentage threatened species

4.3   14   18   25   60   100

Source: IUCN

The loss of coral reef ecosystems will have devastating effects on a wide spectrum of marine species, as well as for people and nations that depend on reef resources for their livelihoods and economic security. © Jerker Tamelander

conducted to seek proper protection of the species and to introduce much-needed monitoring and management measures. More than anything, a greater awareness is needed on the plight of this species.

The objectives of the IUCN SSC Grouper and Wrasse Specialist Group's work, after determining the conservation status of each grouper species, is to focus on those species that are most threatened, address major threatening factors, fill gaps in information and raise awareness of the problems these species face. In addition to Red List assessments, ongoing projects provide support and information to enable spawning aggregations to be managed and considered in protected marine area planning, and seek sustainable practices in the LRFFT and grouper mariculture in Southeast Asia. The need for regional as opposed to national-level management and conservation initiatives for groupers should

be highlighted, as many grouper species are highly mobile as adults and all have a widely dispersive pelagic larval phase.

Corals

The world's known 845 species of reef-building zooxanthellate corals (order Scleractinia plus the families Helioporidae, Tubiporidae, and Milleporidae) have also been assessed for the first time (Carpenter *et al.* 2008). These reef-building corals are essential habitat for many species of fish and invertebrates making them the most biologically diverse ecosystems in the ocean. More than one-quarter of these corals (27%) have been listed in threatened categories, representing an elevated risk of extinction (Figure 2). Over 20% of species are listed as Near Threatened, and are expected to join a threatened category in the near future. Although approximately 17% of reef-building corals are listed as Data Deficient, more than half of these

are in the family Acroporidae, which is characterized by species with high susceptibility to bleaching and disease. Primary threats to these reef-building corals are increased frequency and duration of bleaching and disease events that have been linked to the increase in sea temperatures, a symptom of global climate change. The impacts of these oceanographic environmental changes are also compounded by anthropogenic threats including coastal development, coral extraction, sedimentation and pollution. A further sinister threat to corals is ocean acidification as a result of increasing levels of atmospheric carbon dioxide. This is reducing ocean carbonate ion concentrations and the ability of corals to build skeletons.

Globally, the Indo-Malay-Philippine Archipelago or the "Coral Triangle" has the highest number of species in threatened

Magellanic Penguins Spheniscus magellanicus *at Punta Tumbo (Argentina). This colony has decreased by nearly 30% since 1987. The species is globally listed as Near Threatened.* © Jean-Christophe Vié

categories. This region is also known as the epicenter of marine biodiversity, and has the highest coral species richness (Figure 3). Coral reefs in the Caribbean region have been impacted by recent, rapid population declines of two key species: Staghorn Coral *Acropora cervicornis* and Elkhorn Coral *Acropora palmata*, both of which have been listed as Critically Endangered. Although they have been impacted by localized warming events, coastal development, and other human activities, coral reefs in the South and Eastern Pacific have lower numbers of threatened species, but have relatively higher levels of endemism in some areas such as the Hawaiian Islands. In any region, the potential loss of these coral ecosystems will have huge cascading effects for reef-dependent species, and on the large number of people and nations that depend on coral reef resources for economic and food security.

Marine mammals

Marine mammals represent a diverse group of species and include whales, dolphins, porpoises (order Cetartiodactyla), seals (family Phocidae), Sea Otter (family Mustelidae), Polar Bear (family Ursidae), Walrus (family Odobenidae), manatees and the Dugong (order Sirenia). Almost 35% of marine mammal species are listed as Data Deficient, most of which are Cetaceans that are mainly known from individuals that have been stranded on beaches, or have been captured in fishing nets. One-quarter of marine mammal species are in threatened categories. Major threats to these species include entanglement in fishing gear, directed harvesting, the effects of noise pollution from military and seismic sonar, and boat strikes. In many regions, marine mammals are also threatened by water pollution, habitat loss from coastal development, loss of prey or other food sources due to poor fisheries

management, and intensive hunting both historically and in place today. The Polar Bear *Ursus maritimus* currently listed as Vulnerable, is primarily threatened by the accelerated loss of habitat and food resources that have been associated with climate change as large tracts of ice within the Arctic zone are rapidly disappearing. Two marine mammals have gone extinct in the past 50 years, the Japanese Sea Lion *Zalophus japonicus*, and the Caribbean Monk Seal *Monachus tropicalis*, both primarily due to intensive persecution.

Seabirds

With less than 1% of species listed as Data Deficient, seabirds (class Aves) are one of the best known groups of marine species. However, almost one-third of these species (27.5%) are threatened, and four species have gone Extinct in the past 500 years. Major threats to seabirds include mortality in long-line fisheries and gill-nets, oil spills,

The Sulawesi Coelacanth *Latimeria menadoensis*: A Living Fossil

A new addition to the 2008 Red List is the Sulawesi Coelacanth *Latimeria menadoensis*. Coelacanths are considered to be "living fossils," as they were thought to be extinct since the end of the Cretaceous period, until a specimen was found off the coast of South Africa in 1938. The Sulawesi Coelacanth was first recorded in 1997 when it was captured off the coast of Manado, Indonesia in the Sulawesi Sea. It is a relative of the Critically Endangered African Coelacanth *Latimeria chalumnae* which occurs in the Indian Ocean, and is known from Grand Comoro and Anjouan islands, the coast of South Africa, Madagascar, and Mozambique. Although the two Coelacanths from the two regions are outwardly identical, genetics show that they are actually separate species. The Sulawesi Coelacanth is only currently known from three locations and a small number of specimens, the most recent being caught in May 2007. Although the population status and trends of this species is unknown, it is believed to be a naturally small population. The Coelacanth in both regions live in caves and rocky slopes between 150 and 200 meters deep, is rarely captured, and very difficult to observe in its natural habitat.

Not much is specifically known about the biology and ecology of the Sulawesi Coelacanth, but what is known suggests that its life history traits are similar to the African coelacanth. Coelacanths are at high risk for extinction when subjected to threats because they are slow-growing and late to mature, and long-lived. They also typically produce a small number of eggs at one time. The Sulawesi Coelacanth that was caught in May 2007 in Bunaken National Marine Park was a pregnant female and had a number of large, orange-sized eggs. These large eggs are thought to hatch within the oviduct before the female gives birth to live young. Scientists in Indonesia, France and Japan are currently conducting research to better understand their reproductive biology.

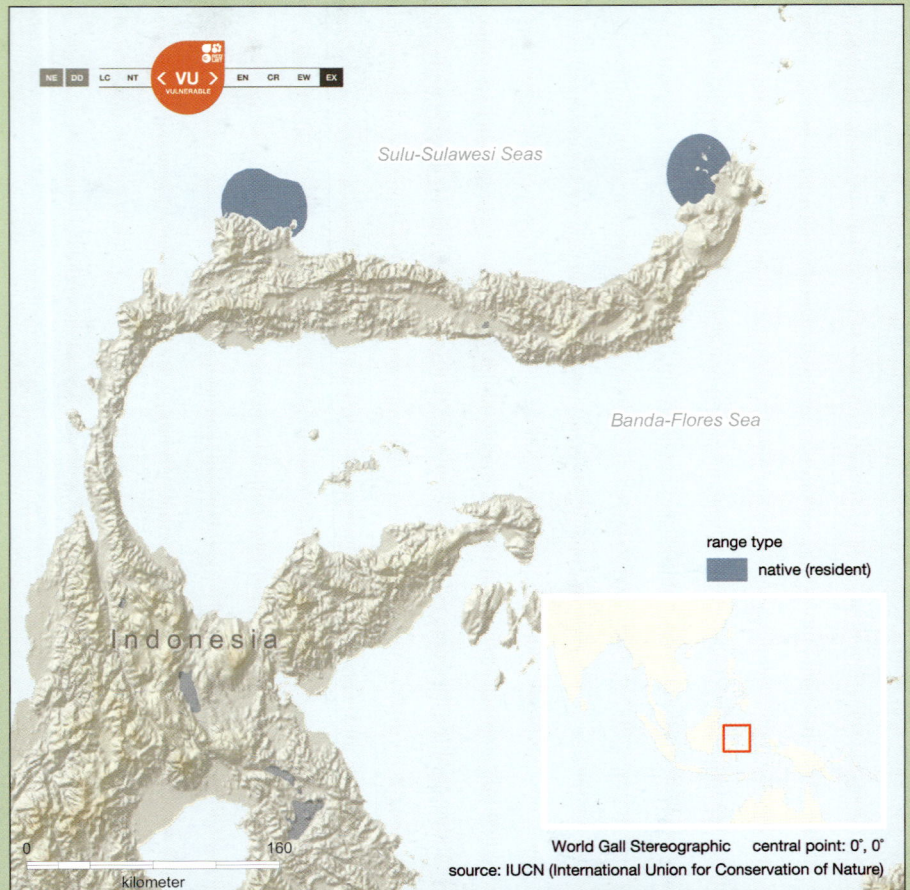

The Sulawesi Coelacanth lives in deep-water caves and rocky slopes, and is only known from a few locations along the northern Sulawesi coast, Indonesia.

Although the Sulawesi Coelacanth is poorly known, it is listed as Vulnerable given its life history, predicted small population size, and susceptibility to several threats, including capture as bycatch in deep shark nets, and by hook and line fisheries that target deepwater snapper. The Coelacanth is also sought after for large aquarium display, although no specimen has ever been successfully kept alive for this purpose. The African Coelacanth assessment is in need of updating due to new information since the last assessment made in 2000.

As awareness of the Sulawesi Coelacanth is increasing, more information is being collected about these mysterious fish. Now that the fishermen know that these fish are unique, there is a better chance of a catch being reported, and specimens being kept for further investigation. Better reporting may also give more insight into the size of the population, and the effects that bycatch may be having on the population of the Sulawesi Coelacanth. This Coelacanth is currently protected locally by Indonesian fishing regulations, and also internationally by the Convention on International Trade in Endangered Species (CITES Appendix I).

The Sulawesi Coelacanth Latimeria menadoensis *is considered a "living fossil," and has recently been added to the IUCN Red List of Threatened Species as Vulnerable. © Mark Erdmann*

The Critically Endangered Balearic Shearwater Puffinus mauretanicus only breeds in the Balearic Islands, Spain where it is threatened by predation by introduced cats and rats, incidental capture in long-line fisheries, loss of habitat from urbanization and coastal development, and water pollution.
© Ben Lascelles / BirdLife

and the impact of invasive alien species (in particular predation by rodents and cats) at the breeding colonies. Additional threats to breeding sites of seabirds are habitat loss and degradation from coastal development, logging and pollution. In many cases, seabirds are subjected to a number of these different threats at the same time. The Critically Endangered Balearic Shearwater *Puffinus mauretanicus* for example, only breeds in the Balearic Islands, Spain, where it is threatened by predation by introduced cats and rats, incidental capture in long-line fisheries, loss of habitat from urbanization and coastal development, and water pollution from high hydrocarbon and mercury levels in nearby areas. Albatrosses belong to one of the most threatened families of birds with 86% (19 species) facing extinction. Among these, the Tristan Albatross *Diomedea dabbenena* was uplisted to Critically Endangered in 2008 owing to its extremely small breeding range and a projected population decline. Modeled population declines of at least 80% over three generations (70 years) are a consequence of very low adult survival owing to incidental mortality in long line fisheries, compounded by low fledging success caused by predation of chicks by introduced mice.

Marine turtles

As of 2008, six of the seven species of marine turtle (Order Testudines) are listed in threatened categories. Only the Flatback *Natator depressus* is currently listed as Data Deficient, as there has been insufficient data

in the past to apply the criteria. Threats to marine turtles occur at all stages of their life cycle. Marine turtles lay their eggs on beaches, which are subject to threats such as coastal development and sand mining. The eggs and hatchlings are threatened by pollution and predation by introduced predators such as pigs and dogs, and eggs are collected by humans for food in many parts of the world. Once at sea, marine turtles are faced with threats from targeted capture in small-scale subsistence fisheries, bycatch largely by long-line and trawling activities, entanglement in marine debris, and boat strikes. Their life history characteristics of being long-lived, late to mature and with a long juvenile stage, combined with the many threats from human activities in the sea and on land that affect at all stages of their life cycle are among the reasons for their high risk of extinction. In addition, global climate change is now considered to be a serious, if not entirely understood threat.

Given their long generation times, global distributions, and the paucity of long-term data, assessing the risk of extinction for marine turtle species is challenging. In light of these complexities, the IUCN SSC Marine Turtle Specialist Group has pledged to complete global assessments for every species as one of its principle outcomes, and to renew them every five years to

reflect improved data and new thinking on how to apply the IUCN Red List Criteria most effectively. An Assessment Steering Committee (ASC) was established in 2006 to take on an ambitious plan for completing this mandate. Since the birth of the ASC, two species have been re-assessed. A status of Vulnerable for the Olive Ridley *Lepidochelys olivacea* has been approved at the end of a long process that included responding to a 2006 petition against the former Endangered listing. The assessment for the Hawksbill *Eretmochelys imbricata* was approved in May, 2008 with a status of Critically Endangered. A draft assessment for the currently Data Deficient Flatback *Natator depressus* is currently under review by the ASC, and will be included on the 2009 Red List. Updated assessments for the Loggerhead *Caretta caretta*, currently listed as Endangered, the Leatherback *Dermochelys coriacea*, currently listed as Critically Endangered, and the Kemp's Ridley *Lepidochelys kempii*, currently listed as Critically Endangered, are forthcoming. The Green Turtle *Chelonia mydas* assessment of Endangered was accepted by the IUCN in 2004, and will be up for review again in 2009.

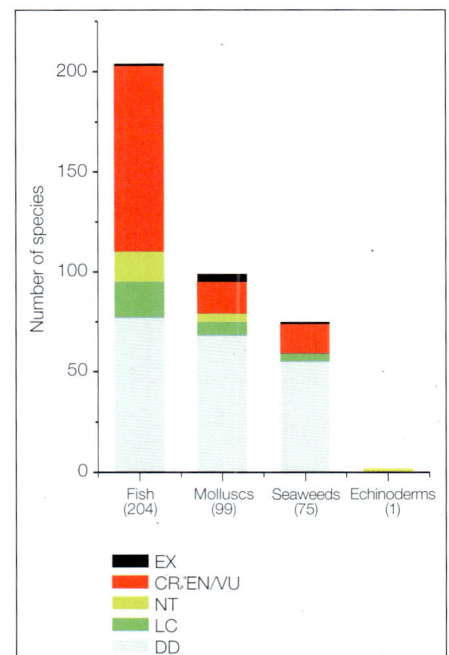

Figure 4. *Summary of 2008 Red List Categories for uncompleted clades of marine species. Number of species assessed in each group in parentheses.*

Short-beaked Common Dolphins Delphinus delphis *around the island of Kalamos, in western Greece. These animals used to be abundant until the mid 1990s. Prey depletion caused by overfishing resulted in a decline from 150 to only 15 animals in ten years. The species is globally listed as Least Concern but Endangered at the Mediterranean level. © Giovanni Bearzi / Tethys Research Institute*

Members of the IUCN SSC Marine Turtle Specialist Group are also discussing the potential for regional-scale assessments, as has been done with other taxonomic groups like sharks, and have recently completed regional assessments for Mediterranean and Hawaiian Islands turtles. The IUCN SSC Marine Turtle Specialist Group also continues to pioneer new methods for finer scale conservation priority setting for marine turtles through regular meetings that since 2003 have generated a number of useful priority setting tools including the Hazards List of the anthropogenic pressures that prevent marine turtle recovery; a Top Ten List of most threatened marine turtle populations worldwide; and a list of the Unsolved Mysteries of Marine Turtles or great unknowns that answers to which will vastly improve the ability to conserve marine turtles. The IUCN SSC Marine Turtle Specialist Group is also a founding member of the State of the World's Marine Turtles Initiative (SWOT), a network of global-scale date providers that is perfecting a mechanism for collecting, managing and disseminating information on marine turtle abundance and conservation status.

Key Messages

- The preservation and protection of our ocean resources, not only for the marine species they contain, but also for the food, products, and ecosystem services that they provide for billions of people around the globe needs to become an urgent priority. Many of the threats listed for marine species are overlapping. The development of sustainable fisheries, including the elimination of harmful fishing or harvesting practices, the enforcement of current fishery regulations, and implementation of improved fishery technology, are essential for reducing the extinction risks for marine species. Similarly, more attention needs to be aimed at reducing pollution and destructive development of coastal areas. The need to slow or reverse global climate change is becoming more important to protect our planet's resources and quality of life, not only for the survival of the plants and animals living in the ocean, but for those that live on land or in freshwater as well. The continued assessment of the status of marine species is essential for monitoring the impact of threats to the ocean's health and survival. It is only in the last 20 years that scientists have begun to worry seriously about the effects that human activities have on the marine realm, and they are discovering that the loss of biodiversity in the oceans is taking place at a similar rate to that of terrestrial areas. Climate change, in fact, may prove to have a more serious affect on marine species than those on land.

- The conservation status of the vast majority of marine species has not yet been investigated on a global scale. Other than the completed groups described here, there are fewer than 400 other marine species that have been assessed for The IUCN Red List (Figure 4). Of these, approximately 200 are marine fishes, 100 are marine molluscs, and 75 are seaweeds. Only one species of echinoderm, the edible European Sea Urchin *Echinus esculentus* has been assessed, although in many parts of the world there have been huge declines in population for commercially important echinoderms such as sea cucumbers and sea urchins. The marine species that have been assessed so far are very unlikely to be representative of the overall risk of extinction in the marine environment, as they have not been done in any systematic way. Completing entire groups of species gives a much clearer view of the status of marine species.

- To address the marine gap on the Red List, plans to compile data on geographic distributions, ecology, population numbers and trends, and threats for the world's marine species are well underway. Priority groups include all marine vertebrates (approximately 15,000 marine fishes and reptiles), as well as important habitat-forming primary producers such as the remaining corals, mangroves, seagrasses and certain seaweeds. The conservation status of species in several important invertebrate groups such as gastropod molluscs, bivalve molluscs and echinoderms (such as starfishes, sea urchins and sea cucumbers) will also be assessed. It is the largest effort to compile marine species threat data ever attempted, and will provide essential information for the protection and conservation of the world's vital marine resources.

References

Carpenter, K.E., Abrar, M., Aeby, G., Aronson, R.B., Banks, S., Bruckner, A., Chiriboga, A., Cortés, J., Delbeek, J.C., DeVantier, L., Edgar, G.J., Edwards, A.J., Fenner, D., Guzman, H.M., Hoeksema, B.W. *et al.* 2008. One-third of reef-building corals face elevated extinction risk from climate change and local impacts. *Science* 321: 560-563.

Dulvy, N.K., Sadovy, Y. and Reynolds, J.D. 2003. Extinction and vulnerability in marine populations. *Fish and Fisheries* 4: 25-64.

Edgar, G.J., Banks, S., Bensted-Smith, R., Calvopiña, M., Chiriboga A., Garske, L.E., Henderson, S., Miller, K.A. and Salazar, S. 2008. Conservation of threatened species in the Galapagos Marine Reserve through identification and protection of marine Key Biodiversity Areas. *Aquatic Conservation: Marine and Freshwater Ecosystems* doi: 10.1002/aqc.

Hoffmann, M., Brooks, T.M., da Fonseca, G.A.B., Gascon, C., Hawkins, A.F.A., James, R.E., Langhammer, P., Mittermeier, R.A., Pilgrim, J.D., Rodrigues, A.S.L. and Silva, J.M.C. 2008. Conservation planning and the IUCN Red List. *Endangered Species Research* doi: 10.3354/esr00087.

Roberts, C.M. and Hawkins, J.P. 1999. Extinction risk in the sea. *Trends in Ecology and Evolution* 14: 241-246.

Rodrigues, A.S.L., Pilgrim, J.D., Lamoreux, J.F., Hoffmann, M. and Brooks, T.M. 2006. The value of the IUCN Red List for conservation. *Trends in Ecology and Evolution* 21: 71-76.

Newly hatched Leatherback Turtles Dermochelys coriacea *that survive predation and pollution on their way to the sea will then face a myriad of other threats, including capture in fisheries, entanglement in marine debris, and boat strikes. The Leatherback Turtle is listed as Critically Endangered.* © Suzanne Livingstone

Broadening the coverage of biodiversity assessments

Ben Collen, Mala Ram, Nadia Dewhurst, Viola Clausnitzer, Vincent J. Kalkman, Neil Cumberlidge and Jonathan E.M. Baillie

While species coverage in The IUCN Red List of Threatened Species™ has increased in number each year since the inception of the Red Data Book in the 1960s, assessments have in general been restricted to the better known taxonomic groups. The number of described species still lags a long way behind the estimated global total species richness; even describing biodiversity remains a significant challenge, and so defining its status is larger still (Hilton-Taylor et al. this volume, Vié et al. this volume). However, a new initiative is being employed to broaden the taxonomic coverage of The IUCN Red List in order to better represent biodiversity, provide increased data coverage, enable a better understanding of biodiversity status, and to identify key regions and taxa that require greater conservation attention. Importantly, this will supply a broader range of species groups whose conservation status can be tracked over time. This will enhance the accuracy of key indicators of biodiversity change, and improve the breadth of information provided to inform key targets like Convention on Biological Diversity 2010 target and the UN Millennium Development Goals.

A broader view of biodiversity

The conservation status of about 2.7% of the world's described biodiversity is currently known. Clearly this limits understanding of the impact of humans on biodiversity, and with it the ability to make informed decisions on conservation planning and action. One of the major challenges for The IUCN Red List is assessing the larger groups that represent the majority of the world's biodiversity. With these larger groups of less well-known

The Usambara Eyelash Viper Atheris ceratophora (Vulnerable) pictured here eating a frog (Afrixalus sp.) gets its name from the Usambara Mountains, part of the Eastern Arc range where it is found in Tanzania. This species is threatened by high rates of deforestation and habitat degradation due to agriculture and increasing human population, which is occurring throughout the Eastern Arc range. © Michele Menegon

organisms, a comprehensive survey of extinction risk for the whole group is not feasible. To illustrate the problem, consider the estimated 298,506 plant species of which roughly 85% have been described but of those around 4% have had their conservation status assessed (Hilton-Taylor et al. this volume). Producing conservation assessments for the remaining 96% within a reasonable timeframe is not possible. However the need for a broader view of the status of biodiversity is urgent. It has become increasingly clear that taxa differ in the relative level of threat they face, with certain groups at higher risk of extinction

than others (Baillie et al. 2004, Purvis et al. 2000). Determining the reasons for these differences is one of the key actions needed for proactive conservation. As a first step, increased coverage of baseline data is required for The IUCN Red List. A truly global picture of biodiversity requires coverage of all major taxonomic groups. We can no longer afford to base conservation decisions on a restricted and non-representative subset of species.

A new approach has been developed (Baillie et al. 2008) that takes a large random sample of species groups – just as when

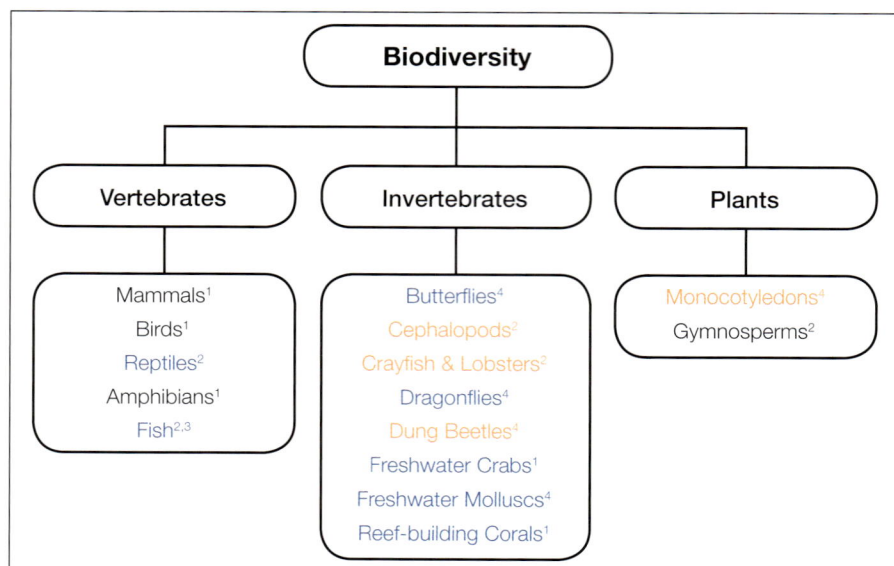

Figure 1. Groups being assessed using the sampled approach and comprehensively. 1 = Comprehensive assessment complete; 2 = Comprehensive assessment underway; 3 = Sample of Freshwater fish complete, marine fish underway; 4 = sampled approach group. Blue are phase 1 groups (2007-2009), orange are phase 2 groups (2008-2010).

forecasting election results, a poll of voters is taken. Using a random sample of 1,500 species from a group, this approach allows for the identification of the general level of threat to each group, the mapping of areas likely to contain the most threatened species, the identification of the main drivers of threat and helps pinpoint what key actions are required to address declines in the group. Results from this new approach are set to revolutionize our understanding of the status of the world's species. It has enabled an understanding of the conservation status of reptiles for the first time and the status of the world's terrestrial vertebrates (mammals, birds, amphibians, and reptiles) can be defined. In addition, it is now possible to describe and therefore address the threat faced by a number of the megadiverse groups of invertebrates. In turn, a greatly expanded understanding of the impact that humans are having on the world's species will be used to feed information into internationally important agreements aiming to address biodiversity loss.

The first results of this new approach to IUCN Red Listing, presented here, provides new insights into our understanding of the status of the world's species, and attempts to do so in a systematic way that can be built upon in time. Further assessments will be delivered in 2009 and 2010 (Figure 1).

Status of the world's terrestrial vertebrates

Understanding where threatened species are found aids conservation action, informs key biodiversity targets and allows

us to better evaluate human impact on biodiversity. Previously, the true proportion of threatened species amongst the under-studied vertebrate classes (reptiles and fish) has only been partially known. Combining the new assessments of reptiles with the new global mammal assessment data (Schipper *et al.* 2008), and updated datasets on amphibians (IUCN 2008) and birds (BirdLife International 2008), a more accurate picture of globally threatened terrestrial vertebrates can be drawn. These new data will aid in the efforts to move from monitoring, to prediction and pre-emptive action. Almost one in four (24%) terrestrial vertebrates within non Data

Deficient categories are threatened with extinction (Figure 2a). The exact threat level is unknown, as the status of 2,601 Data Deficient species is undetermined, but ranges between 21% (assuming no Data Deficient species threatened) and 32% (assuming all Data Deficient species threatened).

Terrestrial vertebrates play a key role in the provisioning of many ecosystem services. Terrestrial ecosystems supply humans with raw materials, food and support livelihoods, so their healthy function is of great importance. Terrestrial vertebrates face several major threats that must be tackled in order to maintain terrestrial systems. The overwhelming threat to terrestrial species is loss and degradation of habitat, primarily through conversion of land to agriculture and the impact of logging (Figure 2b - note that birds are excluded from this analysis, see figure legend). Of secondary importance is the impact of pollution and intrinsic factors such as limited dispersal and restricted ranges faced by small populations, particularly on islands. It is difficult to estimate the impact that climate

Figure 2. Proportion of terrestrial vertebrates in each Red List Category **(a)** and major threats to globally threatened terrestrial vertebrates **(b)**. Note that birds are excluded from this analysis as they are coded against the new threat classification scheme (Salafsky et al. 2008).

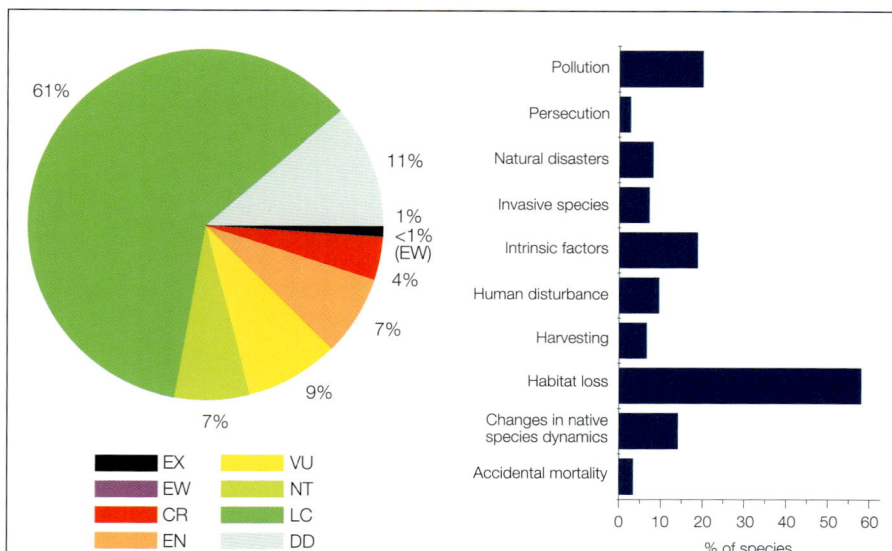

The Lyre Head Lizard Lyriocephalus scutatus (Vulnerable) is endemic to the southwest wet zone of Sri Lanka. Habitat degradation caused by human activities including the clearing of land for agriculture, conversion to plantations, logging, mining, and the pressures linked with human settlements, have resulted in severe deforestation. It is estimated that only 5% of the original wet zone rainforest remains in Sri Lanka. The international pet trade may also pose a threat as individuals are harvested from the wild.
© Ruchira Somaweera

change has already had, but it is clear that in the future this will be a dominant driver of extinction (Foden et al. this volume).

Neither species richness, nor threats to these species are evenly distributed across the planet (Sanderson et al. 2002). Threats to vertebrates vary across realms, according to the intensity and history of the threats. Unfortunately the regions that maintain the highest diversity also tend to be the most threatened, and usually the least well understood (Collen et al. 2008; Mace et al. 2005). The Indo-Malayan and Neotropical realms are consistently home to the greatest proportion of threatened species of terrestrial vertebrates. Oceanic islands also contain very high proportions of threatened species, for example Oceania, while having a comparatively low species richness maintains a highly at risk terrestrial vertebrate fauna (Figure 3). The final step in evaluating vertebrates is to produce assessments for fishes. This will be completed in the spring of 2009. This will enable, for the first time, the status of the world's fishes, and the status of the world's vertebrates to be reported.

The geography of reptile threat

Reptiles are an ancient, diverse and versatile group, present on all continents except Antarctica, having colonized many of the earth's habitats. The first evaluation of a representative sample of reptiles shows that over one in five (22%) reptiles in non Data Deficient categories are threatened with extinction (Figure 4a). The exact threat level is unknown, as the status of 284 Data Deficient species is undetermined, but ranges between 18% (assuming no Data Deficient species threatened) and 37% (assuming all Data Deficient species threatened). The great majority of species show a negative

response to anthropogenic manipulation of habitat worldwide. The principle threats to reptile species are habitat loss and degradation. Additionally, overexploitation, principally through uncontrolled pet trade is a problem in certain families (Figure 4b).

The proportion of species threatened varies across reptile groups. For example, 43% of crocodilians are threatened. This contrasts with 12% of snakes, and 20% of lizards. Broad species level differences are likely to reflect differences in geography, range size, habitat specificity and biology, as well as threat intensity. Identifying the reasons for these differences among groups is a major goal for conservation biology. It is important to understand where threat impact is most intense, and work to reverse the process. This, in combination with identifying those attributes that predispose species to a higher risk of

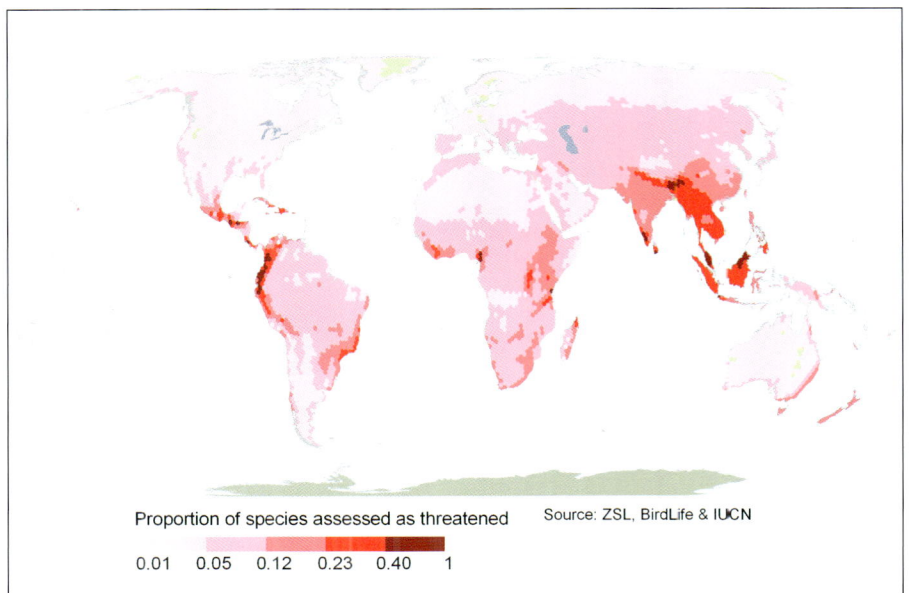

Proportion of species assessed as threatened

Source: ZSL, BirdLife & IUCN

0.01   0.05   0.12   0.23   0.40   1

Figure 3. Threatened species richness map for terrestrial vertebrates (n = 9,606 species).

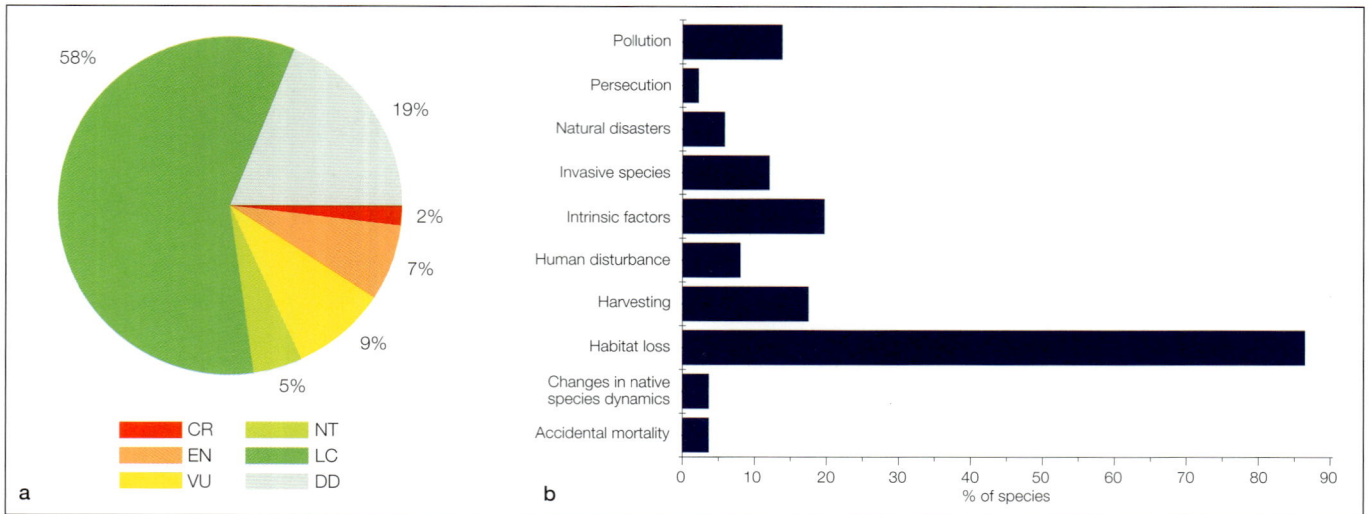

Figure 4. Proportion of reptiles in each Red List Category (a) and the major threats to globally threatened reptiles (b).

extinction will help predict outcomes of different future scenarios and therefore facilitate proactive conservation efforts. This will help to minimize human impact on biodiversity.

Indo-Malaya is the most species rich biogeographic realm for reptiles, as it is for many other species groups. The Indo-Malayan realm also has the greatest density of threatened (CR, EN, and VU) species (Figure 5). High levels of deforestation and over-exploitation are prevalent throughout the area, and are thought to be responsible for these elevated levels of threat. However, it is in the Neotropical realm where the most species with an extremely high risk of extinction can be found. Almost half of the Critically Endangered reptiles are endemic to the Caribbean, Central or South America (43%), a percentage more than double that for any other realm. While the threats to reptiles in the Neotropical realm are not unique, predation by introduced mammals and habitat loss (primarily due to conversion of land for agriculture, urban development and tourism) are common problems. Approximately one in five Neotropical species are distributed in the Caribbean. These species are more likely to have narrower ranges, smaller populations, and limited genetic diversity because the distribution of these species

is restricted to islands. Therefore, in the presence of threats, these island species are at a higher risk of extinction.

Although not all reptile species have been assessed, a random sample reveals new details about the threats faced by reptiles. Others have been updated from old 1996 assessments. In the well-known crocodile group, the Cuban Crocodile *Crocodylus rhombifer* has been uplisted to Critically Endangered (Box 1). New assessments of the IUCN Red List status for some of the more poorly known groups have also been possible. For example, amphisbaena (worm lizards) are little studied, due to their burrowing lifestyle (Box 2). Many of the group that can be assigned a category are Least Concern, protected from human impact by their burrowing habits.

However, a number of species are listed in threatened categories, principally due to having restricted distributions in threatened habitats. Time will tell whether those classified as Data Deficient (DD) turn out to be threatened or not.

Assessments of species in nine families that have never before been red listed allow a more confident appraisal of the status of the world's reptiles. The threats to reptiles and regions where threatened species are concentrated that have been identified in this random sample pave the way for a comprehensive assessment of all reptile species, while providing timely status information for the CBD 2010 target. This approach allows identification of key attributes from which comprehensive coverage will allow finer scale analysis.

Figure 5. Threatened species richness map for reptiles (n = 244 species).

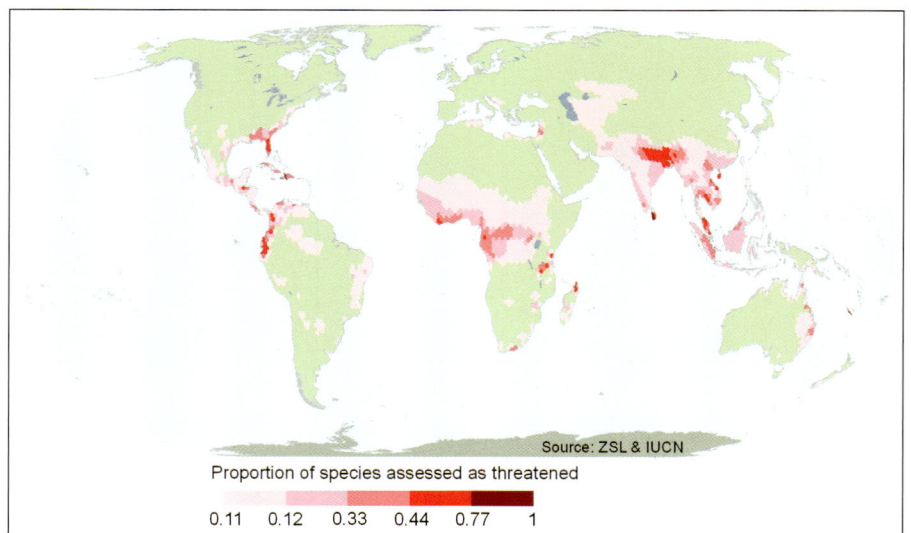

Box 1: Uplisting the Cuban Crocodile *Crocodylus rhombifer* (Cuvier, 1807)

The Cuban Crocodile is a species of freshwater crocodile renowned for its leaping ability, allowing it to prey on arboreal mammals. This species is a relict from the Pleistocene Era and has the narrowest distribution of any extant crocodilian due to its preference of peat swamp habitats. This unusual crocodile once had a wider distribution in the Caribbean but is now confined to the Zapata Swamp in Cuba with a smaller subpopulation

in Lanier Swamp on the Isla de la Juventud off the southwest coast of Cuba. Like many other threatened species, the Cuban Crocodile is negatively impacted by habitat degradation. In the past, intensive charcoal burning was the main cause of habitat loss, but now development of the tourism industry, land transportation, and agriculture are degrading the areas in which this species inhabits. Possible sea level rise because of global warming in the future could also degrade its swamp habitat. As well as these environmental threats to the Cuban Crocodile, this species is directly threatened by poaching for its meat. In 1996, the Cuban Crocodile was assessed as Endangered. However, recent genetic analysis has shown that extensive hybridization with the American Crocodile *Crocodylus acutus* (Cuvier, 1807) has taken place, probably over many centuries. Hybridization is a newly recognized major threat to the Cuban Crocodile because it is decreasing the genetic purity of this highly threatened species, which has just a small natural distribution remaining. A reassessment of the conservation status of the Cuban Crocodile has now listed this species as Critically Endangered.

The Cuban Crocodile Crocodylus rhombifer *(Critically Endangered) is a species of freshwater crocodile renowned for its leaping ability, allowing it to prey on arboreal mammals. Hybridization with the American Crocodile,* Crocodylus acutus *is a newly recognized major threat to the Cuban Crocodile because it is decreasing the genetic purity of this highly threatened species, which has just a small natural distribution remaining. © John White*

First globally representative groups of invertebrates on The IUCN Red List

Existing biodiversity information is strongly biased toward the terrestrial megafauna and megaflora, and to temperate rather than tropical areas. However the highest extinction risk and therefore greatest loss of biodiversity is expected to be suffered by invertebrates (Dunn 2005, Thomas *et al.* 2004). These first invertebrate group assessments show a great range in extinction risk, with dragonflies and damselflies (Odonata) that could be assigned a conservation status the least threatened (14%) and corals the most threatened (32.8%) (Polidoro *et al.* this volume, Carpenter *et al.* 2008). The main message drawn from these invertebrate assessments is that current data suggest invertebrates may be just as threatened as vertebrates, and in certain taxa, more so. The status of different invertebrate groups is extremely variable; therefore generalising across invertebrates is not particularly meaningful.

Box 2: The secret lives of fossorial reptiles

There are two main groups of burrowing reptiles – Amphisbaenia and Scolecophidia. Amphisbaenia, known as worm lizards, are mostly limbless species, uniquely adapted to a burrowing habitat, and occur on both sides of the Atlantic. The blind snakes belonging to the infraorder Scolecophidia, consist of snakes in the Typhlopidae, Leptotyphlopidae, and Anomalepididae families. These burrowing snakes are distributed in tropical and subtropical areas around the world. Burrowing reptiles are ecologically important because they aerate soil and help regulate the populations of insects such as ants and termites. There are over 600 species of burrowing reptiles within these two groups and an estimated half of these species are Data Deficient. Clearly, the fossorial nature of these species means that surveying is more problematic than for ground-dwelling or even arboreal species. Even though ranges are often narrow, the distribution of these species is still not fully understood, and this further highlights that too little effort is being put into researching these unusual and important reptiles. This situation is more worrying as it has been shown that habitat loss and degradation have a negative impact on burrowing reptiles because they are unlikely to occupy or re-colonize fragmented landscapes. Ground compaction is also a threat because it reduces the burrowing ability of these species. One in five of burrowing reptiles within non Data Deficient categories are considered threatened. Therefore more research is urgently needed into these

The Speckled Worm Lizard Amphisbaena fuliginosa *(Not Evaluated) is a burrowing reptile. This species was thought to be restricted to the rainforests of northern South America but has recently been discovered in Brazilian Cerrado, a dry savanna habitat. © Laurie Vitt*

species to better understand, not only the conservation status of them, but also their biology and ecology.

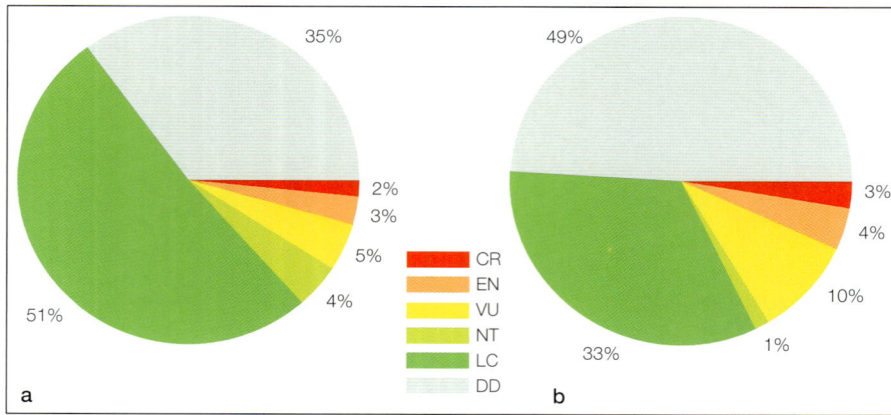

Figure 6. *Proportion of dragonflies and damselflies* **(a)** *and proportion of freshwater crabs in each Red List Category* **(b)**.

Freshwater ecosystems provide raw materials, food and support livelihoods; they perform many important environmental functions and contribute to general human well-being. In spite of only covering about 1% of the total land surface, inland waters are home to around 126,000 aquatic species (Darwall *et al.* this volume). The results from the first freshwater invertebrate group assessed, the Odonata, show that 14% of species in non Data Deficient categories are threatened with extinction (Figure 6a). The exact threat level is unknown, as the status of 526 Data Deficient species is undetermined, but ranges between 9% (assuming no Data Deficient species threatened) and 44% (assuming all Data Deficient species threatened). However, Odonata only comprise a small invertebrate order, with above average dispersal ability and wide distribution ranges, so results are likely not to be indicative of many other invertebrate groups. The majority of the threatened species in this group inhabit lotic (flowing) waters (Clausnitzer *et al.* 2009). A combination of a more specialised ecology and the higher environmental pressure on these waters may provide the explanation for the increased risk to species in these habitats.

Almost one fifth of the world's crabs are restricted to freshwater, a total of 1,281 species. Overlooked in comparison to their more speciose marine counterparts, they are distributed throughout almost all freshwater habitats in tropical regions. Traits such as low reproductive output in combination with fragmentation caused of human impact to freshwater habitats has resulted in relatively high threat levels in this group that is defined by high levels of endemism.

A total of 32% of freshwater crabs in non Data Deficient categories are threatened with extinction (Figure 6b) (Cumberlidge *et al.* 2009). The exact threat level is unknown, as the status of 629 Data Deficient species is undetermined, but ranges between 16% (assuming no Data Deficient species threatened) and 65% (assuming all Data Deficient species threatened). The majority of the threatened species have highly restricted ranges. This exposes them to the impacts of human disturbance from habitat loss, particularly in forested regions, with alteration of water regimes and pollution being most frequently cited as cause of threat.

A map of the distribution of threatened species from the crabs and dragonflies reveals some centres of threat for freshwater systems (Figure 7). Marked concentrations of threatened species exist in Viet Nam, Thailand, Cambodia, Malaysia and the Philippines in Southeast Asia; Sri Lanka (Box 3) and the Indian Western Ghats in South Asia; and Colombia and Mexico in central and South America. These patterns are heavily influenced by the distribution of restricted-range species. Healthy freshwater invertebrate populations are indicative of freshwater systems that are able to provide critical services, such as flood control, economic and livelihood benefits. However, river basin and wetland management is complex, as they are open systems with ill-defined boundaries.

Comparing invertebrates to vertebrates

Rather than differences between vertebrates and invertebrates *per se*, the assessments highlighted in this chapter, suggest that key differences

exist between system and habitat, regardless of whether the species is a vertebrate or not. Freshwater groups are consistently at higher risk than their terrestrial counterparts, yet it is a system about which we still know very little. Restricted-range species tied to particular habitats are at greater risk in all systems (Box 4a) in comparison to wide ranging species with more generalist ecological requirements. While some of the major threatening processes differ between taxa (e.g., broadly speaking, over-exploitation is less of a threat for Odonata and crabs than for terrestrial vertebrates, particularly mammals), habitat loss and degradation represent major threats across all groups.

In situations where habitat loss is the primary cause of decline, it is reasonable to assume that there might be a positive correlation between declines in vertebrate and non-vertebrate populations. However, where threats such as exploitation or pollution are the cause of a decline, the expectation might be that trends observed in one set of species will not necessarily be indicative of population trends in other species in the same ecosystem. The impacts of climate change remain complex, though an increased understanding of species biology may provide some clues (Box 4b) (Foden *et al.* this volume). Addressing the

Box 3: Sri Lankan freshwater crabs under threat

Sri Lanka is rich in freshwater crabs, with 49 out of its 50 species endemic to the island. The biodiversity assessment presented in this chapter listed 80% of all Sri Lankan freshwater crabs as threatened (Critically Endangered, Endangered, or Vulnerable), with a shocking 50% of all species CR, and possibly on the brink of extinction. The elevated levels of endemism and threat in Sri Lanka are surprisingly high considering the moderate size of the island (<65, 000 km²) and its close proximity to the Asian continental mainland. Threats to the Sri Lankan freshwater crabs include deforestation, pollution from excessive pesticide use, and the impact of alien invasive species on native species. Many of the threatened species of freshwater crabs have a limited distribution in the rainforests of the wet zone in the southwest part of the island, where they are increasingly under pressure from the rising human population density.

Johora singaporensis (Critically Endangered) is a freshwater crab that is endemic to Singapore. It was previously known from two locations, one of which is a nature reserve. However, despite intensive surveys this species has not been sighted in the nature reserve since the early 1990s, suggesting that it no longer exists in this location. Acidification of water is suspected to be the cause of this extirpation. The surviving population of this species is under severe threat due to recent development of the land, which has lowered the water table, and continuing decline in the quality of its habitat. © Peter Ng

lack of invertebrate coverage on the IUCN Red List to date, is particularly pressing in view of the ecosystem services that they provide; therefore, it is important that they be assessed, inventoried, monitored, and protected (Rohr *et al*. 2007).

Evaluating trends in biodiversity

In order to mitigate biodiversity loss effectively, greater investment of conservation attention is required in tropical regions where there is the most to lose in terms of species richness, and where species groups have to date been largely ignored (Collen *et al*. 2008). Broad-reaching global legislation may provide an impetus for such investment. One important example is the CBD, under which the 190 signatory nations have ambitiously committed themselves to actions to "achieve by 2010, a significant reduction of the current rate of biodiversity loss at the global, regional and national levels" (UNEP 2002). Assessing progress towards this important goal requires data on the status and trends in biodiversity for a given group, county or region.

IUCN and its partners are working towards this by evaluating change in conservation status with The IUCN Red List Index (RLI) (Butchart *et al*. 2004, 2005, 2007, Vié *et al*. this volume). By conducting conservation assessments at regular intervals, changes in the threat status of species in a taxonomic group can be used to monitor trends in extinction risk. As exemplified by the RLIs calculated for birds (Butchart *et al*. 2004), amphibians (Baillie *et al*. 2004), mammals and corals

Figure 7. *Threatened species richness map for freshwater crabs (n = 210 species) and dragonflies and damselflies (n = 136 species).*

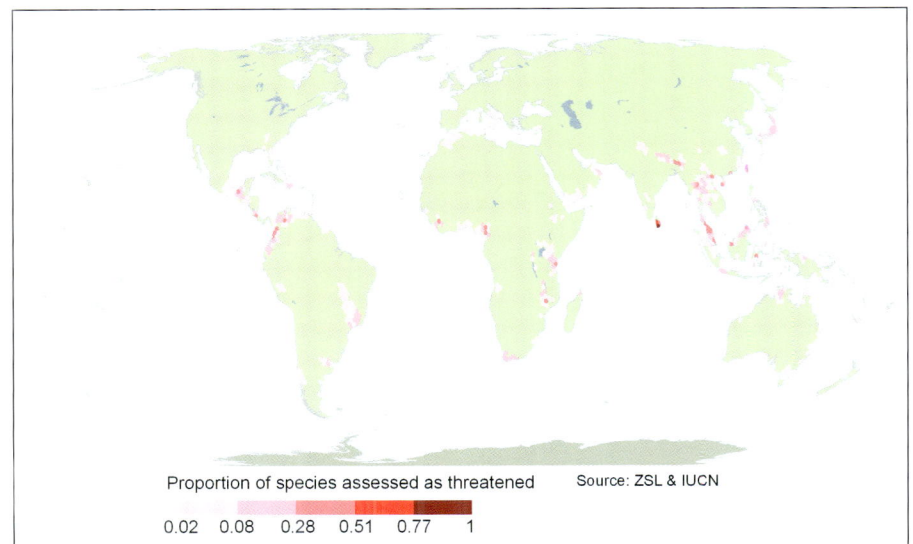

Proportion of species assessed as threatened   Source: ZSL & IUCN

0.02   0.08   0.28   0.51   0.77   1

Viridithemis viridula (Data Deficient) is a dragonfly endemic to western Madagascar. Only the female of this species is known to science and its habitat preferences have not yet been determined. Further research is needed before a more accurate assessment can be made. © Klemens Karkow

(Hilton-Taylor *et al.* this volume), robust trends in change in conservation status is achievable with regular assessment, and with retrospective assessment. Assessments can realistically take place every four to five years for the vertebrates and some plant groups, and at least every 10 years for all other groups (Baillie *et al.* 2008).

Delivery by 2010

The choice of groups for which this approach is being applied is currently limited by the presence of active networks of specialists, up-to-date species lists, and available data. However, by the year 2010, the coverage of the IUCN Red List will have expanded to include eight groups of invertebrates, doubling invertebrate coverage on the Red List in a systematic manner (Figure 1). Using the approach outlined here, this will not only provide insight into the conservation status of invertebrates, but also plant groups such as the monocots and other groups such as fungi. The results, when used to calculate the Red List Index, will provide a

species-based biodiversity indicator that is considerably more broadly representative of all biodiversity than anything hitherto available. Finally it will provide a data set that will enable a broad range of trend indices to be generated ranging from specific taxonomic groups, to functional groups, to species trends in biomes, or biogeographic regions.

References

Baillie, J.E.M., Collen, B., Amin, R., Akçakaya, H.R., Butchart, S.H.M., Brummitt, N., Meagher, T.R., Ram, M., Hilton-Taylor, C. and Mace, G.M. 2008. Towards monitoring global biodiversity. *Conservation Letters* 1: 18-26.

Baillie, J.E.M., Hilton-Taylor, C. and Stuart, S.N. (editors). 2004. *2004 IUCN Red List of Threatened Species: a Global Species Assessment*. IUCN, Gland, Switzerland and Cambridge, UK.

BirdLife International. 2008. *State of the World's Birds: Indicators for our Changing World*. BirdLife International, Cambridge, UK.

Butchart, S.H.M., Akçakaya, H.R., Chanson, J.S., Baillie, J.E.M., Collen, B., Quader, S., Turner, W.R., Amin, R., Stuart, S.N. and Hilton-Taylor, C. 2007. Improvements to the Red List Index. *PLoS ONE* 2: e140. doi:110.1371/journal.pone.0000140.

Butchart, S.H.M., Stattersfield, A.J., Baillie, J.E.M., Bennun, L.A., Stuart, S.N., Akçakaya, H.R., Hilton-Taylor, C. and Mace, G.M. 2005. Using Red List Indices to measure progress towards the 2010 target and beyond. *Philosophical Transactions of the Royal Society of London* 360: 255-268.

Butchart, S.H.M., Stattersfield, A.J., Bennun, L.A., Shutes, S.M., Akçakaya, H.R., Baillie, J.E.M., Stuart, S.N., Hilton-Taylor, C. and Mace, G.M. 2004. Measuring global trends in the status of biodiversity: Red List Indices for birds. *PLoS Biology* 2: e383.

Carpenter, K.E., Abrar, M., Aeby, A., Aronson, R.B., Banks, S., Bruckner, A., Chiriboga, A., Cortes, J., Delbeek, J.C., DeVantier, L., Edgar, G.J., Edwards, A.J., Fenner, D., Guzman, H.M., Hoeksema, B.W. *et al.*. 2008. One-third of reef-building corals face elevated risk from climate change and local impacts. *Science* 321: 560-563.

Clausnitzer, V., Kalkman, V.J., Ram, M., Collen, B., Baillie, J.E.M., Bedjani, M., Darwall, W.R.T., Dijkstra, K.-D.B, Dow, R., Hawking, J., Karube, H., Malikova, E., Paulson, D., Schütte, K., Suhling, F., Villanueva, R., von Ellenrieder, N. and Wilson, K. 2009. Odonata enter the biodiversity crisis debate: the first global assessment of an insect group. *Biological Conservation* 142(8): 1864-1869.

Collen, B., Ram., M, Zamin, T. and McRae, L. 2008. The tropical biodiversity data

Box 4: Threats to dragonflies and damselflies

a. Restricted range damselflies

The Pemba Featherleg *Platycnemis pembipes*, a fragile black-and-white damselfly was first discovered in 2001 on the island of Pemba off the Tanzanian coast. Remarkably its nearest relatives occur on Madagascar, separated by 1,000 km of ocean. Although the species might have reached Pemba aided by strong monsoon winds recent studies suggest it may be the survivor of an ancient African fauna that is now largely confined to Madagascar. The species only inhabits the single stream flowing through Pemba's last remnant of forest and is listed as Critically Endangered. The Pemba Featherleg shares this fate with two other East African damselflies of unknown origin. *Amanipodagrion gilliesi* (Critically Endangered) survives on a single stream in Tanzania's Usambara Mountains. It shares no similarities with any other known species. Equally unique is *Oreocnemis phoenix* (Critically Endangered), named for its bright red males. Streams on the high plateau of Mount Mulanje in Malawi, known aptly as 'the island in the sky' and a mere 24 km across, are its only known habitat. The plateau is made up of bauxite deposits: mining these would significantly impact the habitat.

b. Climate change impact on the Ancient Greenling *Hemiphlebia mirabilis*

The Australian endemic damselfly *Hemiphlebia mirabilis* (Endangered), the Ancient Greenling, is notable for its apparent archaic characters, its male mating displays and its biogeography. Originally thought to have been a Victorian endemic, the species was subsequently found in northeastern Tasmania and then on Flinders Island. This suggests that the species would have occupied the Bassian Ridge when it was exposed during glacial times and this may have been a dispersal route at some time. The species is cryptic within its reed habitat except when the males in particular display by waving their expanded, white anal appendages. The species breeds in open, sedge marshes with a low water level and seems to be capable of recolonizing habitats when they have become dry, probably surviving in the egg stage. In recent times, however, dry spells are longer and more frequent due to climate change and they pose a severe threat to this already rare species. The Ancient Greenling is not the only Australian dragonfly to be affected by climate change and it seems likely that dry spells will become a major driver for decline in the near future.

gap: addressing disparity in global monitoring. *Tropical Conservation Science* 1: 75-88. Available online: http://tropicalconservationscience.mongabay.com/content/v1/08-06-09-Ben_Collen_et_al.pdf.

Cumberlidge, N., Ng, F.K., Yeo, D.C.J., Magalhães, C., Campos, M.R., Alvarez, F., Naruse, T., Daniel S.R., Esser, L.J., Attipoe, F.Y.K., Clotilde-Ba, F.-L., Darwall, W., Mcivor, A., Baillie, J.E.M., Collen, B. and Ram, M. 2009. Freshwater crabs and the biodiversity crisis: importance, threats, status, and conservation challenges. *Biological Conservation* 142(8): 1665-1673.

Dunn, R.R. 2005. Modern insect extinctions, the neglected majority. *Conservation Biology* 19: 1030-1036.

IUCN. 2008. The IUCN Red List of Threatened Species. Available online at: www.iucnredlist.org.

Mace, G.M., Masundire, H. and Baillie, J.E.M. 2005. Ecosystems and human well-being: Biodiversity. *Millennium Ecosystem Assessment.* Island Press, Washington DC.

Purvis, A., Agapow, P.-M., Gittleman, J.L. and Mace, G.M. 2000. Nonrandom extinction and the loss of evolutionary history. *Science* 288: 328-330.

Rohr, J.R., Mahan, C.G. and Kim, K.C. 2007. Developing a monitoring program for invertebrates: guidelines and a case study. *Conservation Biology* 21: 422-433.

Salafsky, N., Salzer, D., Stattersfield, A.J., Hilton-Taylor, C., Neugarten, R., Butchart, S.H.M., Collen, B., Cox, N., Master, L.L., O'Connor, S. and Wilkie, D.. 2008. A standard lexicon for biodiversity conservation: unified classifications of threats and actions. *Conservation Biology* 22: 897-911.

Sanderson, E.W., Jaiteh, M., Levy, M.A., Redford, K.H., Wannebo, A.V. and Woolmer, G. 2002. The human footprint and the last of the wild. *BioScience* 52: 891-904.

Schipper, J., Chanson, J., Chiozza, F., Cox, N., Hoffmann, M., Katariya, V., Lamoreux, J., Rodrigues, A., Stuart, S.N., Temple, H.J., Baillie, J.E.M., Boitani, L., Lacher, T.E., Mittermeier, R.A., Smith, A.T., et al. 2008. The status of the world's land and marine mammals: diversity, threat and knowledge. *Science* 322(5899): 225-230.

Thomas, C.D., Cameron, A., Green, R.E., Bakkenes, M., Beaumont, L.J., Collingham, Y.C., Erasmus, B.F.N., Siqueira, M.F.D., Grainger, A., Hannah, L., Hughes, L., Huntley, B., Jaarsveld, A.S.V., Midgley, G.F., Miles, L. et al. 2004. Extinction risk from climate change. *Nature* 427: 145-148.

UNEP. 2002. Report on the sixth meeting of the Conference of the Parties to the Convention on Biological Diversity (UNEP/CBD/COP/20/Part 2) Strategic Plan Decision VI/26.

Geothelphusa ancylophallus (Least Concern) is a freshwater crab that is endemic to Taiwan. Although localized habitat loss and degradation, as well as problems associated with pollution are occurring in parts of its range, the current rate of decline that this species is experiencing is not significant enough to warrant listing in a threatened category. However, if the quality of this species' habitat continues to decline, this species will become threatened in the future. © Hsi-Te Shih

Species susceptibility to climate change impacts

Wendy B. Foden, Georgina M. Mace, Jean-Christophe Vié, Ariadne Angulo, Stuart H.M. Butchart,
Lyndon DeVantier, Holly T. Dublin, Alexander Gutsche, Simon N. Stuart and Emre Turak

Background

There is growing evidence that climate change will become one of the major drivers of extinction in the 21st century. An increasing number of published studies have documented a variety of changes attributable to climate change, for example changes in species breeding times and shifts in distributions (Figure 1). The Intergovernmental Panel on Climate Change concludes that approximately 20-30% of plant and animal species are likely to be at increasingly high risk of extinction as global mean temperatures exceed warming of 2-3°C above preindustrial levels (Fischlin et al. 2007). Another synthesis study predicts 15-37% 'commitment to extinction' by 2050 of the wide range of regionally endemic and near-endemic species examined (Thomas et al. 2004). How can we predict which species will be most threatened by climate change, and how best can we mitigate the impacts?

To date, most assessments of species extinctions under climate change have been based on either isolated case studies or large-scale modelling of species' distributions. These methods depend on broad and possibly inaccurate assumptions, and generally do not take account of the biological differences between species. As a result, meaningful information that could contribute to conservation planning at both fine and broad spatial scales is limited. Conservation decision-makers, planners and practitioners currently have few tools and little technical guidance on how to incorporate the differential impacts of climate change into their plans and actions.

IUCN is developing assessment tools to identify the potential effects of climate change on species. The IUCN Red List Categories and Criteria were developed before climate change impacts on species were widely recognized, and although they

Figure 1. A summary of some of the predicted aspects of climate change and examples of the effects that these are likely to have on species.

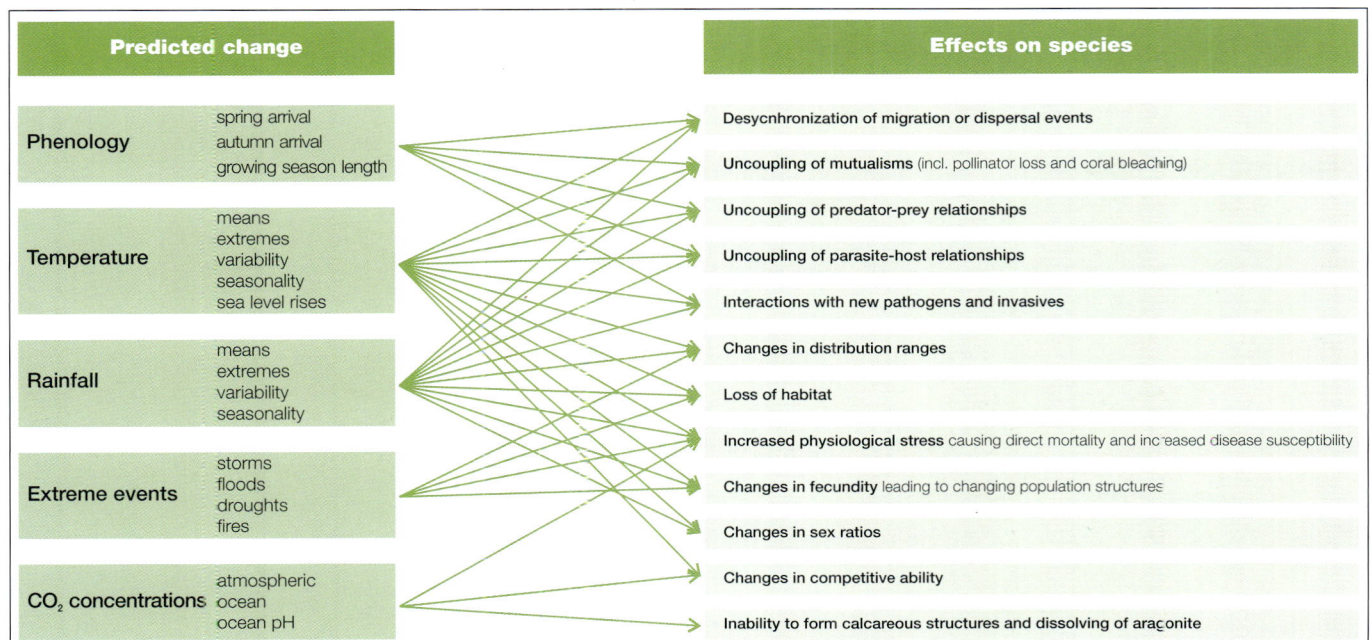

Predicted change		Effects on species
Phenology	spring arrival autumn arrival growing season length	Desynchronization of migration or dispersal events
		Uncoupling of mutualisms (incl. pollinator loss and coral bleaching)
Temperature	means extremes variability seasonality sea level rises	Uncoupling of predator-prey relationships
		Uncoupling of parasite-host relationships
		Interactions with new pathogens and invasives
Rainfall	means extremes variability seasonality	Changes in distribution ranges
		Loss of habitat
		Increased physiological stress causing direct mortality and increased disease susceptibility
Extreme events	storms floods droughts fires	Changes in fecundity leading to changing population structures
		Changes in sex ratios
CO_2 concentrations	atmospheric ocean ocean pH	Changes in competitive ability
		Inability to form calcareous structures and dissolving of aragonite

Climate change is causing population declines of the Quiver Tree Aloe dichotoma, a long-lived giant tree aloe from the Namib Desert region. Growing evidence suggests that desert ecosystems may be more sensitive to climate change than previously suspected. © Wendy Foden

remain effective for identifying species that are undergoing declines in ranges or population sizes, they may need further refinement in order to identify the full suite of species at risk from climate change. A new initiative aimed at examining how the IUCN Red List Criteria can be used for identifying the species most at risk from climate change is underway. This study, although it forms part of the overall project looking at the impacts of climate change on species, is not discussed further here.

Methodological approach

General Circulation Models (GCMs) predict that climate change will affect different areas of the world to different degrees. But it is also widely recognized that not all species will respond in the same way, even to similar levels of climatic change. A species' individual susceptibility to climate change depends on a variety of biological traits, including its life history, ecology, behaviour, physiology and genetic makeup. Species exposed to large climatic changes in combination with intrinsic susceptibility to climate change face the greatest risk of extinction due to climate change (Figure 2).

We assessed susceptibility to climate change according to taxon-specific biological traits and present an analysis of the potential impacts of climate change on species based on an analysis of these traits. Using expert assessments for birds (9,856 species), amphibians (6,222 species) and warm-water reef-building corals (799 species), we examined the taxonomic and geographical distributions of the species most susceptible to climate change and compared these to the existing assessments of threatened species in The 2008 IUCN Red List of Threatened Species™ (herein The IUCN Red List; IUCN 2008). Specifically we address the following questions:

- What are the biological traits that make species potentially susceptible to climate change?

- How common are these traits in birds, amphibians and warm-water reef-building corals?

- Are the species that are potentially susceptible to climate change the same as those already identified as threatened on The IUCN Red List?

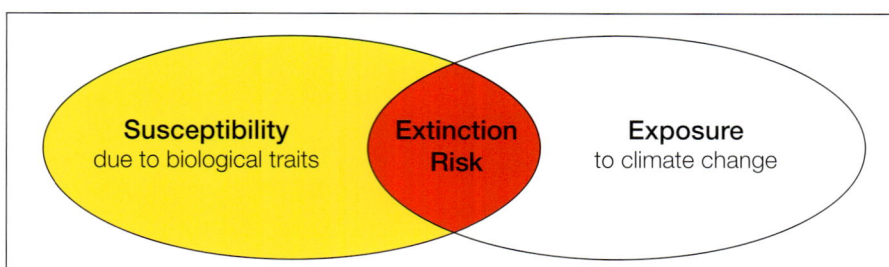

Figure 2. Increased risk of extinction due to climate change occurs when species possess biological traits or characteristics that make them particularly susceptible to change, and simultaneously occur in areas where climatic changes are most extreme.

- How do taxonomic and geographic concentrations of species that are potentially susceptible to climate change compare with those of threatened species?

What are the biological traits that make species most susceptible to climate change?

Through detailed consultations with a wide range of experts, we identified over 90 biological traits that may be associated with enhanced susceptibility to climate change. These were consolidated into five groups of traits (Table 1 and Box), and each trait within these groups, was assessed using a range of biological information. Specific trait combinations were developed for each of the three taxonomic groups covered in this study. Table 1 shows the groups, traits and the number of bird, amphibian and coral species that met each of them either singly or in combination.

There were a number of challenges in selecting traits. These included the scarcity of key species-level data (e.g., population sizes, temperature-tolerance thresholds, prey species), as well as defining traits in quantifiable, objective and replicable ways. Although not always possible, we aimed to represent each of the trait groups with at least one specific trait for each taxonomic group. Even though many species possess multiple susceptibility traits, for this analysis we defined "susceptible species" to be those that were recorded as having any one or more susceptibility traits. While this method allows for very broad comparability between taxonomic groups, accurate quantification of the contribution of each trait to extinction risk is necessary before cross-taxonomic group comparisons can be made with confidence. Our approach to assessing species' susceptibility to climate change assesses relative susceptibility within each taxonomic group only.

The IUCN Red List and BirdLife International's World Bird Database provided essential information such as taxonomy, distribution maps, habitats and

Which traits or characteristics make species susceptible to climate change?

In October 2007, Imperial College London, IUCN and the Zoological Society of London hosted a four-day workshop to identify the traits associated with elevated extinction risk, particularly due to climate change. Thirty-one biologists, whose expertise spanned a broad range of taxonomic groups and geographic regions, identified, discussed and eventually reached consensus on a list of over 90 traits that are generally indicative of species' vulnerability to extinction across most taxonomic groups. These traits were subsequently refined and form the basis of IUCN's ongoing assessment of species susceptibility to climate change. The traits fall into the following five trait groups:

A. *Specialized habitat and/or microhabitat requirements.* Species with generalized and unspecialized habitat requirements are likely to be able to tolerate a greater level of climatic and ecosystem change than specialized species. Where such species are able to disperse to new climatically suitable areas, the chances of fulfillment of all their habitat requirements are low (e.g., plants confined to limestone outcrops; cave-roosting bats). Susceptibility is exacerbated where a species has several life stages, each with different habitat or microhabitat requirements (e.g., water-dependent larval-developing amphibians), or when the habitat or microhabitat to which the species is specialized is particularly vulnerable to climate change impacts (e.g., mangroves, cloud forests or polar habitats). In some cases (e.g., deep sea fish), extreme specialization may allow species to escape the full impacts of competition from native or invading species so the interaction of such traits with climate change must be considered carefully for each species group assessed.

B. *Narrow environmental tolerances or thresholds that are likely to be exceeded due to climate change at any stage in the life cycle.* The physiology and ecology of many species is tightly coupled to very specific ranges of climatic variables such as temperature, precipitation, pH and carbon dioxide levels, and those with narrow tolerance ranges are particularly vulnerable to climate change. Even species with broad environmental tolerances and unspecialized habitat requirements may already be close to thresholds beyond which ecological or physiological function quickly breaks down (e.g., photosynthesis in plants, protein and enzyme function in animals).

C. *Dependence on specific environmental triggers or cues that are likely to be disrupted by climate change.* Many species rely on environmental triggers or cues for migration, breeding, egg laying, seed germination, hibernation, spring emergence and a range of other essential processes. While some cues such as day length and lunar cycles will be unaffected by climate change, others such as rainfall and temperature (including their interacting and cumulative effects) will be heavily impacted upon by climate change. Species become vulnerable to changes in the magnitude and timing of these cues when they lead to an uncoupling with resources or other essential ecological processes e.g., early spring warming causes the emergence of a species before their food sources are available. Climate change susceptibility is compounded when different stages of a species' life history or different sexes rely on different cues.

D. *Dependence on interspecific interactions that are likely to be disrupted by climate change.* Many species' interactions with prey, hosts, symbionts, pathogens and competitors will be affected by climate change either due to the decline or loss of these resource species from the dependent species' ranges or loss of synchronization in phenology. Species dependent on interactions that are susceptible to disruption by climate change are at risk of extinction, particularly where they have high degree of specialization for the particular resource species and are unlikely to be able to switch to or substitute other species.

E. *Poor ability to disperse to or to colonise a new or more suitable range.* In general, the particular set of environmental conditions to which each species is adapted (its 'bioclimatic envelope') will shift polewards and to increasing altitudes in response to climate change. Species with low rates or short distances of dispersal (e.g., land snails, ant and rain drop splash dispersed plants) are unlikely to migrate fast enough to keep up with these shifting climatic envelopes and will face increasing extinction risk as their habitats become exposed to progressively greater climatic changes.

Even where species could disperse to newly suitable bioclimatic areas, several other factors may affect colonization success. Species' phenotypic plasticity and genetic diversity will determine the likelihood of adaptation over different time scales. Where they exist, direct measures of genetic variability can be supplemented with information on naturalization outside species' native ranges and on the success of any past translocation efforts. Extrinsic factors likely to decrease dispersal success include the presence of any geographic barriers such as mountain ranges, oceans, rivers, and for marine species, ocean currents and temperature gradients. Anthropogenic transformation of migration routes or destination habitats increases species' susceptibility to negative impacts from climate change.

Trait Group	Biological Trait	Number of species qualifying		
		Birds	Amphibians	Corals
A. Specialized habitat and/or microhabitat requirements	Altitudinal range narrow and at high elevation	224		
	Restricted to habitats susceptible to climate change	820	757	15
	High degree of habitat specialization	693		28
	Dependence on a particular microhabitat	438	889	
	Contribution of trait group	*46%*	*42%*	*5%*
B. Narrow environmental tolerances or thresholds that are likely to be exceeded due to climate change at any stage in the life cycle	Global temperature tolerances likely to be exceeded			61
	Larvae particularly susceptible to heat stress			108
	Sensitive to increased sedimentation			143
	Vulnerable to physical damage from storms and cyclones			183
	Contribution of trait group	*0%*	*0%*	*68%*
C. Dependence on specific environmental triggers or cues that are likely to be disrupted by climate change	Environmental trigger/cue disruption observed or likely	316	315	
	Contribution of trait group	*9%*	*10%*	*0%*
D. Dependence on interspecific interactions which are likely to be disrupted by climate change	Dependent on very few prey or host species	27		
	Dependent on an interspecific interaction that is likely to be impacted by climate change	44		
	Susceptible to chytridiomycosis and/or enigmatic decline		1,034	
	Susceptible to breakdown of coral-zooxanthellae interaction			144
	Contribution of trait group	*2%*	*32%*	*25%*
E. Poor ability or limited opportunity to disperse to or colonize a new or more suitable range	Low maximum dispersal distances	1,500		73
	Geographic barriers limit dispersal opportunity	709	744	117
	Limited opportunity to establish at new locations	769	602	55
	Low genetic diversity or known genetic bottleneck	63		
	Contribution of trait group	*69%*	*40%*	*40%*
Number of climate change susceptible species		3,438	3,217	566
Number of species assessed		9,856	6,222	799
Climate change susceptible species (%)		35%	52%	71%

Table 1. *A summary of the trait groups, biological traits and numbers of bird, amphibian and warm-water reef-building coral species that qualify as having the trait in question. Trait group summary rows (italic) show the relative contribution of each trait group to the total number of climate change susceptible species for each taxonomic group. The sum of these values is >100% because many qualifying species have multiple traits. Detailed descriptions of trait groups are given in the Box.*

threats, and additional information was gathered from published and unpublished data, online resources, literature and expert knowledge. While we attempted to address data gaps with experts' inferences and assumptions, numerous uncertainties remain. In summary, our results are based on the following assumptions: that species' susceptibility to climate change is associated with the possession of specific biological traits that we have identified; that the possession of any one of these traits increases the susceptibility of a species to climate change; and that our classification of each species according to these traits is accurate.

How common are these traits in the amphibians, birds and warm-water reef-building corals?

Birds

Eleven traits were selected for this relatively information-rich group. 3,438 of the world's 9,856 extant bird species (35%) possess traits that make them potentially susceptible to climate change. Of these, 1,288 species have between two and seven such traits with the majority

of species qualifying due to specialized habitat and microhabitat requirements, and poor or limited opportunity to establish at new locations, particularly due to low maximum dispersal distances. We also examined any evidence of impacts of changing seasonal cues, confinement to narrow altitudinal ranges at high elevations, and dependence on five or fewer prey or host species.

Susceptibility to climate change in birds shows strong taxonomic and geographic patterns with all species considered susceptible within the Diomedeidae (albatross), Spheniscidae (penguin), Procellariidae, Pelecanoididae and Hydrobatidae (petrel and shearwater) families. Large families with particularly high levels of susceptibility include Turdidae (thrushes, 60%), Thamnophilidae (antbirds, 69%), Scolopacidae (sandpipers and allies, 70%), Formicariidae (antthrushes and antpittas, 78%) and Pipridae (manakins, 81%). In contrast, large families showing low levels of climate change susceptibility include Ardeidae (herons and egrets, 3%), Accipitridae (osprey, kites, hawks and eagles, 10%), Estrildidae (waxbills, grass

finches and munias, 12%), Cuculidae (cuckoos, 15%), Picidae (woodpeckers, 21%) and Columbidae (doves and pigeons, 27%).

Amphibians

We found that 3,217 of the 6,222 global amphibian species (52%) are potentially susceptible to climate change, and 962 species possess two to four climate change susceptibility traits. Within three small families in order Caudata (salamanders), namely the Amphiumidae (amphiumas, three species), Sirenidae (sirens, four species) and Proteidae (mudpuppies and waterdogs, six species), all species are "climate-change-susceptible". The low numbers of susceptible species in the order Gymnophiona (caecilians) (18%), might be due to the scarcity of global knowledge of the group. The Sooglossidae (Seychelles frogs and Indian Burrowing Frog), Myobatrachidae and Limnodynastidae (Australian ground frogs), Ceratophryidae (horned toads), and Centrolenidae (glassfrogs) families have 80-100% of species assessed as "climate-change-susceptible". Large families with more than 50% "climate-change-susceptible" species include Strabomartidae, Bufonidae (toads and true toads), Hylidae (treefrogs) and Plethodontidae (lungless salamanders).

Of the six traits used to assess amphibian susceptibility to climate change, those relating to specialized habitat requirements, poor dispersal and colonization ability, and disruption of interspecific interactions identified the majority of susceptible species. These included species occurring exclusively in habitats vulnerable to climate change; those with water-dependant larvae occurring exclusively in unbuffered habitats; those unable to disperse due to barriers such as large water bodies or unsuitable habitat; and those with small ranges in combination with very low population densities.

Emerging infectious diseases, such as chytridiomycosis, caused by the chytrid fungus *Batrachochytrium dendrobatidis*, and 'enigmatic' or unexplained declines play an increasingly large role in driving amphibians towards extinction (Bosch and Rincon 2008; Corey and Waite 2008; Lips *et al.* 2003; Navas and Otani 2007; Pounds *et al.* 2006; Stuart *et al.* 2004). While the direct contribution of climate change to these threats remains disputed, at best, species particularly susceptible to or already experiencing such declines are more likely to fare poorly in the face of increasing climate change.

Under the trait group covering 'dependence on interspecific interactions that are likely to be disrupted by climate change', amphibians were regarded as "climate-change-susceptible" where chytrid infection has been recorded in wild populations, where experts have deemed infection highly likely, or where enigmatic declines have been recorded but not directly linked to chytridiomycosis. In many cases, susceptibility to chytrid infection shows a taxonomic and ecological bias (Corey and Waite 2008; Stuart *et al.* 2004), so we have provisionally included in this trait group all species in genera containing infected members (e.g., members of genus *Atelopus*) that also occur exclusively in subtropical or tropical montane environments and are dependent on water bodies. Species with chytrid presence but no symptoms of clinical disease (e.g., members of genus *Xenopus*) did not qualify as susceptible under this trait.

Restricted to a very small range in north-west Ecuador, the Black-breasted Puffleg Eriocnemis nigrivestis *has been assessed as both Critically Endangered according to The IUCN Red List and "climate-change-susceptible" based on its biological traits. These "climate-change-susceptibility" traits include its habitat specialization, restriction to a climate change susceptible habitat, a narrow and high altitude range, very short typical dispersal distances and an extremely small population size. The species is suffering ongoing declines from deforestation. © Francisco Enriquez*

Corals

To date, most climate change studies have focussed on reef-level impacts and few have attempted to distinguish individual species' responses to climate change impacts. Here we assess only the warm-water reef-building corals, including 789 species (those either *zooxanthellate or hermatypic*) of order Scleractinia (stony corals), eight Milleporina species (fire or stinging corals), one Helioporaceae (blue coral) and one Stolonifera species (organ pipe coral). Due to insufficient information and taxonomic uncertainties, we were unable to assess the 46 other species in the group.

We found that 566 of 799 global warm-water reef-forming coral species (71%) are potentially susceptible to the impacts of climate change, while 253 species possess between two and six susceptibility traits. Families Acroporidae (including staghorn corals), Agariciidae and Dendrophylliidae had particularly high numbers of susceptible species, while Fungiidae (including mushroom corals), Mussidae (including some brain corals) and Pocilloporidae (including cauliflower corals) possess relatively few.

Coral susceptibility assessments were based on 10 traits and most species qualified due to their sensitivity to increases in temperature both by adult polyps as well as free-living larvae; sedimentation; and physical damage from storms and cyclones. Poor dispersal ability and colonization potential proved a further important trait group and included larval longevity (as a proxy for maximum dispersal distance) and the presence of currents or temperatures as barriers to dispersal. Although climate change related ocean acidification is likely to become a serious threat to coral survival in the future (Kleypas et al. 1999; Royal Society 2005), we did not include it in susceptibility assessments due to sparse information about differentiation in species' aragonite decalcification rates. We plan, however, to include acidification impacts in the climate change exposure component of overall climate change vulnerability assessments.

Are the "climate-change-susceptible" species the same as those already identified as threatened on The IUCN Red List, or are they different?

For each taxonomic group, we assigned all species into the following four categories: (i) threatened (according to The IUCN Red List) and "climate-change-susceptible"; (ii) threatened but not "climate-change-susceptible"; (iii) not threatened but "climate-change-susceptible"; and (iv) neither threatened nor "climate-change-susceptible". A summary of the results is shown in Table 2.

The summaries in Table 2 and Figure 3 show that each taxonomic group faces

different challenges in response to climate change. At 32%, the amphibians already have a very high number of threatened species. Seventy-five percent of these are also susceptible to climate change, greatly exacerbating their extinction risk. In addition, 41% of currently non-threatened species are "climate-change-susceptible".

The overall percentage of threatened birds is lower than those of the other groups assessed (12%), but most threatened birds (80%) are also susceptible to the impacts of climate change. In addition, a quarter of all bird species and nearly 30% of all non-threatened species are susceptible to climate change.

At 51%, corals have the greatest proportion of not threatened but "climate-change-susceptible" species of the groups assessed, while a further 19% of species are both susceptible and threatened. Corals are the only group in which non-threatened but susceptible species outnumber those that are neither threatened nor susceptible (21%), and they do so by more than two-fold. This suggests that if climate change becomes extreme globally, more than three quarters

of all warm-water reef-building coral species could be at risk of extinction.

The large overlap between threatened and "climate-change-susceptible" amphibian and bird species means that, ideally, they may already be included in conservation prioritization strategies. However, the question above has more complex implications. Species that already face a high risk of extinction, irrespective of the threat type, are far less likely to be resilient to environmental and climatic changes. A large overlap between threatened and "climate-change-susceptible" species may therefore mean that climate change may cause a sharp rise in both the extinction risk and extinction rate of already threatened species. It is also important to identify susceptible species which, while currently not threatened, are likely to become so in the future as climate change impacts intensify. By highlighting such species before they decline, we hope to promote preemptive and more effective conservation actions.

Data Deficient Species

While Data Deficient species (i.e., those with insufficient information to conduct

Table 2. *The numbers and percentages of species assessed for "climate-change-susceptibility" and in the 2008 IUCN Red List for birds, amphibians and warm-water reef-building corals. These values fall into categories: (i) threatened and "climate-change-susceptible" (red); (ii) threatened but not "climate-change-susceptible" (orange); (iii) not threatened but "climate-change-susceptible" (yellow); and (iv) neither threatened nor "climate-change-susceptible" (green).*

Birds		Threatened			
			YES	NO	TOTAL
Climate Change Susceptible	YES		976 / 10%	2,462 / 25%	35%
	NO		246 / 2%	6,172 / 63%	65%
	TOTAL		12%	88%	9,856

Amphibians		Threatened			
			YES	NO	TOTAL
Climate Change Susceptible	YES		1,488 / 24%	1,729 / 28%	52%
	NO		503 / 8%	2,502 / 40%	48%
	TOTAL		32%	68%	6,222

Corals		Threatened			
			YES	NO	TOTAL
Climate Change Susceptible	YES		155 / 19%	411 / 51%	71%
	NO		68 / 9%	165 / 21%	29%
	TOTAL		28%	72%	799

With a small range in Southern Africa, the Spotted Snout Burrower Hemisus guttatus *is already considered Vulnerable due to habitat loss from afforestation, agriculture, the introduction of alien fishes and lowering of the water table by invasive alien plants. The species' reliance on seasonal rainfall events to break periods of hibernation, as well as tadpoles' dependence on temporary water bodies, make it particularly vulnerable to negative impacts due to climate change.*
© Marius Burger

Red List assessments) represent only one per cent of bird species, 25% and 14% of amphibians and corals respectively fall into this Red List Category. Because a trait-based assessment of species susceptibility to climate change requires different information to Red List assessments, we were able to infer that 38 (58%), 679 (44%) and 94 (81%) of Data Deficient bird, amphibian and coral species respectively are potentially susceptible to climate change. For corals, these susceptibility assessments were based on traits inferred from knowledge of close taxonomic relatives (e.g., similar reproductive modes), while

Figure 3. *The proportion of bird, amphibian and coral species falling in one of the 6 following categories: (i) threatened (according to The 2008 IUCN Red List) (orange); (ii) threatened and "climate-change-susceptible" (red); (iii) not threatened but "climate-change-susceptible" (yellow); (iv) Data Deficient and "climate-change-susceptible" (grey); (v) Data Deficient and not "climate-change-susceptible" (dark green); and (vi) neither threatened, Data Deficient nor "climate-change-susceptible" (light green).*

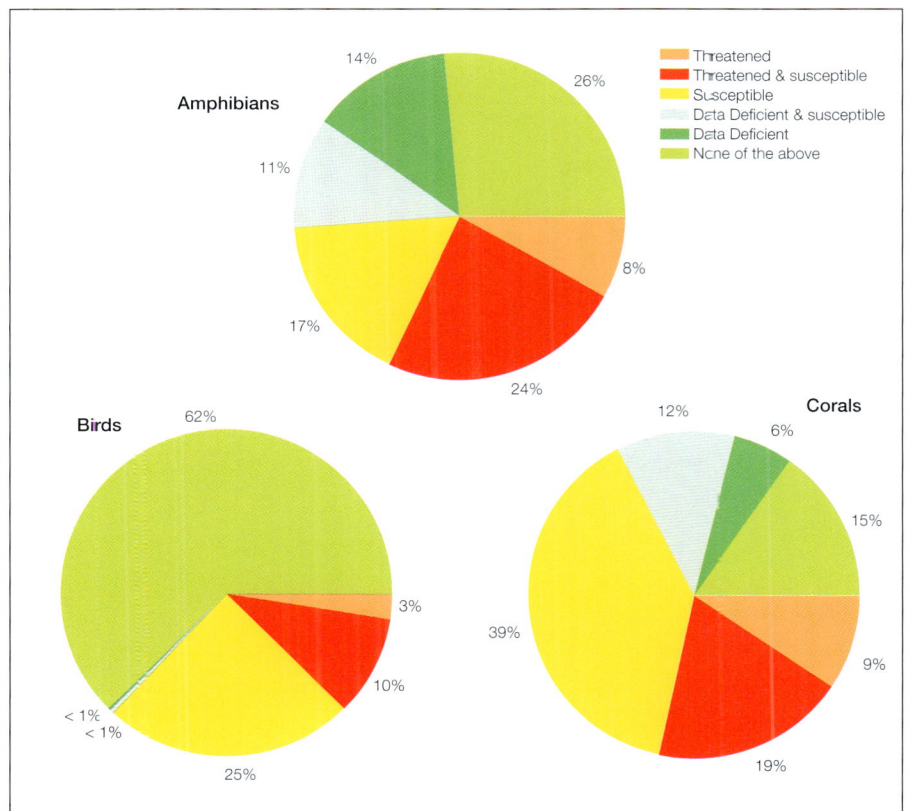

83

inferences were made based largely on habitats (e.g., disease susceptibility) for amphibians. Due to particularly poor distribution information for most Data Deficient species, they were not included in the geographic analyses.

Where are the areas of highest concentrations of "climate-change-susceptible" species?

Although birds are generally a data rich group, range maps are not currently available for many of the non-threatened species, making meaningful analysis of global geographic trends in "climate-change–susceptible" species impossible. For this reason we are only able to present global geographical trends for amphibians and corals.

Amphibians

We identified high concentration areas by selecting areas with the top 10%, 5% and 2.5% of species richness (or nearest appropriate percentages when these were not distinguishable). For amphibians assessed as threatened and "climate-change-susceptible" (Figure 4a), the areas of greatest richness span Mesoamerica and northwestern South America. Smaller areas of high richness include various Caribbean Islands; south-eastern Brazil; Sri Lanka, the Western Ghats of India; northern Borneo (Malaysia); and the eastern coast of Australia. As expected, this shows strong correspondence with areas of both overall and threatened species richness (Stuart *et al.* 2004), although interesting exceptions are the lower levels of susceptibility in the Amazon Basin; central to southern Africa;

and the southeastern USA. While each of these areas has moderate to high species richness and many threatened species, they did not meet the thresholds for inclusion as high concentration areas.

High concentration areas for amphibians assessed as not threatened but "climate-change-susceptible" complement the threatened and "climate-change-susceptible" species' coverage of the regions of overall species richness. The largest and most dominant high concentration areas are southern Brazil and its neighbouring countries, and a large region from east to central and southern Africa. We also identified smaller high concentration areas in West Africa, New Guinea and eastern and northern Australia. High concentration areas of not threatened and susceptible species co-occur with those of threatened and susceptible species in Mesoamerica, northwestern South America and western Australia.

High concentration areas for threatened and "climate-change-susceptible" amphibians cover a relatively small geographic extent. This is due to the typically extremely small ranges of most threatened amphibians, particularly in Mesoamerica, the northern Andes and the Caribbean, where threatened species richness is greatest. That relatively small areas contain such high amphibian richness, particularly of priority species, further highlights their extreme importance for amphibian conservation. In contrast with threatened and "climate-change-susceptible" species, several that are not threatened but susceptible have much larger ranges. In combination with the larger number of these species, this results in much larger high concentration areas for this group.

In order to identify areas of disproportionately high threat or susceptibility, we compared

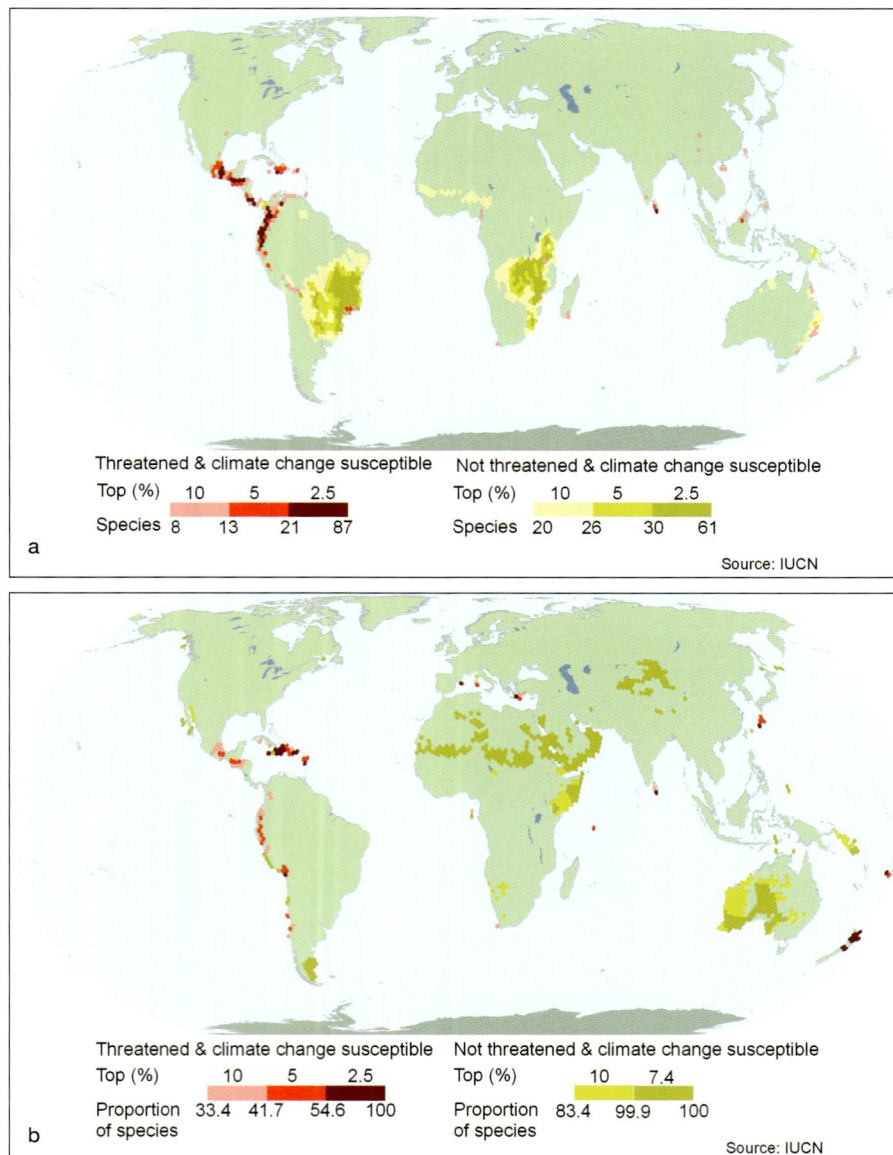

Figure 4. Areas of high concentration of amphibian species assessed as **(a)** threatened and "climate-change-susceptible" (reds), and not threatened but "climate-change-susceptible" (yellows). **(b)** Shows areas containing high proportions of threatened and "climate-change-susceptible" (reds), and not threatened and "climate-change-susceptible" amphibian species (yellows) (expressed as the percentage of species in these categories relative to the total number of species occurring there). High concentration areas indicate those with the top 10%, 5% and 2.5% of values, and when these were not distinguishable, the nearest appropriate percentages were used.

the number of threatened and susceptible species relative to the total number of species in any one area (expressed as the percentage of species of interest relative to the total species number). This information complements high concentration areas of overall species richness and is particularly important for conservation planning at regional and global scales.

For amphibians, mapping the relative richness of threatened and "climate-change-susceptible" species (Figure 4b) once again highlights Mesoamerica, the northern Andes and the Caribbean, but for this group, the area of high concentration continues intermittently through southwestern North America and the Andes as far south as central Chile. Additional areas of high concentration include several Mediterranean islands and south-western Turkey; the Seychelles; the southern Japanese islands; New Zealand's North Island; and Fiji. Areas of high concentration of relative numbers of species assessed as not threatened but "climate-change-susceptible" include western and central Australia; the Solomon Islands; south eastern South America; north-western

Mexico; the arid region extending from the Western Sahara through the Red Sea basin, south to the Horn of Africa and along the coastal regions of the Arabian Peninsula; and the foothills surrounding the northern Himalayan Plateau.

Corals

Based on high concentration area analysis, a single high concentration area is identified for warm-water reef-building corals based on all assessment categories and their combinations, namely the 'Coral Triangle' bordered by the Philippines, Malaysia and Indonesia (Figure 5a). This is the high concentration area of threatened and "climate-change-susceptible", not threatened but "climate-change-susceptible", as well as for overall coral species richness and threatened coral species richness (Carpenter *et al.* 2008). The 'Coral Triangle' is already being negatively impacted by climate change (Carpenter *et al.* 2008) and our results reinforce the extreme importance of effective conservation in this region.

Mapping areas with high proportions of threatened and "climate-change-

susceptible" coral species (Figure 5b) also highlights the 'Coral Triangle', though additional high concentration areas include the northern parts of the Great Barrier Reef; the south-western coast of Australia; the Yellow Sea; the East China Sea and the Sea of Japan; and various areas along the coastlines of Pakistan, India and Bangladesh. Although not particularly species rich on a global scale and therefore not appearing as high concentration areas in Figure 5a, these regions clearly face an extremely high level of threat.

Areas with high concentrations of non-threatened but "climate-change-susceptible" corals show a markedly different pattern to those of other coral groups (Figure 5b). The species-rich 'Coral Triangle' is not highlighted, but in several areas of generally low species richness, more than 90% of all species are not threatened but "climate-change-susceptible". High concentration areas of these species include the Mediterranean and extending to north-west Africa; the east coast of the United States; the southern United States coast; north-

Emperor Penguin Aptenodytes forsteri and man meet in front of Mount Discovery (McMurdo Sound, Antarctica). Although currently listed as Least Concern, Emperor Penguins' poor dispersal potential and low reproductive rate make them likely to be susceptible to negative impacts of climate change. © Colin Harris

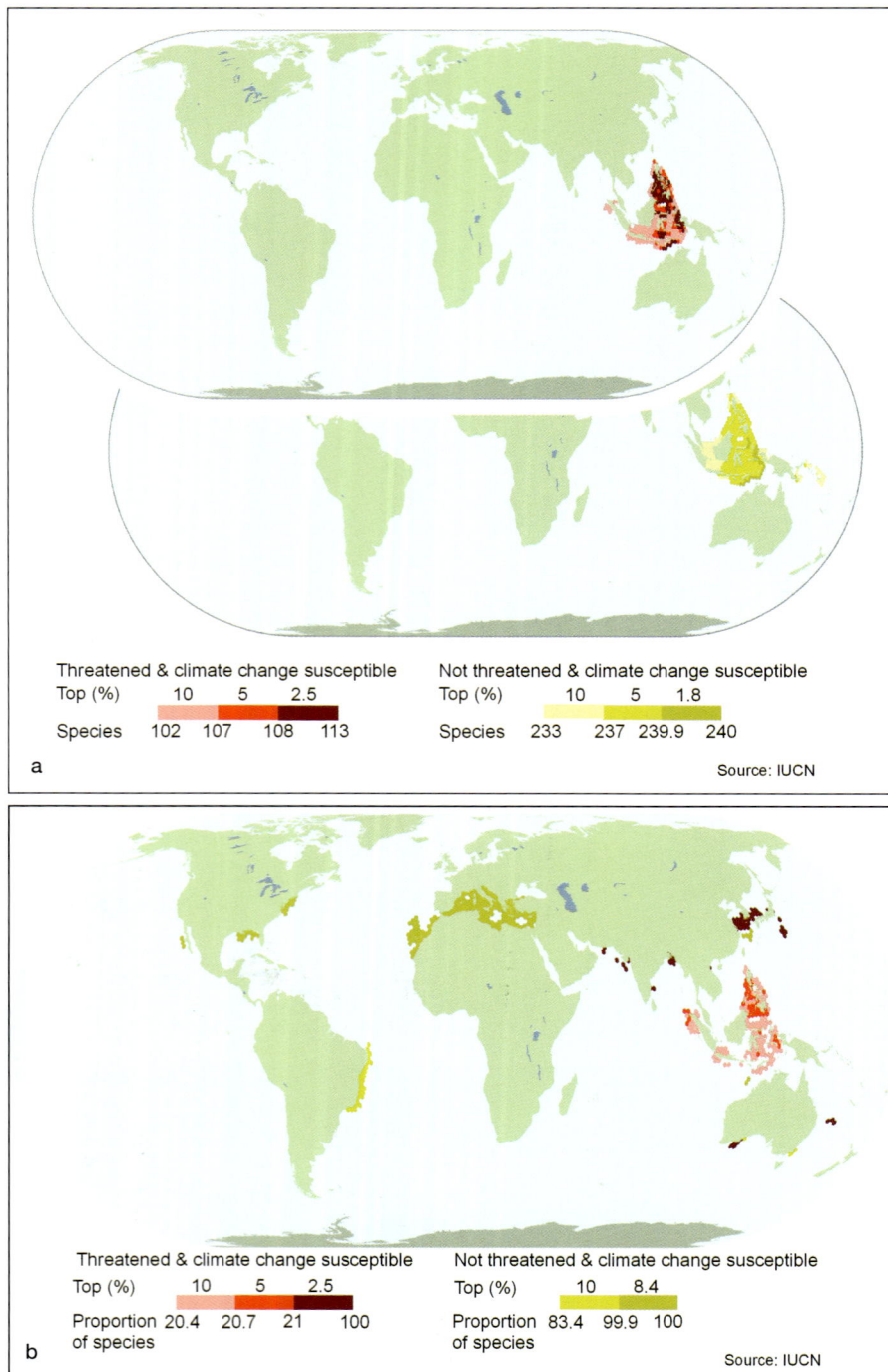

Figure 5. *Areas of high concentration of warm-water reef-building coral species assessed as* **(a)** *threatened and "climate-change-susceptible" (reds), and not threatened but "climate-change-susceptible" (yellows).* **(b)** *Shows areas containing high proportions of threatened and "climate-change-susceptible", and not threatened and "climate-change-susceptible" coral species (yellows) (expressed as the percentage of species in these categories relative to the total number of species occurring there). High concentration areas indicate those with the top 10%, 5% and 2.5% of values, and when these were not distinguishable, the nearest appropriate percentages were used.*

are reliable predictors within and across species groups.

Key messages

- Some species are much more susceptible to climate change impacts than others due to inherent biological traits related to their life history, ecology, behaviour, physiology and genetics.

- High risks of extinction occur when species experience both high susceptibility to climate change and large climatic changes.

- IUCN has conducted assessments of susceptibility to climate change for the world's birds, amphibians and warm-water reef-building coral species. Based on a range of taxon-specific traits, we found that 35%, 52% and 71% of these groups respectively have traits that render them particularly susceptible to climate change impacts.

- 70-80% of birds, amphibians and corals that are already threatened are also "climate-change-susceptible". Given exposure to large climatic changes, these species which also have least resilience to further threat, already face the greatest risk of extinction. Of species that are not considered threatened, 28-71% are "climate-change-susceptible". We identify the taxonomic groups and geographic regions harbouring the greatest concentrations of the above species and recommend that they are given high conservation priority.

- Assessments of "climate-change-susceptibility" complement IUCN Red List assessments of extinction risk and serve as a 'warning flag' highlighting the need for intensive monitoring and

western Mexico; the east and south-east of Brazil; the East China Sea; and smaller areas around Australia. These areas are likely to be subject to rapid coral declines if they are exposed to large climatic changes.

In the long term we plan to compare the distribution of "climate-change-susceptible" species with areas of large climatic change exposure, based on General Circulation Model projections, which will allow us to identify species,

taxonomic groups and areas where species potentially face the highest risk of extinction due to climate change. However, first we propose to examine the traits and their distribution across species in order to evaluate the extent to which they can be shown to be predictive of climate change impacts, as well as to examine the inter-relationships and possible redundancies in the trait set. This process will contribute to the validation and testing of our methods in order to provide reassurance that the traits used

These corals of the Solomon Islands are healthy and none is currently threatened, yet 4 of the 5 species photographed have traits that are likely to make them susceptible to climate change impacts. These susceptible corals include Acropora digitifera (Near Threatened), A. gemmifera (Least Concern), A. robusta (Least Concern) and Pocillopora eydouxi (Near Threatened), which are more vulnerable to bleaching because their symbiont algae have low temperature tolerances, while those of the pink coral shown (Pocillopora verrucosa – Least Concern) may be more robust. © Emre Turak

potentially conservation action for affected species.

References

Bosch, J. and Rincon, P.A. 2008. Chytridiomycosis-mediated expansion of *Bufo bufo* in a montane area of Central Spain: an indirect effect of the disease. *Diversity and Distributions* 14(4): 637-643

Carpenter, K.E., Abrar, M., Aeby, G., Aronson, R.B., Banks, S., Bruckner, A., Chiriboga, A., Cortes, J., Delbeek, J.C., DeVantier, L., Edgar, G.J., Edwards, A.J., Fenner, D., Guzman, H.M., Hoeksema, B.W. *et al.* 2008. One-Third of reef-building corals face elevated extinction risk from climate change and local impacts. *Science* 321(5888): 560-563.

Corey, S.J. and Waite, T.A. 2008. Phylogenetic autocorrelation of extinction threat in globally imperilled amphibians. *Diversity and Distributions* 14: 614-629.

Fischlin, A., Midgley, G.F., Price J.T., Leemans, R., Gopal, B., Turley, C., Rounsevell, M.D.A., Dube, O.P., Tarazona, J. and Velichko, A.A. 2007. Ecosystems, their properties, goods, and services. In: M.L. Parry, O.F. Canziani, J.P. Palutikof, P.J. van der Linden and C.E. Hanson (eds). Climate Change 2007: Impacts, Adaptation and Vulnerability. Contribution of Working Group II to the Fourth Assessment Report of the Intergovernmental Panel on Climate Change, Cambridge University Press, Cambridge, UK, pp 211-272.

IUCN 2008. *2008 Red List of Threatened Species.* Available at: http://www.iucnredlist.org.

Kleypas, J.A., Buddemeier, R.W., Archer, D., Gattuso, J.P., Langdon, C. and Opdyke, B.N. 1999. Geochemical consequences of increased atmospheric carbon dioxide on coral reefs. *Science* 284(5411): 118-120.

Lips, K.R., Reeve, J.D. and Witters L.R. 2003. Ecological traits predicting amphibian population declines in Central America. *Conservation Biology* 17(4): 1078-1088.

Navas, C.A. and Otani, L. 2007. Physiology, environmental change, and anuran conservation. *Phyllomedusa* 6(2): 83-103.

Pounds, J.A., Bustamante, M.R., Coloma, L.A., Consuegra, J.A., Fogden M.P.L., Foster, P.N., La Marca, E., Masters, K.L., Merino-Viteri, A., Puschendorf, R., Ron, S.R., Sanchez-Azofeifa, G.A., Still, C.J. and Young B.E. 2006. Widespread amphibian extinctions from epidemic disease driven by global warming. *Nature* 439(7073): 161-167.

Royal Society 2005. Ocean acidification due to increasing atmospheric carbon dioxide. Report nr 12/05.

Stuart, S.N., Chanson, J.S., Cox, N.A., Young, B.E., Rodrigues, A.S.L., Fischman, D.L. and Waller, R.W. 2004. Status and trends of amphibian declines and extinctions worldwide. *Science* 306(5702): 1783-1786.

Thomas, C.D., Cameron, A. Green, R.E., Bakkenes, M., Beaumont, L.J., Collingham, Y.C., Erasmus, B.F.N., de Siqueira, M.F., Grainger, A., Hannah, L., Hughes, L., Huntley, B., van Jaarsveld, A.S., Midgley G.F., Miles, L., Ortega-Huerta, M.A., Townsend Peterson A., Phillips, O.L. and Williams, S.E. 2004. Extinction risk from climate change. *Nature* 427(6970): 145-148.

The Mediterranean: a biodiversity hotspot under threat

Annabelle Cuttelod, Nieves García, Dania Abdul Malak, Helen J. Temple and Vineet Katariya

The diverse Mediterranean

The Mediterranean Basin is one of the world's richest places in terms of animal and plant diversity. This diverse region, with its lofty mountains, ancient rivers, deserts, forests, and many thousands of islands, is a mosaic of natural and cultural landscapes, where human civilization and wild nature have coexisted for centuries (Figure 1). The unique conjunction of geography, history, and climate has led to a remarkable evolutionary radiation that continues to the present day, as animals and plants have adapted to the myriad of opportunities for life that the region

presents. The Mediterranean is particularly noted for the diversity of its plants – about 25,000 species are native to the region, and more than half of these are endemic – in other words, they are found nowhere else on earth. This has led to the Mediterranean being recognized as one of the first 25 Global Biodiversity Hotspots (Myers et al. 2000).

Besides this great richness of plants, a high proportion of Mediterranean animals are unique to the region: 2 out of 3 amphibian species are endemic, as well as half of the crabs and crayfish, 48% of

the reptiles, a quarter of mammals, 14% of dragonflies, 6% of sharks and rays and 3% of the birds. The Mediterranean is also hosting 253 species of endemic freshwater fish. Although the Mediterranean Sea makes up less than 1% of the global ocean surface, up to 13% of the world's macroscopic marine species are found there, of which 25 to 30% are endemic – an incredibly rich biodiversity for such a small area (Bianchi and Morri 2000). The Mediterranean's importance for wildlife is not limited to the richness or uniqueness of its resident fauna and flora: millions of migratory birds from the far reaches

Figure 1. Map of the Mediterranean Sea and surrounding countries.

Box 1: Why is species conservation important?

Species provide us with essential services: not only food, fuel, clothes and medicine, but also purification of water and air, prevention of soil erosion, regulation of climate, pollination of crops by insects, and many more. In the Mediterranean, they provide a vital resource for the tourism and fishing industries, as well as having significant cultural, aesthetic and spiritual values. Consequently the loss of species diminishes the quality of our lives and our basic economic security. From an ethical point of view, species are part of our natural heritage and we owe it to future generations to preserve and protect them.

of Europe and Africa use Mediterranean wetlands and other habitats as stopover or breeding sites.

The Human Factor

In addition to its thousands of species of fauna and flora, the Mediterranean region is home to some 455 million human inhabitants, from a wide variety of countries and cultures. Considerable economic disparities exist within the region, with the GNI per capita of the Mediterranean EU countries (USD 20,800) being ten times that of the North African ones (USD 2,100) (World Bank 2006). Poor people depend heavily on natural resources and the loss of biodiversity is undermining the potential for economic growth, affecting the security of populations (food, health, etc.) and limiting their options. On the other hand, economic development increases the pressures on the environment and hence conservation challenges and options in the region are driven by these economic inequities.

The region also receives a large number of visitors: in 2005, 246 million people –

31% of all international tourists – visited the Mediterranean, particularly its coastal areas (Blue Plan 2008). Many visitors to the region are drawn by its natural beauty, but heavy pressure from visitors and residents alike is causing severe environmental degradation. Urbanization, coastal development, pollution, and unsustainable exploitation of natural resources such as marine fish are just

"An outstanding centre of biodiversity but also one of the most threatened, mainly by human activity"

some of the many human activities that are leading to an ever-increasing number of Mediterranean species to be facing a high risk of extinction.

Assessing Mediterranean Species

Assessing the conservation status of species at the Mediterranean regional level is particularly relevant to regional policy

instruments such as the Convention for the Protection of the Marine Environment and the Coastal Region of the Mediterranean or Barcelona Convention. It gives a timely overview of the status of biodiversity, and provides sound scientific data to decision-makers for policy development and management of natural resources. These assessments will help Mediterranean countries to determine whether or not they have met their obligations, commitments and targets under international agreements, such as the target to reduce the rate of biodiversity loss by 2010 under the Convention on Biological Diversity (CBD 2002). Hence IUCN is coordinating a process to evaluate the conservation status of all vertebrates and selected invertebrate and plant groups in the Mediterranean region, including terrestrial, freshwater and marine species. In total, 1,912 species have been assessed to date (Table 1). Some taxonomic groups have been assessed at the global level (amphibians, birds, mammals and reptiles), while others have been evaluated regionally (cartilaginous fishes, cetaceans, crabs

Table 1. *Numbers of species from Mediterranean countries assigned to each IUCN Red List category, by taxonomic group. Assessments carried out between 2004 and 2008 by IUCN and its partners. Data Deficient means that there is not enough information to assign the species to one of the other Categories, and it does not imply that the species is not threatened.*

IUCN Red List Categories	Amphibians[1]	Birds[1]	Cartilaginous fishes[2]	Cetaceans[2,3]	Crabs and Crayfish[2,3]	Endemic Freshwater fishes[1,4]	Mammals[1]	Dragonflies[2,4]	Reptiles[1]	TOTAL
Extinct[5]	1	1	0	0	0	8	2	4	0	16
Critically Endangered	4	6	13	1	0	45	5	5	14	93
Endangered	13	9	8	2	3	46	15	13	22	131
Vulnerable	16	13	9	2	2	51	27	13	11	144
Near Threatened	17	29	13	0	4	10	20	27	36	156
Least Concern	63	543	10	0	5	52	231	96	253	1253
Data Deficient	1	0	18	4	0	41	30	6	19	119
TOTAL	115	601	71	9	14	253	330	164	355	1912
Endemic	71 (62%)	16 (3%)	4 (6%)	0 (0%)	7 (50%)	253 (100%)	87 (26%)	23 (14%)	170 (48%)	631 (33%)

[1] Species assessed at the global level.
[2] Species assessed at the regional level.
[3] Preliminary data; still to be confirmed by the IUCN Red List Authority.
[4] Only the species occurring in river basins flowing into the Mediterranean Sea and adjacent Atlantic waters were included in the assessment (Smith and Darwall 2006).
[5] "Extinct" includes the categories Extinct, Extinct in the Wild and Regionally Extinct.

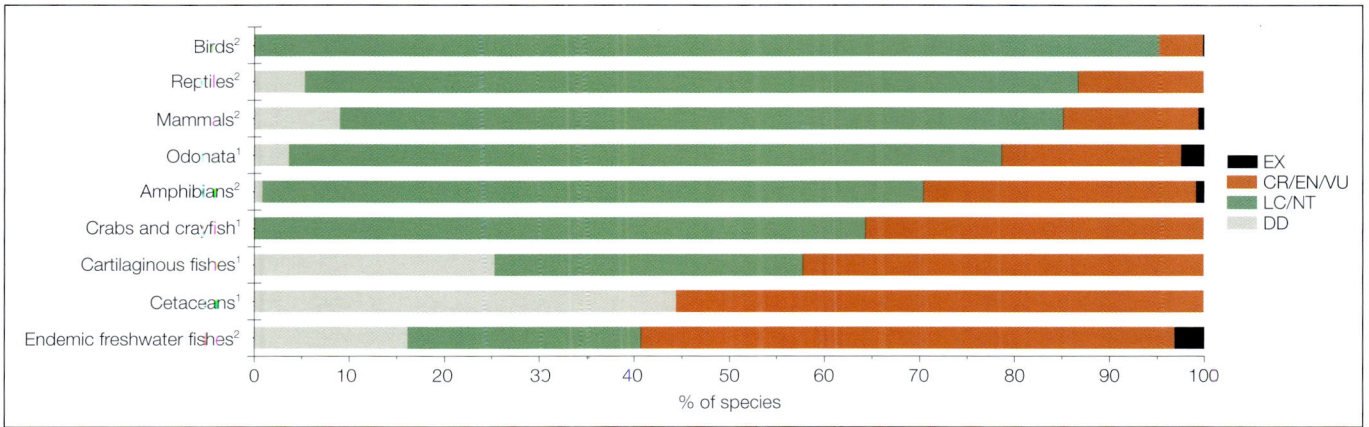

¹ Species assessed at the regional level
² Species assessed at the global level

Figure 2. Percentages of Extinct, threatened, non-threatened and Data Deficient species in each major taxonomic group assessed.

and crayfish, endemic freshwater fishes and Odonata (dragonflies and damselflies, later referred to collectively as dragonflies)). Although the global and regional assessments are not directly comparable with each other, they do give an indication of the different levels of threat faced by each taxonomic group.

A closer look at the different groups shows that at least 56% of endemic freshwater fishes, 56% of dolphins and whales, 42% of sharks and rays, 36% of crabs and crayfish, 29% of amphibians, 19% of dragonflies and damselflies, 14% of mammals, 13% of reptiles and 5% of birds are threatened with extinction. Overall, the proportion of

threatened species in the Mediterranean (those classified as Critically Endangered, Endangered or Vulnerable), either at the global or at the regional level, is about one fifth (19%) and about 1% of the species is already extinct in the region. These percentages will be higher if some of the currently Data Deficient species prove

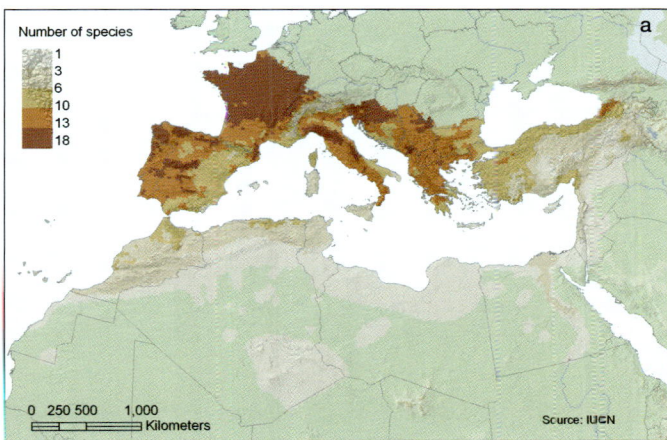

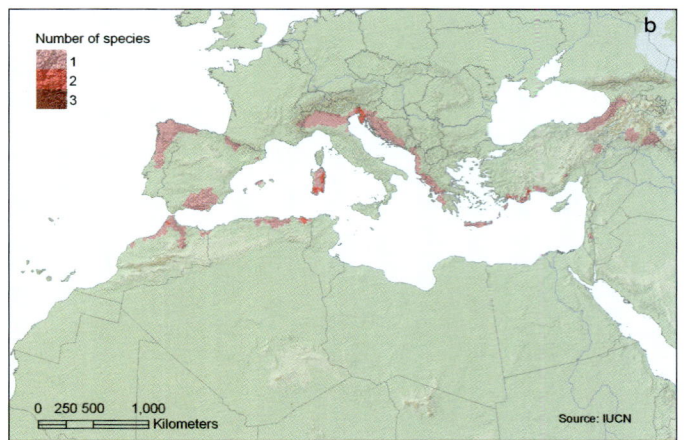

Figure 3. Species richness of amphibians in the Mediterranean basin. **(a)** and species richness of globally threatened amphibians in the Mediterranean basin **(b)**.

Figure 4. Species richness of globally threatened birds in the Mediterranean basin. (Map of bird species richness is not available).

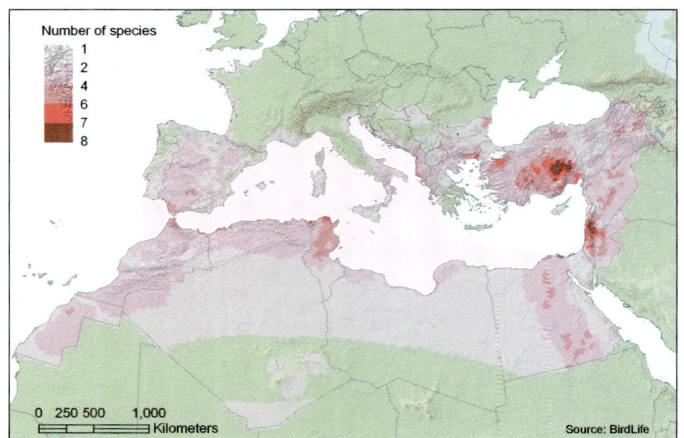

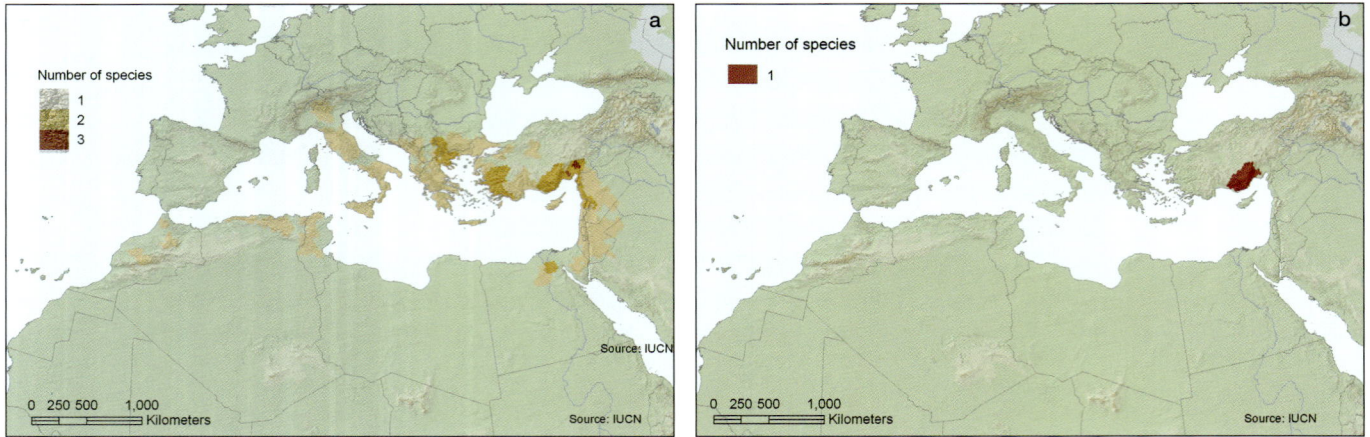

Figure 5. Species richness of crabs in the Mediterranean basin **(a)** and species richness of regionally threatened crabs in the Mediterranean basin **(b)**.

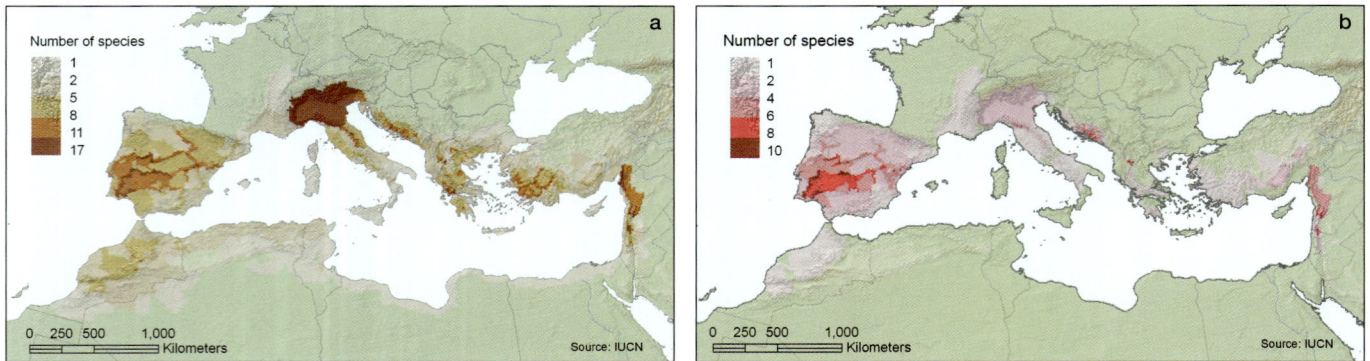

Figure 6. Species richness of endemic freshwater fish in the Mediterranean basin **(a)** and species richness of threatened endemic freshwater fish in the Mediterranean basin **(b)**.

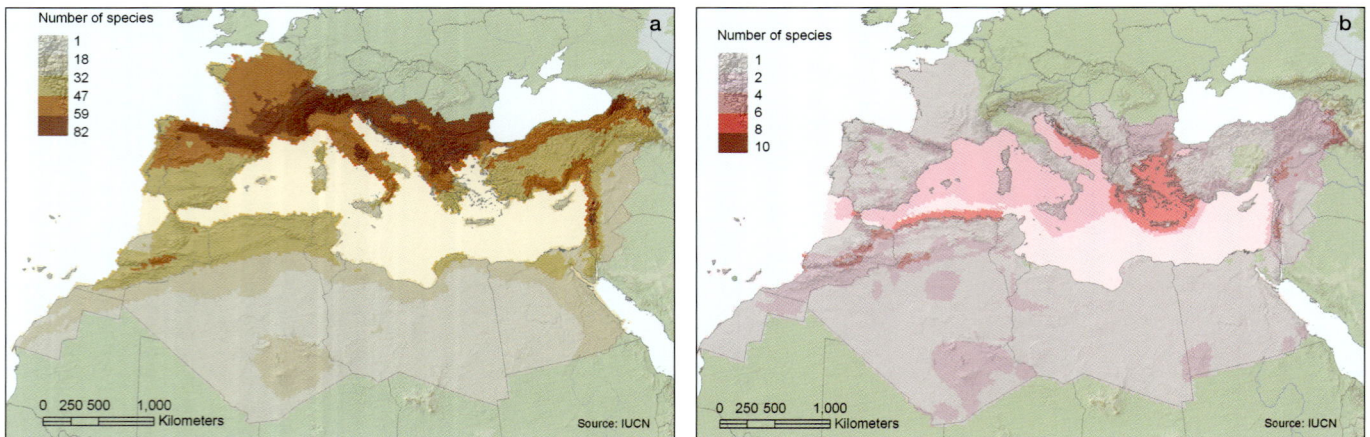

Figure 7. Species richness of mammals (including cetaceans) in the Mediterranean basin **(a)** and species richness of globally threatened mammals (including cetaceans) in the Mediterranean basin **(b)**.

to be threatened. Sixteen species are already extinct in the region, including some endemics such as the Hula Painted Frog *Discoglossus nigriventer*, the Canary Islands Oystercatcher *Haematopus meadewaldoi* and seven endemic freshwater fishes: *Tristramella intermedia*, *Tristramella magdelainae*, *Alburnus akili*, *Chondrostoma scodrense*, *Mirogrex hulensis*, *Telestes ukliva* and *Salmo pallaryi*. These extinctions signify the definitive loss of an important part of the world's biological heritage.

The geographic distribution of species richness and threatened species richness, highlighting regions with greater concentrations of species at risk that should be given particular attention, is presented for each taxonomic group in Figures 3 to 9.

Freshwater habitats

In addition to invaluable "ecosystem services" such as food, water purification, flood and pollution control, and fertile sediment for agriculture, rivers and wetlands also provide irreplaceable habitats for thousands of species. But freshwater habitats are facing major threats: Mediterranean rivers contain more than 3,500 dams, sediment discharge is drastically reduced and water

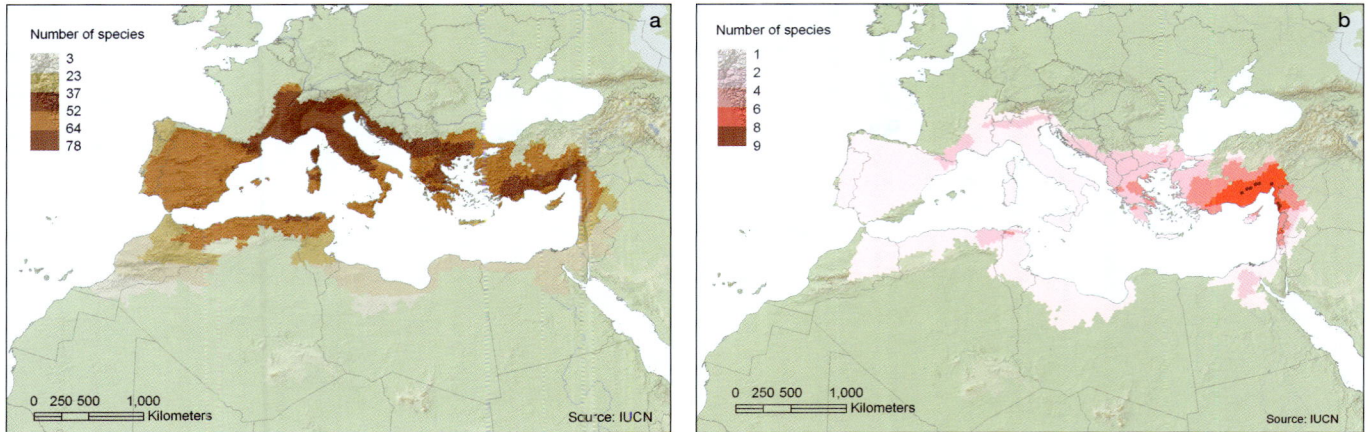

Figure 8. *Species richness of dragonflies in the Mediterranean basin* **(a)** *and species richness of regionally threatened dragonflies in the Mediterranean basin* **(b).**

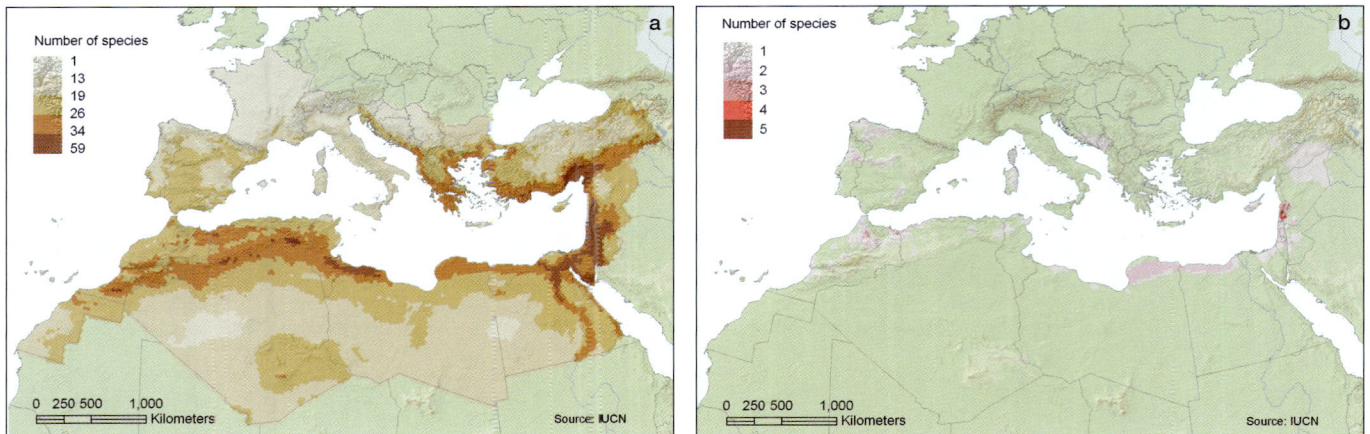

Figure 9. *Species richness of reptiles in the Mediterranean basin* **(a)** *and species richness of globally threatened reptiles in the Mediterranean basin* **(b).**

is diverted for energy production, irrigation or water supply, reducing therefore the original basin drainage area by about 78% (Poulos and Collins 2002). In most Mediterranean countries, water-use is approaching the limit of available resources (Blue Plan 2005) and several rivers are now seasonally dry.

Of the species assessed, 547 amphibians, crabs, freshwater fishes, dragonflies,

"Freshwater species contribute significantly to the economy, environment and livelihoods of Mediterranean societies"

reptiles and mammals are dependant of freshwater habitat for at least some part of their life cycle. The fact that 38% of them are threatened gives an indication of the

worrying status of Mediterranean wetlands and rivers. Freshwater species have been mapped based on river basins flowing into the Mediterranean Sea and adjacent Atlantic Ocean river basins. Figure 10a provides an indication of patterns of high freshwater species richness, based on the species assessed to date, while Figure 10b indicates concentrations of species at risk, in particular in the Iberian Peninsula,

Figure 10. *Species richness of freshwater amphibians, crabs, endemic fishes, mammals, dragonflies and reptiles in the Mediterranean basin* **(a)** *and species richness of freshwater threatened amphibians, crabs, endemic fishes, mammals, dragonflies and reptiles in the Mediterranean basin* **(b).**

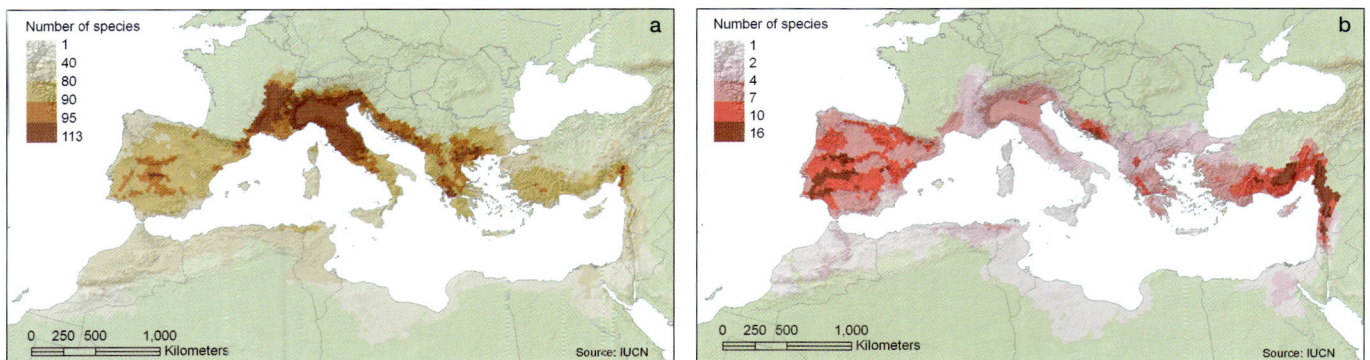

Mediterranean freshwater-dependent species: Green Gomphid Ophiogomphus cecilia *– Least Concern © Jean-Pierre Boudot. Pond Water-crowfoot* Ranunculus peltatus *– Not Evaluated © Serge Müller.* Economidichthys pygmaeus *– Least Concern © Ioannis Rousopoulos. Pyrenean Frog* Rana pyrenaica *– Endangered © Lars Bergendorf*

the Balkans, the western part of Greece and the area from Turkey down to Israel. More information about the conservation status of amphibians and endemic freshwater fish are detailed in Cox et al. (2006) and Smith and Darwall (2006).

Some species, such as various amphibians and dragonflies, are particularly sensitive to water quality and considered to be good indicators of the health of freshwater systems. Monitoring the status

of these freshwater species is therefore a key tool in the conservation of important Mediterranean wetlands.

Terrestrial habitats

The Mediterranean region is made up of a mosaic of different terrestrial habitats, containing a diverse range of species, including 355 species of reptiles (Cox *et al.* 2006), 330 species of mammals, 106 species of amphibians, 158 species of dragonflies, about half of the species in

these groups being endemic. There is also a high diversity of birds, invertebrates and plants. The initial results show that about 16% of the assessed terrestrial species are threatened with extinction.

Based on these results, terrestrial species richness is shown in Figure 11a. It's interesting to note the Hoggar mountain region, in the south of Algeria, which is an important refuge for numerous species. However, this map is only indicative, as

Figure 11. *Species richness of terrestrial amphibians, mammals, dragonflies and reptiles in the Mediterranean basin* **(a)** *and species richness of threatened terrestrial amphibians, mammals, dragonflies and reptiles in the Mediterranean basin* **(b)**.

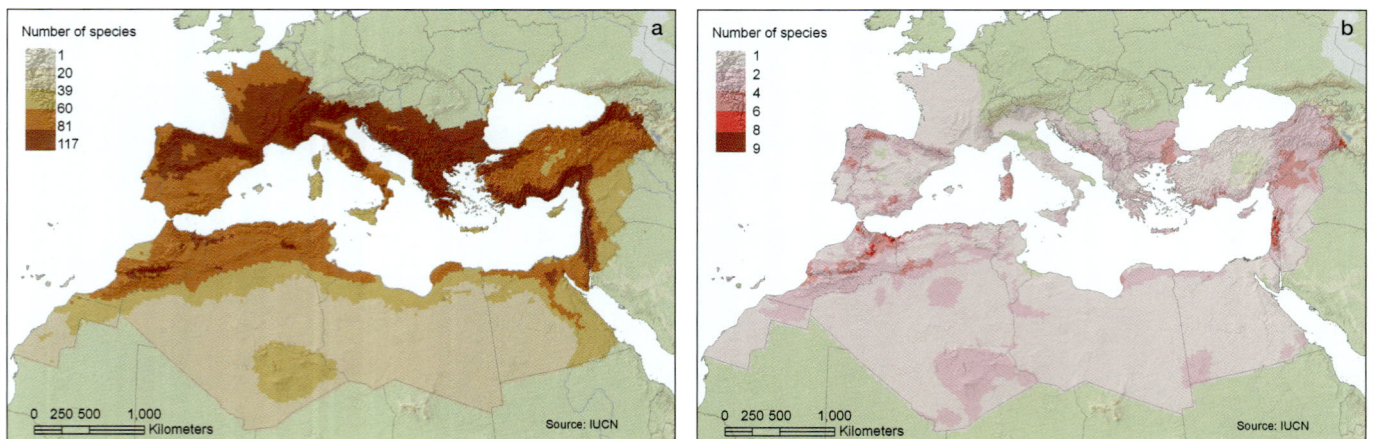

Box 2: Mediterranean island plants

With almost 5,000 islands and islets, the Mediterranean comprises one of the largest groups of islands in the world. Mediterranean islands display extraordinary features, with high rates of endemism, and act as a natural laboratory for evolutionary studies. Their particularities give rise to specific conservation challenges. Thus many of the endemic island plant species are confined to single small locations, they are extremely vulnerable to habitat destruction, overgrazing, and urban expansion. The *Top 50 Mediterranean Island Plants* highlights some of the most threatened plant species of the Mediterranean islands, stressing particular situations and conservation needs (Montmollin and Strahm 2005).

Box 3: Medicinal plants: Biodiversity that saves lives

North African people have ancient and rich traditions associated with the use of medicinal plants. Plant-derived products are used in the production of traditional medicines, cosmetics and perfumes. They are particularly important for people of the region, as they are sometimes the only source of medicine readily available. Mediterranean plants have been used in the development of modern pharmaceutical products and crop varieties, and about 70% of the North African wild plants in the Mediterranean are known to be of potential value in fields such as medicine, biotechnology and crop improvements (UNEP 2006). Increased demand, coupled with unsustainable collection from the wild has led a number of important plant species to become scarce in areas where they were previously abundant. The regulation of their collection is therefore essential, to ensure that these valuable species continue to be available in future.

---

plants and invertebrates, which account for most of the terrestrial species, have not yet been assessed. Figure 11b indicates some areas of particular concern, due to the high numbers of threatened species, in particular Morocco, the eastern rim of the Mediterranean basin and Turkey.

Marine habitats

The Mediterranean Sea contains an immense diversity of life despite its small area. Of the world's 85 cetacean species, 23 are known to occur in the Mediterranean and the Black Seas. Some are just visitors, but nine species are known to be year-round residents in the Mediterranean (Reeves and Notarbartolo 2006). An additional marine mammal species is encountered in the Mediterranean Sea: the Mediterranean Monk Seal *Monachus monachus* – the world's most endangered pinniped.

Cartilaginous fishes (sharks, rays and chimaeras) are also present, with 71 species living and breeding in the Mediterranean Sea (Cavanagh and Gibson 2007).

Despite the general impression of homogeneity, under-sea ecosystems are very diverse, with submarine mountains, canyons and other specific hydrological features. As a consequence, marine species are not evenly distributed (Figures 12a and 12b) and some areas are of critical importance for the conservation of these species as they provide unique nursery and feeding sites.

Sharks are the top predators of the Mediterranean Sea food chain: they regulate species abundance, distribution and diversity, and they help maintain the marine ecosystem's health and limit disease dispersal by taking sick or weak prey. Nonetheless, they are facing a particularly high risk of extinction. Accidental killing, intensive fishing activities and pollution are severe threats for these species, and the situation in the Mediterranean Sea is much bleaker than it is worldwide: 42% of the shark species are threatened in the region in comparison with 17% globally (Polidoro *et al.* this volume), which has led to the Mediterranean being described as the most dangerous

Mediterranean terrestrial species: Kythrean Sage Salvia veneris – Critically Endangered © Yiannis Christofides. Desert Horned Viper Cerastes cerastes – Least Concern © Wolfgang Böhme. Egyptian Vulture Neophron percnopterus – Endangered © Pedro Regato. Spanish Ibex Capra pyrenaica – Least Concern © Pedro Regato

Mediterranean marine species: Short-Beaked Common Dolphin Delphinus delphis *– Endangered © Giovanni Bearzi/Tethys. Long-snouted Seahorse* Hippocampus guttulatus *– Data Deficient © TUDAV. Goose Foot Star* Peltaster placenta *– Not Evaluated © TUDAV. Giant Devilray* Mobula mobular *– Endangered © Maurizio Würtz*

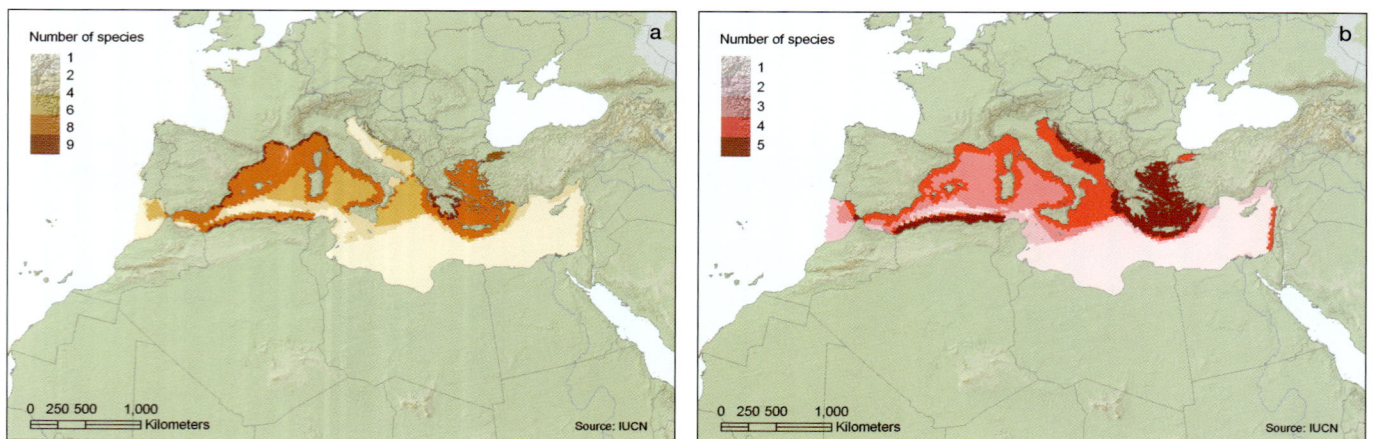

Figure 12. *Species richness of marine mammals in the Mediterranean Sea* **(a)** *and species richness of threatened marine mammals in the Mediterranean Sea* **(b).**

sea in the world for cartilaginous fishes (Cavanagh and Gibson 2007).

The analysis of marine mammals and shark species, as well as the first results for other marine fish species, displays a particularly striking feature of the marine ecosystem: about one third of species are assessed as Data Deficient; in other words, there is insufficient information to determine which Red List Category a species should be placed in. Research at sea is logistically more difficult and more expensive than on dry land, even in a sea as much used as the Mediterranean. This

means that the real number of threatened species could well be much higher and that species could be declining or perhaps even disappearing from our waters without us even noticing.

The main causes of threat: why are so many species in peril?

Habitat Loss and Degradation

As Figure 13 clearly shows, the loss, fragmentation and degradation of habitats as a direct or indirect result of human activities is the main threat to Mediterranean species. This applies to

Box 4: The most important causes of threat for Mediterranean species (by order of importance):

- Habitat loss and degradation
- Pollution
- Overexploitation (unsustainable harvesting, hunting and fishing)
- Natural disasters
- Invasive alien species
- Human disturbance
- Bycatch

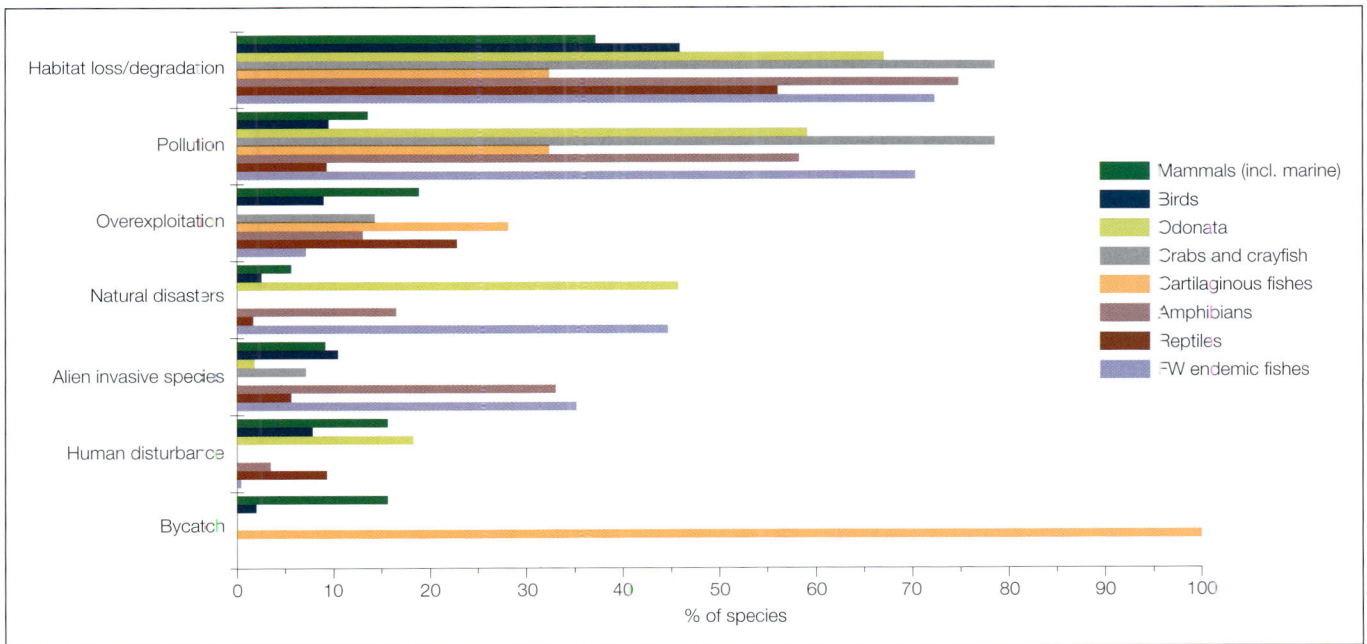

Figure 13. Breakdown of the major threats to amphibians, birds, cartilaginous fishes, crabs and crayfish, dragonflies, endemic freshwater fishes, mammals (including marine mammals) and reptiles in the Mediterranean.

Dam in Northern Spain. © Kevin Smith

all taxonomic groups and to all parts of the Mediterranean region. Changes in land-use patterns, such as intensification or abandonment of agricultural practices, urbanization, industrialization or tourism development are some of the main causes of this degradation.

Infrastructure development is strongly affecting some of the most fragile habitats. For example, 32% of freshwater fishes are threatened by dam construction (Smith and Darwall 2006), which drastically alters hydrological processes, reduces the amount of water available downstream, blocks migratory routes and can impair reproduction (McAllister et al. 2001).

Pollution

For the groups of species assessed so far, the second most important cause of threat is pollution. For example, chemical pollutants such as polychlorinated biphenyls (PCBs) are well known to affect the immune system, increasing sensitivity to illness, and causing increased mortality and impaired reproductive success. Mediterranean populations of Striped Dolphins *Stenella coeruleoalba* have suffered population declines due to morbillivirus infection, and it is believed that PCBs played an important role by compromising the immune system of the affected animals.

Another type of pollution is noise pollution: in the Mediterranean Sea, the increasing levels of noise due to marine traffic are harming cetaceans by impairing their ability to communicate and to locate their prey.

Solid waste is also a serious problem: discarded plastic bags have caused the death of many marine animals such as turtles, birds or dolphins that mistake the bags for jellyfish and die from ingesting them. Runoff of agricultural fertilizers causes eutrophication of coastal waters and can result in the formation of "dead zones", where no oxygen is available and fish and crustaceans cannot survive (Diaz and Rosenberg 2008).

Overexploitation (Unsustainable Harvesting, Hunting and Fishing)

Overexploitation is a serious problem for Mediterranean species, affecting many threatened plants, reptiles, fishes, and other species. Overexploitation is driven by several causes: for example, demand for traditional medicines is threatening some plants, seahorses and mammals species. Illegal trade is also of major concern in the Mediterranean: the Critically Endangered Egyptian Tortoise *Testudo kleinmanni* is, for example, heavily affected by the illegal national and international pet trade. Increased fishing activities and more efficient fishing boats and gear have resulted in the overfishing and consequent decline of some fish species. Overexploitation is likely to be of major importance for some Mediterranean species groups (e.g., marine fishes and medicinal plants) for which comprehensive assessments have not yet been completed.

Natural Disasters

Many Mediterranean species are threatened by natural disasters or extreme climatic events, notably forest fires and droughts. The frequency of such events is expected to increase as a result of global climate change. Climate change models indicate that the Mediterranean region will experience decreasing rainfall and increasing sea temperatures (Bates

et al. 2008), which will have an impact on the distribution and survival of species. Information collected during the Red List assessment process shows that populations of North African freshwater species such as molluscs and dragonflies are already shifting their ranges northwards in response to rising temperatures and decreasing availability of water – and there is a limit as to how far north they can move, given that the Mediterranean Sea presents a major barrier to dispersal. Through a combination of climate change, increased water abstraction, and the construction of dams, some rivers in North Africa are now completely dry for parts of the year when previously they flowed year-round, and some springs have completely dried out. A number of range-restricted molluscs are already feared to have gone extinct.

Invasive Alien Species

Invasive alien species are alien species which become established in natural or semi-natural ecosystems or habitat, are an agent of change, and threaten native biological diversity (IUCN 2000). Their introduction can be deliberate, to satisfy

human needs (food, pest control) or accidental (often as a result of increased globalization of transport). These invasive species can cause enormous damage to ecosystems, livelihoods and human health, and they are one of the most important causes of biodiversity loss, especially on Mediterranean islands. Due to a lack of information and awareness, the issue of invasive species and their effects is often underestimated and adequate prevention and mitigation measures are lacking.

Human Disturbance

The Mediterranean is a densely populated region that receives large numbers of visitors each year, and direct disturbance by humans is an important threat to some animals and plants, including iconic Mediterranean species such as the Northern Bald Ibis *Geronticus eremita* and the Mediterranean Monk Seal *Monachus monachus*, both listed as Critically Endangered. Disturbance at breeding sites can be particularly problematic, as species may abandon their young.

Bycatch

Most of the Mediterranean marine species assessed so far are affected by accidental capture in fishing gear, also called "bycatch". This is considered to be a major threat for sharks, rays, dolphins and marine turtles. All shark species are considered to be threatened by bycatch,

---

Box 5: The Water Hyacinth

The Water Hyacinth *Eichornia crassipes*, originally from the Amazon basin in South America and introduced for ornamental use in garden ponds, is now widely distributed in the Mediterranean basin. Doubling its population every two weeks, it cover quickly rivers and water bodies, impeding boat traffic, competing with endemic aquatic plants and, by reducing the light available under the water, threatening the whole ecosystem (Lowe *et al.* 2000). Its control costs millions of Euros each year just in the European Mediterranean countries.

© Geoffrey Howard

Box 6: The Mediterranean Monk Seal

The Mediterranean Monk Seal *Monachus monachus* is classified as Critically Endangered and is the most threatened pinniped species in the world. Only 350-450 animals survive, with the largest remaining population sited in the eastern Mediterranean, in Greece, western Turkey and some islands in the Ionian and Aegean Seas. Remnant populations are fragmented and declining. The main threats are linked to human activities and include exploitation, bycatch and persecution. More recently, tourism has grown to become one of the most significant hazards faced by monk seals, particularly in the eastern Mediterranean: as well as causing significant disturbance to individuals and breeding colonies, tourist activities increase the risk of vessel accidents, spills, transmission of disease, and the discharge of pollutants and waste near the seals (IUCN 2007).

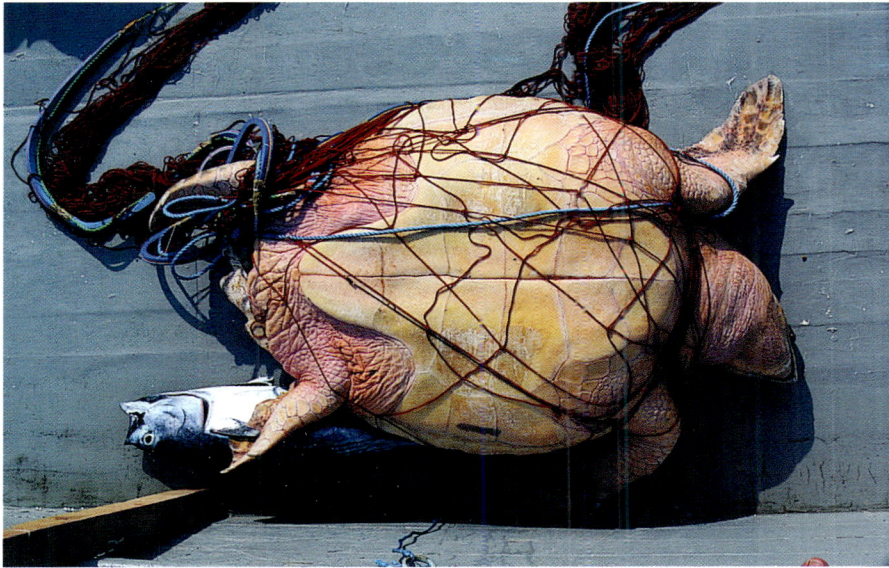

A Loggerhead Turtle Caretta caretta *trapped in a swordfish driftnet. This species is listed as Endangered.* © TUDAV

has significantly improved the status of some Mediterranean species, such as the Dalmatian Pelican *Pelecanus crispus* (Nagy and Crockford 2004). The Barcelona Convention (1976) defined action plans for key Mediterranean threatened species (such as the Monk Seal, the sharks or the marine vegetation). Mediterranean assessments can support this process by providing crucial information on threatened species.

The primary goal of species conservation is the preservation of viable populations of wild species in their original native range. However, in certain circumstances, in particular for the most threatened species, intensive management such as captive breeding may be necessary to ensure the survival of species that are close to extinction.

but the large coasta species, which are more exposed to intensive fishing pressure, are thought to be the most seriously affected (Cavanagh and Gibson 2007).

Conservation actions to prevent extinction

Numerous national, regional and international actions have been put in place to enhance species survival in the Mediterranean. Through the Mediterranean Species Assessment process participating scientists propose recommendations for targeted conservation actions that are needed to reduce species extinction risk and which can help Mediterranean states to monitor whether or not they are meeting their obligations under the regional and global conventions and multi-lateral agreements.

Species protection

Improvement and enforcement of legal protection for threatened species and their habitats is the most urgent conservation action to be taken at both regional and national levels. Many threatened species currently have no legal protection: for

example, little more than a quarter (27%) of threatened sharks benefit from any form of protective legislation.

Species Action Plans can be an effective means of determining specific conservation actions that are needed and for promoting coordinated activities. The implementation of action plans

Site protection

Site protection, for example through the designation of protected areas, is one of the most effective means of reducing global biodiversity loss. However, the goal of protecting at least 10% of each of the world's ecological regions by 2010 is still far out of reach, especially the commitment to create and sustain a coherent network of Marine Protected Areas (MPAs) by 2012 (Word Sustainable Development Summit, Johannesburg 2002). In order to effectively safeguard all threatened species, especially threatened endemics, protected areas need to be designed as representative networks and to integrate a gap analysis assessing the adequacy of species coverage. In this respect, Mediterranean marine protected areas are notably under-developed and the southern and eastern parts of the area, despite their importance for the marine biodiversity, are insufficiently protected.

"Integrating species data and spatial distributions into the planning and designation of protected areas allows the establishment of more efficient and representative networks."

Box 8: The Mallorcan Midwife Toad

The Mallorcan Midwife Toad *Alytes muletensis* is a very rare species endemic to the island of Mallorca (Spain). It is threatened by predation by the introduced Viperine Snake *Natrix maura*, and competition for space with Perez's Frog *Pelophylax perezi*, as well as by habitat loss owing to the development of tourism and human settlements. A conservation and recovery plan is being undertaken by the Species Conservation Service of the Conselleria de Medi Ambient in Mallorca, with captive breeding and reintroductions taking place. Thanks to these measures, the species, initially assessed as Critically Endangered, has now been downlisted to Vulnerable.

Male Mallorcan Midwife Toad carrying eggs. © Richard Griffiths

Box 7: Main conservation actions for Mediterranean species:

- Species protection
- Site protection
- Conservation of the wider environment
- Communication and education
- Monitoring and research

Conservation of the wider environment

Although protected areas are a key tool for protecting species, many plants and animals live outside these areas, often in semi-natural or man-made environments where coexistence with humans is the only option for survival, and where wildlife conservation is just one of many competing land-uses. For species to prosper in the wider environment, it is essential that biodiversity conservation is integrated into public policy in other sectors that impact on species and their habitats, notably agriculture, fisheries, forestry, urban planning, transport, water management, and so on.

Integrated River Basin Management (IRBM) takes an ecosystem approach to the management of rivers, associated wetlands and groundwater systems. River basins are dynamic systems, and any single management intervention has implications for the system as a whole. IRBM is the process of coordination, through stakeholder participation, of biodiversity conservation, management and water resource allocation decisions across the river basin as a whole to ensure that freshwater ecosystems are maintained whilst ensuring that human development needs are equitably met.

Communication and Education

Effective conservation cannot be achieved without the support of those people dependent on natural resources. Communicating about the status of their environment, its importance for humans, the main threats and the

Tajo National Park (Guadalajara, Spain). © Pedro Regato

actions that could be taken to mitigate them is an essential part of sustainable development. The species assessments provide timely and reliable information on which such communication can be based. They promote synergies and collaboration between regional actors to enhance conservation efforts to halt biodiversity loss, as exemplified by the establishment of the first Intercontinental Biosphere Reserve between Spain and Morocco.

Monitoring and research

A solid evidence base is necessary to determine conservation priorities and take appropriate action. The Mediterranean assessments provide a baseline against which future progress can be assessed, and provide a wealth of information on species status, population size and trends, distribution, habitat requirements, threats, conservation actions in place and needed, and other factors that will be of use to policymakers, conservation practitioners, natural resource managers and others. However, scientific information, especially in the marine ecosystem, is still lacking and research programmes are crucial to further develop the understanding and knowledge needed to underpin sound natural resource management.

Key findings

- The regional assessments confirm the high diversity and endemism of Mediterranean plants and animals, but also underline the severe threats that these species face. Nine species groups have been comprehensively assessed to date (amphibians, birds, cartilaginous fishes, cetaceans, crabs and crayfish, endemic freshwater fishes, mammals, dragonflies and reptiles), and almost a fifth of these species are threatened with extinction, with 5% Critically Endangered, 7% Endangered and 7% Vulnerable.

- Freshwater ecosystems are under particularly severe pressure - over 56% of endemic freshwater fish species are threatened with extinction.

- The marine ecosystem is poorly known, with around one third of marine species assessed to date listed as Data Deficient.

- Assessment of additional groups of species in this biodiversity hotspot is ongoing, and results will soon be available for marine fishes, freshwater molluscs, butterflies and endemic plants.

- Mediterranean species are threatened with extinction as a result of human

---

Box 9: The Gizani

The Gizani *Ladigesocypris ghigii* is a fish species endemic to Rhodes, Greece, which is threatened by water abstraction and is classified as Vulnerable. It has been the subject of a LIFE-Nature project involving the study of its geographic range, life history, reproduction, nutrition, habitat preferences, genetics and threats. Conservation actions designed include an Action Plan focusing on the sustainable management of the island's water resources. This project has helped to ensure the future survival of this species, showing that good conservation practices can reduce a species' risk of extinction (Stoumboudi 2000).

activities. Habitat destruction, pollution, unsustainable exploitation, and other threats are taking a heavy toll on the region's biodiversity. Climate change, which is predicted to cause increasing droughts in this already arid region, is set to be an increasingly significant threat in the future.

- Urgent action is needed to preserve the future of the Mediterranean. Sustainable management and legal protection of species and their habitats are the key conservation actions to be promoted in the Mediterranean region, but education and research are also needed.

- Conservation actions applied to date have had positive results and some species have already been saved from extinction. However, in a region like the Mediterranean, where biodiversity is so strongly influenced by human activities, biodiversity loss is a constant reality that will only be stopped when humans realize how much their present and future health and prosperity can be damaged when species disappear.

References

Bates, B.C., Kundzewicz, Z.W., Wu, S. and Palutikof, J.P. (eds.). 2008. *Climate Change and Water*. Technical Paper of the Intergovernmental Panel on Climate Change, IPCC Secretariat, Geneva, 210 pp.

Bianchi, C.N. and Morri, C. 2000. Marine Biodiversity of the Mediterranean Sea: Situation, Problems and Prospects for Future Research. *Marine Pollution Bulletin* 40(5): 367-376.

Blue Plan. 2005. *A Sustainable Future for the Mediterranean. The Blue Plan's Environment and Development Outlook*. UNEP Blue Plan Activity Centre, Sophia Antipolis, France.

Blue Plan. 2008. *The Blue Plan's Sustainable Development Outlook for the Mediterranean*. UNEP Blue Plan Activity Centre, Sophia Antipolis, France.

Cavanagh, R.D. and Gibson, C. 2007. *Overview of the Conservation Status of Cartilaginous Fishes (Chondrichthyans) in the Mediterranean Sea*. IUCN, Gland, Switzerland and Malaga, Spain.

CBD Secretariat. 2002. *Strategic Plan for the Convention on Biological Diversity*. Publication of the Decision VI/26. Sixth meeting of the Convention on Biological Diversity 2002.

Cox, N., Chanson, J. and Stuart, S. (compilers). 2006. *The Status and Distribution of Reptiles and Amphibians of the Mediterranean Basin*. IUCN, Gland, Switzerland and Cambridge, UK.

Diaz, R.J. and Rosenberg, R. 2008. Spreading dead zones and consequences for marine ecosystems. *Science* 321: 926-929.

IUCN. 2000. *IUCN Guidelines for the Prevention of Biodiversity Loss Caused by Alien Invasive Species*. IUCN, Gland, Switzerland.

IUCN. 2007. *Monachus monachus*. In: IUCN. 2007. European Mammal Assessment. Downloaded on 11 August 2008, from http://ec.europa.eu/environment/nature/conservation/species/ema/.

Lowe, S., Browne, S., Boudjelas, S. and De Poorter, M. 2000. *100 of the World's Worst Invasive Alien Species A selection from the Global Invasive Species Database*. International Union for Conservation of Nature and Natural Resources.

McAllister, D., Craig, J.F., Davidson, N., Delany, S. and Seddon, M. 2001. Biodiversity Impacts of Large Dams. Background Paper Nr.1 Prepared for IUCN/ UNEP/ WCD. International Union for Conservation of Nature and Natural Resources and the United Nations Environmental Programme.

Montmollin, B. de and Strahm, W. (eds.). 2005. *The Top 50 Mediterranean Island Plants: Wild plants at the brink of extinction, and what is needed to save them*. IUCN/ SSC Mediterranean Islands Plant Specialist Group. IUCN, Gland, Switzerland and Cambridge, UK.

Myers, N., Mittermeier, R.A., Mittermeier, C.G., da Fonseca, G.A.B. and Kent, J. 2000. Biodiversity hotspots for conservation priorities. *Nature* 403: 853-858.

Nagy, S. and Crockford, N. 2004. *Implementation in the European Union of Species Action Plans for 23 of Europe's Most Threatened Birds*. BirdLife International, Wageningen, The Netherlands.

Poulos, S.E. and Collins, M.B. 2002. Fluviatile sediment fluxes to the Mediterranean Sea: a quantitative approach and the influence of dams. In: S.J. Jones and L.-E. Frostick (eds), *Sediment Flux to Basins: Causes, Controls and Consequences, pp. 227-245*. Geological Society (Special Publications), London.

Reeves, R. and Notarbartolo di Sciara, G. (compilers and editors). 2006. *The Status and Distribution of Cetaceans in the Black Sea and Mediterranean Sea*. IUCN Centre for Mediterranean Cooperation, Malaga, Spain.

Smith, K.G. and Darwall, W.R.T. (compilers). 2006. *The Status and Distribution of Freshwater Fish Endemic to the Mediterranean Basin*. IUCN, Gland, Switzerland and Cambridge, UK.

Stoumboudi, M.T. 2000. Conservation measures for the endangered fish *Ladigesocypris ghigii*. A Life-Nature project. Symposium on "Freshwater Fish Conservation - Options for the future". Algarve, Portugal.

UNEP. 2006. *Africa Environment Outlook 2 - Our Environment, Our Wealth*. United Nations Environment Programme, Nairobi, Kenya.

World Bank. 2006. *Key development data and statistics*. Downloaded on August 25, 2008, from http://web.worldbank.org/WBSITE/EXTERNAL/DATASTATISTICS/0,,content MDK:20535285~menuPK:1192694~pag ePK:64133150~piPK:64133175~theSite PK:239419,00.html.

Educational activity in Salum, Egypt. © Nature Conservation Egypt

Appendices

Appendix 1: Summary of the five criteria (A–E) used to evaluate if a taxon belongs in a threatened category (Critically Endangered, Endangered or Vulnerable)

Use any of the criteria A–E	Critically Endangered	Endangered	Vulnerable
A. Population reduction	Declines measured over the longer of 10 years or 3 generations		
A1	≥ 90%	≥ 70%	≥ 50%
A2, A3 & A4	≥ 80%	≥ 50%	≥ 30%

A1. Population reduction observed, estimated, inferred, or suspected in the past where the causes of the reduction are clearly reversible **AND** understood **AND** have ceased, based on and specifying any of the following:

- **(a)** direct observation
- **(b)** an index of abundance appropriate to the taxon
- **(c)** a decline in area of occupancy (AOC), extent of occurrence (EOO) and/or habitat quality
- **(d)** actual or potential levels of exploitation
- **(e)** effects of introduced taxa, hybridization, pathogens, pollutants, competitors or parasites.

A2. Population reduction observed, estimated, inferred, or suspected in the past where the causes of reduction may not have ceased **OR** may not be understood **OR** may not be reversible, based on (a) to (e) under A1.

A3. Population reduction projected or suspected to be met in the future (up to a maximum of 100 years) based on (b) to (e) under A1.

A4. An observed, estimated, inferred, projected or suspected population reduction (up to a maximum of 100 years) where the time period must include both the past and the future, and where the causes of reduction may not have ceased **OR** may not be understood **OR** may not be reversible, based on (a) to (e) under A1.

	Critically Endangered	Endangered	Vulnerable
B. Geographic range in the form of either B1 (extent of occurrence) AND/OR B2 (area of occupancy)			
B1. Extent of occurrence (EOO)	< 100 km²	< 5,000 km²	< 20,000 km²
B2. Area of occupancy (AOO)	< 10 km²	< 500 km²	< 2,000 km²

AND at least 2 of the following:

- **(a)** Severely fragmented, **OR**

	Critically Endangered	Endangered	Vulnerable
Number of locations	= 1	≤ 5	≤ 10

- **(b)** Continuing decline in any of: **(i)** extent of occurrence; **(ii)** area of occupancy; **(iii)** area, extent and/or quality of habitat; **(iv)** number of locations or subpopulations; **(v)** number of mature individuals.
- **(c)** Extreme fluctuations in any of: **(i)** extent of occurrence; **(ii)** area of occupancy; **(iii)** number of locations or subpopulations; **(iv)** number of mature individuals.

	Critically Endangered	Endangered	Vulnerable
C. Small population size and decline			
Number of mature individuals	< 250	< 2,500	< 10,000
AND either C1 or C2:			
C1. An estimated continuing decline of at least: (up to a max. of 100 years in future)	25% in 3 years or 1 generation	20% in 5 years or 2 generations	10% in 10 years or 3 generations
C2. A continuing decline **AND** (a) and/or (b):			
(a i) Number of mature individuals in each subpopulation: or	< 50	< 250	< 1,000
(a ii) % individuals in one subpopulation =	90–100%	95–100%	100%
(b) Extreme fluctuations in the number of mature individuals.			

	Critically Endangered	Endangered		Vulnerable
D. Very small or restricted population				
Either:				
Number of mature individuals	< 50	< 250	**D1.**	< 1,000
				AND/OR
			D2.	typically: AOO < 20 km² or number of locations ≤ 5
Restricted area of occupancy				

	Critically Endangered	Endangered	Vulnerable
E. Quantitative Analysis			
Indicating the probability of extinction in the wild to be:	≥ 50% in 10 years or 3 generations (100 years max.)	≥ 20% in 20 years or 5 generations (100 years max.)	≥ 10% in 100 years

Appendix 2. The IUCN Red List of Threatened Species™ threat categories scale

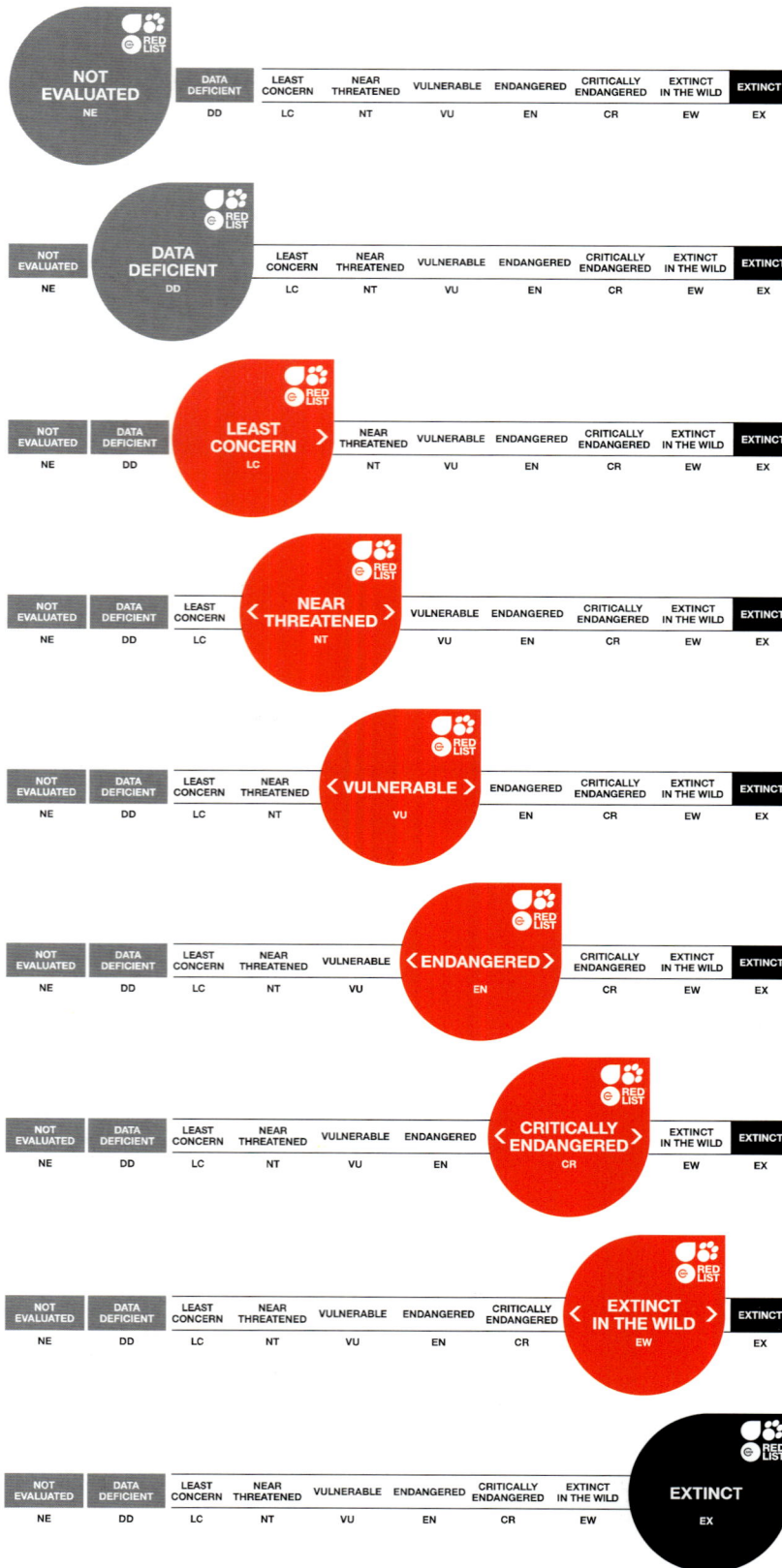

NOT EVALUATED	DATA DEFICIENT	LEAST CONCERN	NEAR THREATENED	VULNERABLE	ENDANGERED	CRITICALLY ENDANGERED	EXTINCT IN THE WILD	EXTINCT
NE	DD	LC	NT	VU	EN	CR	EW	EX

The IUCN Red List Categories and Criteria are the world's most widely used system for gauging the extinction risk faced by species. Each species assessed is assigned to one of eight different Categories (Extinct, Extinct in the Wild, Critically Endangered, Endangered, Vulnerable, Near Threatened and Data Deficient), based on a series of quantitative criteria.

Species classified as Vulnerable, Endangered and Critically Endangered are regarded as 'threatened'.

The threat category scales are an easy to use graphic element that clearly identifies the threat category of a species.

The scales can be used on signage, posters, in publications, etc. They can only be used in relation to a species that has been assessed and appears on The IUCN Red List. The scale must always be placed next to the name of the species.

There are alternative versions of the scale depending on usage. The preferred option is to use the long scale however where space is limited, the short scale, or a single button can be used. If the short scale or single button is used, then an explanation of the Threat Categories must appear nearby to ensure a clear understanding of the scale or button.

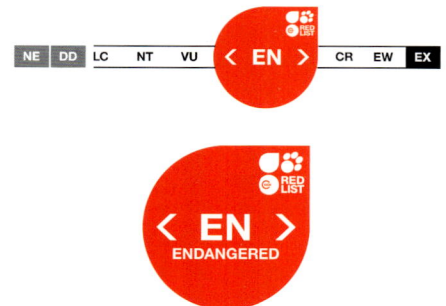

NE	DD	LC	NT	VU	< EN >	CR	EW	EX

< EN >
ENDANGERED

If you are interested in using the threat category scales please email iucnredlist.logo@iucn.org

Appendix 3. Sources for numbers of described species used in Table 1 of the chapter State of the World's Species

Taxonomic Group	Data source used
Vertebrates	
Mammals	From Wilson and Reeder (2005; see http://www.bucknell.edu/msw3/), with deviations based on new revisions and published papers that have appeared since the accounts in Wilson and Reeder (2005) were compiled and largely up until 31 December 2007, but there are a few exceptions where new species published early in 2008 were included. In cases where there are alternative taxonomic treatments, the Global Mammal Assessment coordinating team working with the relevant IUCN SSC Specialist Group has advised on which treatment to follow.
Birds	BirdLife International. 2008. The BirdLife checklist of the birds of the world, with conservation status and taxonomic sources. Version 1. Available from http://www.birdlife.org/datazone/species/downloads/BirdLife_Checklist_Version_1.zip [.xls zipped 1 MB]. Accessed: 28 September 2008.
Reptiles	Based on the figures (as of February 2008) provided by The Reptile Database compiled by Peter Uetz and Jakob Hallermann. Available at: http://www.reptile-database.org. Accessed: 28 September 2008.
Amphibians	From Frost (2008).
Fishes	Based on Froese and Pauly (2008).
Invertebrates	
Insects	Estimates of the number of insects in the world vary from about 751,000 to more than 1 million, but the most commonly cited figure is 950,000 (see discussion in Chapman (2005)).
Molluscs	From Bouchet (2007). (For further discussion on the numbers of molluscs, see Chapman (2005)).
Crustaceans	The estimated number of described species of Crustacea in the world varies from 30,000 to 67,000 but the best estimate is 40,000 (see discussion in Chapman (2005)).
Corals	Corals fall under the Phylum Cnidaria and are primarily in the Class Anthozoa, although there are some in the Class Hydrozoa. The number of described species reported here are for species typically regarded as 'corals' and are largely based on Spalding et al. (2001) (Alcyonarian corals); and Cairns (1999) (Scleractinian corals). The remainder of the cnidarians, anemones, jellyfish, etc., are treated under "Others".
Arachnids (spiders, scorpions, etc.)	Estimates of the number of described arachnids vary from 60,000 to 96,711, but the best estimate of 98,000 is higher than these figures (see discussion in Chapman (2005)).
Velvet Worms	The number of described species of Onychophora (velvet worms) would appear to be around 165 (for further details see discussion in Chapman (2005)).
Horseshoe Crabs	Horseshoe crabs are placed on the Red List under the traditional class "Merostomata" which excludes the fossil sea scorpions; only four species are extant today (see http://en.wikipedia.org/wiki/Merostomata for further details).
Others	This is a miscellaneous group of invertebrate species that have been assessed for the IUCN Red List. The total number of described species is based on the estimated totals from the following groups from which the assessed species come: Annelida - segmented worms (15,000), Cnidaria - anemones, jellyfish, etc. but excluding the corals which are treated separately (6,825), Echinodermata -starfish (7,000 species), Myriapoda - centipedes and millipedes (12,215) and Platyhelminthes - flat worms (20,000). For further details on the numbers in these groups see: Chapman (2005).
Plants	
Mosses	Based on information provided by Chapman (2005).
Ferns and allies	Based on information provided by Chapman (2005).
Gymnosperms	Based on Donaldson (2003), Farjon (2001) and Mabberley (1997). Chapman (2005) also follows this figure, for discussion see http://www.environment.gov.au/biodiversity/abrs/publications/other/species-numbers/03-03-groups-plants.html#gymnosperms.
Dicotyledons and Monocotyledons	Based on Thorne (2002), but see Mabberley (1997); Schmid (1998); Govaerts (2001, 2003); Bramwell (2002); and Scotland and Wortley (2003) for alternative views on the numbers of seed plant species.
Green and Red Algae	Numbers of green (Chlorophyta) and red (Rhodophyta) algae are from Guiry and Guiry (2008).
Others	
Lichens	The figure of 10,000 from Groombridge and Jenkins (2002) appears to be too low, so the number described is now based on information provided by Chapman (2005).
Mushrooms	Number of mushroom-forming fungi is based on Kirk et al. (2001) (see Tree of Life web site: http://tolweb.org/tree/phylogeny.html. Accessed 28 September 2008).
Brown Algae	The number of brown algae (Ochrophyta) is taken from Guiry and Guiry (2008).

References

Bouchet, P. 2007. Inventorying the molluscan fauna of the world: how far to go? In: K. Jordaens, N. van Houtte, J. van Goethem and T. Backlejau (eds), Abstracts of the World Congress of Malacology. Antwerp, Belgium.

Bramwell, D. 2002. How many plant species are there? Plant Talk 28: 32–34.

Cairns, S.D. 1999. Species richness of recent Scleractinia. Atoll Research Bulletin 459: 1–46.

Chapman, A. 2005 (updated April 2007). Numbers of Living Species in Australia and the World. Available at: http://www.environment.gov.au/biodiversity/abrs/publications/other/species-numbers/03-04-groups-others.html#lichens. Accessed: 28 September 2008.

Donaldson, J. (ed.) 2003. Cycads. Status Survey and Conservation Action Plan. IUCN/SSC Cycad Specialist Group. IUCN, Gland, Switzerland and Cambridge, UK.

Farjon, A. 2001. World Checklist and Bibliography of Conifers. 2nd edition. World Checklists and Bibliographies, 3. Royal Botanic Gardens, Kew.

Froese, R. and Pauly, D. (eds). 2008. FishBase. World Wide Web electronic publication. www.fishbase.org. version (07/2008). Accessed: 28 September 2008.

Frost, D.R. 2008. Amphibian Species of the World: an Online Reference. Version 5.2 (15 July, 2008). Available at: http://research.amnh.org/herpetology/amphibia/index.php. Accessed: 28 September 2008.

Govaerts, R. 2001. How many species of seed plants are there? Taxon 50: 1085–1090.

Govaerts, R. 2003. How many species of seed plants are there? – a response. Taxon 52: 583–584.

Groombridge, B. and Jenkins, M.D. 2002. World Atlas of Biodiversity. Prepared by the UNEP World Conservation Monitoring Centre. University of California Press, Berkeley, USA.

Guiry, M.D. and Guiry, G.M. 2008. AlgaeBase. World-wide electronic publication, National University of Ireland, Galway. http://www.algaebase.org. Accessed: 28 September 2008.

Kirk P.M., Cannon P.F., David J.C. and Stalpers J.A. 2001. Ainsworth and Bisby's Dictionary of the Fungi. 9th edition. Cambridge, United Kingdom: CAB International University Press.

Mabberley, D.J. 1997. The Plant-Book. A portable dictionary of the higher plants. Second edition. Cambridge University Press, Cambridge.

Schmid, R. 1998. Statistics for numbers of extant taxa of major groups in Mabberley. Taxon 47: 245.

Scotland, R.W. and Wortley, A.H. 2003. How many species of seed plants are there? Taxon 52: 101–104.

Spalding, M.D., Ravilious, C. and Green, E.P. 2001. World Atlas of Coral Reefs. Prepared at the UNEP World Conservation Monitoring Centre. University of California Press, Berkley, USA.

Thorne, R.F. 2002. How many species of seed plants are there? Taxon 51: 511–512.

Wilson, D.E. and Reeder, D.M. (eds) 2005. Mammal Species of the World. A Taxonomic and Geographic Reference. Third edition. Johns Hopkins University Press, Baltimore.

Appendix 4. Summary of number of animal species in each Red List Category in each taxonomic class

IUCN Red List Categories: **EX** - Extinct, **EW** - Extinct in the Wild, **CR** - Critically Endangered, **EN** - Endangered, **VU** - Vulnerable, **LR/cd** - Lower Risk/conservation dependent, **NT** - Near Threatened (includes **LR/nt** - Lower Risk/near threatened), **DD** - Data Deficient, **LC** - Least Concern (includes **LR/lc** - Lower Risk/least concern).

Class*	EX	EW	Subtotal	CR	EN	VU	Subtotal	LR/cd	NT	DD	LC	Total
Mammalia	76	2	**78**	188	448	505	**1,141**	0	323	836	3,110	**5,488**
Aves	134	4	**138**	190	361	671	**1,222**	0	835	66	7,729	**9,990**
Reptilia	21	1	**22**	86	134	203	**423**	3	123	180	634	**1,385**
Amphibia**	38	1	**39**	475	755	675	**1,905**	0	381	1,578	2,357	**6,260**
Cephalaspidomorphi	1	0	**1**	1	0	1	**2**	0	2	3	10	**18**
Chondrichthyes	0	0	**0**	22	29	75	**126**	1	107	205	152	**591**
Actinopterygii	90	13	**103**	265	240	640	**1,145**	10	135	426	1,051	**2,870**
Sarcopterygii	0	0	**0**	1	0	1	**2**	0	0	0	0	**2**
Echinoidea	0	0	**0**	0	0	0	**0**	0	1	0	0	**1**
Arachnida	0	0	**0**	2	5	11	**18**	0	2	9	3	**32**
Chilopoda	0	0	**0**	0	0	1	**1**	0	0	0	0	**1**
Diplopoda	0	0	**0**	1	6	7	**14**	0	0	7	10	**31**
Crustacea	7	1	**8**	84	127	395	**606**	9	19	663	430	**1,735**
Insecta	60	1	**61**	70	132	424	**626**	3	93	129	347	**1,259**
Merostomata	0	0	**0**	0	0	0	**0**	0	1	3	0	**4**
Onychophora	0	0	**0**	3	2	4	**9**	0	1	1	0	**11**
Hirudinoidea	0	0	**0**	0	0	0	**0**	0	1	0	0	**1**
Oligochaeta	1	0	**1**	1	0	4	**5**	0	1	0	0	**7**
Polychaeta	0	0	**0**	1	0	0	**1**	0	0	1	0	**2**
Bivalvia	31	0	**31**	52	28	15	**95**	5	60	14	13	**218**
Gastropoda	257	14	**271**	216	196	471	**883**	14	186	557	83	**1,994**
Enopla	0	0	**0**	0	0	2	**2**	0	1	3	0	**6**
Turbellaria	1	0	**1**	0	0	0	**0**	0	0	0	0	**1**
Anthozoa	0	0	**0**	6	23	202	**231**	0	175	147	289	**842**
Hydrozoa	0	0	**0**	1	2	2	**5**	0	1	2	8	**16**
Total	**717**	**37**	**754**	**1,665**	**2,488**	**4,309**	**8,462**	**45**	**2,448**	**4,830**	**16,226**	**32,765**

* Mammalia (mammals), Aves (birds), Reptilia (reptiles), Amphibia (amphibians), Cephalaspidomorphi (lampreys and hag fish), Chondrichthyes (sharks, skates, rays and chimaeras), Actinopterygii (bony fishes), Sarcopterygii (coelacanth), Echinoidea (sea urchins, starfish, etc), Arachnida (spiders and scorpions), Chilopoda (centipedes), Diplopoda (millipedes), Crustacea (crustaceans), Insecta (insects), Merostomata (horshoe crabs), Onychopora (velvet worms), Hirudinoidea (leeches), Oligochaeta (earthworms), Polychaeta (marine bristle worms), Bivalvia (mussels and clams), Gastropoda (snails, etc), Enopla (nemertine worms), Turbellaria (flatworms), Anthozoa (sea anemones and corals), Hydrozoa (corals).

** It should be noted that for certain species endemic to Brazil, it has not yet been possible to reach agreement on the Red List Categories between the Global Amphibian Assessment (GAA) Coordinating Team, and the experts on the species in Brazil. The 2004-2008 figures for Amphibians displayed here are those that were agreed at the GAA Brazil workshop in April 2003. However, in the subsequent consistency check conducted by the GAA Coordinating Team, many of the assessments were found to be inconsistent with the approach adopted elsewhere in the world, and a "consistent Red List Category" was also assigned to these species. The "consistent Red List Categories" are yet to be accepted by the Brazilian experts; therefore the original workshop assessments are retained here. However, in order to ensure comparability between results for amphibians with those for other taxonomic groups, the data used in various analyses (e.g. Hilton-Taylor *et al.* (this volume); the amphibian analysis at www.iucnredlist.org/amphibians) are based on the "consistent Red List Categories". Therefore, figures for Amphibians in the Table above will not completely match figures that appear in other analyses. This note also applies to Appendices 6, 8, 9 and 11.

Appendix 5. Summary of number of plant species in each Red List Category in each taxonomic class

IUCN Red List Categories: **EX** - Extinct, **EW** - Extinct in the wild, **CR** - Critically Endangered, **EN** - Endangered, **VU** - Vulnerable, **LR/cd** - Lower Risk/conservation dependent, **NT** - Near Threatened (includes **LR/nt** - Lower Risk/near threatened), **DD** - Data Deficient, **LC** - Least Concern (includes **LR/lc** - Lower Risk/least concern).

Class*	EX	EW	Subtotal	CR	EN	VU	Subtotal	LR/cd	NT	DD	LC	Total
Bryopsida	2	0	2	11	15	11	37	0	0	0	1	40
Anthocerotopsida	0	0	0	0	1	1	2	0	0	0	0	2
Marchantiopsida	1	0	1	12	16	15	43	0	0	0	9	53
Lycopodiopsida	0	0	0	1	2	8	11	0	1	0	1	13
Sellaginellopsida	0	0	0	0	0	1	1	0	1	0	0	2
Isoetopsida	0	0	0	2	0	1	3	0	0	0	0	3
Polypodiopsida	3	0	3	29	37	58	124	0	14	45	7	193
Coniferopsida	0	0	0	21	54	97	172	25	63	26	334	620
Cycadopsida	0	4	4	45	40	65	150	0	67	18	50	289
Ginkgoopsida	0	0	0	0	1	0	1	0	0	0	0	1
Magnoliopsida	78	22	100	1,299	1,847	3,976	7,122	196	810	458	938	9,624
Liliopsida	2	2	4	149	267	366	782	17	109	138	105	1,155
Chlorophyceae	0	0	0	0	0	0	0	0	0	1	0	1
Ulvophyceae	0	0	0	0	0	0	0	0	0	1	0	1
Florideophyceae	1	0	1	6	0	3	9	0	0	44	4	58
Total	**87**	**28**	**115**	**1,575**	**2,280**	**4,602**	**8,457**	**238**	**1,065**	**731**	**1,449**	**12,055**

Status category summary by major taxonomic group (other groups)

Class*	EX	EW	Subtotal	CR	EN	VU	Subtotal	LR/cd	NT	DD	LC	Total
Basidiomycetes	0	0	0	1	0	0	1	0	0	0	0	1
Lecanoromycetes	0	0	0	1	1	0	2	0	0	0	0	2
Phaeophyceae	0	0	0	4	1	1	6	0	0	9	0	15
Total	**0**	**0**	**0**	**6**	**2**	**1**	**9**	**0**	**0**	**9**	**0**	**18**

* **Plants**: Bryopsida (true mosses); Anthocerotopsida (hornworts); Marchantiopsida (liverworts); Lycopodiopsida (club mosses); Sellaginellopsida (spike mosses); Isoetopsida (quillworts); Polypodiopsida (true ferns); Coniferopsida (conifers); Cycadopsida (cycads); Ginkgoopsida (ginkgo); Magnoliopsida (dicotyledons); Liliopsida (monocotyledons); Chlorophyceae and Ulvophyceae (green algae); Florideophyceae (red algae). **Other groups**: Lecanoromyctes (discolichens); Basidiomycetes (club fungi); Phaeophyceae (brown algae).

Appendix 6. Number of species in each Red List Category in each major animal taxonomic group (Class, Order)

IUCN Red List Categories: EX - Extinct, **EW** - Extinct in the Wild, **CR** - Critically Endangered, **EN** - Endangered, **VU** - Vulnerable, **LR/cd** - Lower Risk/conservation dependent, **NT** - Near Threatened (includes **LR/nt** - Lower Risk/near threatened), **DD** - Data Deficient, **LC** - Least Concern (includes **LR/lc** - Lower Risk/least concern).

Class MAMMALIA

Order	EX	EW	Subtotal	CR	EN	VU	Subtotal	LR/cd	NT	DD	LC	TOTAL
Afrosoricida	0	0	0	1	7	9	17	0	3	4	30	54
Carnivora	5	0	5	8	24	39	71	0	27	19	163	285
Cetartiodactyla	7	2	9	14	46	49	109	0	26	62	123	329
Chiroptera	5	0	5	25	53	99	177	0	77	204	687	1,150
Cingulata	0	0	0	0	0	4	4	0	5	3	9	21
Dasyuromorphia	1	0	1	1	6	5	12	0	10	4	47	74
Dermoptera	0	0	0	0	0	0	0	0	0	0	2	2
Didelphimorphia	1	0	1	1	0	7	8	0	2	17	67	95
Diprotodontia	7	0	7	14	15	16	45	0	16	2	76	146
Eulipotyphla	7	0	7	12	41	31	84	0	13	77	269	450
Hyracoidea	0	0	0	0	0	0	0	0	0	0	5	5
Lagomorpha	1	0	1	2	10	5	17	0	6	8	61	93
Macroscelidea	0	0	0	0	1	2	3	0	1	3	9	16
Microbiotheria	0	0	0	0	0	0	0	0	1	0	0	1
Monotremata	0	0	0	3	0	0	3	0	0	0	2	5
Notoryctemorphia	0	0	0	0	0	0	0	0	0	2	0	2
Paucituberculata	0	0	0	0	0	2	2	0	2	0	2	6
Peramelemorphia	3	0	3	0	4	2	6	0	1	3	9	22
Perissodactyla	0	0	0	5	5	3	13	0	1	0	2	16
Pholidota	0	0	0	0	2	0	2	0	4	0	2	8
Pilosa	0	0	0	1	1	0	2	0	1	0	7	10
Primates	2	0	2	37	86	78	201	0	23	56	133	415
Proboscidea	0	0	0	0	1	0	1	0	1	0	0	2
Rodentia	36	0	36	64	144	150	358	0	103	369	1,389	2,255
Scandentia	0	0	0	0	2	0	2	0	0	3	15	20
Sirenia	1	0	1	0	0	4	4	0	0	0	0	5
Tubulidentata	0	0	0	0	0	0	0	0	0	0	1	1
Subtotal (Mammalia)	**76**	**2**	**78**	**188**	**448**	**505**	**1,141**	**0**	**323**	**836**	**3,110**	**5,488**

Class AVES

Order	EX	EW	Subtotal	CR	EN	VU	Subtotal	LR/cd	NT	DD	LC	TOTAL
Anseriformes	6	0	6	6	10	12	28	0	9	0	124	167
Apodiformes	2	0	2	9	15	11	35	0	24	8	374	443
Caprimulgiformes	0	0	0	3	2	3	8	0	10	4	100	122
Charadriiformes	4	0	4	10	11	17	38	0	34	0	278	354
Ciconiiformes	5	0	5	5	11	5	21	0	5	0	90	121

Class AVES

Order	EX	EW	Subtotal	CR	EN	VU	Subtotal	LR/cd	NT	DD	LC	TOTAL
Coliiformes	0	0	0	0	0	0	0	0	0	0	6	6
Columbiformes	15	1	16	9	15	35	59	0	41	1	219	336
Coraciiformes	1	0	1	3	3	19	25	0	28	3	164	221
Cuculiformes	2	0	2	2	2	7	11	0	11	0	143	167
Falconiformes	2	0	2	10	9	30	49	0	37	1	225	314
Galliformes	2	1	3	5	21	46	72	0	38	0	175	288
Gaviiformes	0	0	0	0	0	0	0	0	0	0	5	5
Gruiformes	22	1	23	5	18	32	55	0	20	5	126	229
Passeriformes	42	1	43	77	168	328	573	0	436	34	4,803	5,889
Pelecaniformes	2	0	2	2	4	10	16	0	7	0	42	67
Phoenicopteriformes	0	0	0	0	0	1	1	0	3	0	2	6
Piciformes	0	0	0	4	2	11	17	0	30	2	360	409
Podicipediformes	2	0	2	2	1	2	5	0	1	0	14	22
Procellariiformes	2	0	2	15	18	25	58	0	18	4	48	130
Psittaciformes	19	0	19	17	34	45	96	0	40	0	219	374
Sphenisciformes	0	0	0	0	4	7	11	0	2	0	5	18
Strigiformes	4	0	4	6	11	16	33	0	24	4	137	202
Struthioniformes	2	0	2	0	1	4	5	0	4	0	2	13
Tinamiformes	0	0	0	0	0	5	5	0	3	0	39	47
Trogoniformes	0	0	0	0	1	0	1	0	10	0	29	40
Subtotal (Aves)	134	4	138	190	361	671	1,222	0	835	66	7,729	9,990

Class REPTILIA

Order	EX	EW	Subtotal	CR	EN	VU	Subtotal	LR/cd	NT	DD	LC	TOTAL
Crocodylia	0	0	0	6	1	3	10	2	0	1	10	23
Rhynchocephalia	0	0	0	0	0	1	1	0	0	0	1	2
Squamata	14	0	14	49	91	140	280	0	81	169	604	1,148
Testudines	7	1	8	31	42	59	132	1	42	10	19	212
Subtotal (Reptilia)	21	1	22	86	134	203	423	3	123	180	634	1,385

Class AMPHIBIA*

Order	EX	EW	Subtotal	CR	EN	VU	Subtotal	LR/cd	NT	DD	LC	TOTAL
Anura	36	1	37	398	650	578	1,626	0	320	1,403	2,146	5,532
Caudata	2	0	2	76	104	93	273	0	61	57	159	552
Gymnophiona	0	0	0	1	1	4	6	0	0	118	52	176
Subtotal (Amphibia)	38	1	39	475	755	675	1,905	0	381	1,578	2,357	6,260

Class CEPHALASPIDOMORPHI

Order	EX	EW	Subtotal	CR	EN	VU	Subtotal	LR/cd	NT	DD	LC	TOTAL
Petromyzontiformes	1	0	1	1	0	1	2	0	2	3	10	18
Subtotal (Cephalaspidomorphi)	1	0	1	1	0	1	2	0	2	3	10	18

Class CHONDRICHTHYES

Order	EX	EW	Subtotal	CR	EN	VU	Subtotal	LR/cd	NT	DD	LC	TOTAL
Carcharhiniformes	0	0	0	6	5	14	25	1	37	61	43	167
Chimaeriformes	0	0	0	0	0	0	0	0	3	20	12	35

Class CHONDRICHTHYES

Order	EX	EW	Subtotal	CR	EN	VU	Subtotal	LR/cd	NT	DD	LC	TOTAL
Heterodontiformes	0	0	0	0	0	0	0	0	0	4	3	7
Hexanchiformes	0	0	0	0	0	0	0	0	3	1	0	4
Lamniformes	0	0	0	0	0	5	5	0	2	5	1	13
Orectolobiformes	0	0	0	0	0	7	7	0	8	2	8	25
Pristiophoriformes	0	0	0	0	0	0	0	0	2	1	2	5
Rajiformes	0	0	0	11	20	38	69	0	39	61	55	224
Squaliformes	0	0	0	1	0	4	5	0	9	38	25	77
Squatiniformes	0	0	0	3	4	1	8	0	1	3	2	14
Torpediniformes	0	0	0	1	0	6	7	0	3	9	1	20
Subtotal (Chondrichthyes)	**0**	**0**	**0**	**22**	**29**	**75**	**126**	**1**	**107**	**205**	**152**	**591**

Class ACTINOPTERYGII

Order	EX	EW	Subtotal	CR	EN	VU	Subtotal	LR/cd	NT	DD	LC	TOTAL
Acipenseriformes	0	0	0	6	11	6	23	0	2	0	2	27
Anguilliformes	0	0	0	1	0	0	1	0	0	0	0	1
Atheriniformes	0	0	0	9	7	44	60	0	9	27	14	110
Batrachoidiformes	0	0	0	0	0	6	6	0	0	0	0	6
Beloniformes	0	0	0	2	3	8	13	0	1	2	1	17
Characiformes	0	0	0	2	2	1	5	0	0	9	20	34
Clupeiformes	0	0	0	3	3	4	10	0	0	5	14	29
Cypriniformes	21	2	23	82	93	161	336	6	34	87	238	724
Cyprinodontiformes	12	5	17	26	19	48	93	0	3	22	35	170
Esociformes	0	0	0	0	0	1	1	0	1	0	1	3
Gadiformes	0	0	0	1	0	2	3	0	0	0	1	4
Gasterosteiformes	1	0	1	1	0	1	2	0	0	4	6	13
Gonorynchiformes	0	0	0	1	0	3	4	0	0	4	6	14
Lophiiformes	0	0	0	1	0	0	1	0	0	0	0	1
Mugiliformes	0	0	0	0	1	0	1	0	0	0	7	8
Ophidiiformes	0	0	0	0	0	7	7	0	0	1	0	8
Osmeriformes	0	0	0	0	1	0	1	0	0	1	3	5
Osteoglossiformes	0	0	0	0	2	2	4	0	2	7	14	27
Perciformes	40	5	45	94	72	254	420	4	69	141	554	1,233
Percopsiformes	0	0	0	1	0	3	4	0	0	0	0	4
Pleuronectiformes	0	0	0	0	1	1	2	0	0	0	3	5
Salmoniformes	14	1	15	18	10	42	70	0	6	25	38	154
Scorpaeniformes	1	0	1	3	2	7	12	0	0	5	10	28
Siluriformes	1	0	1	13	11	29	53	0	6	41	70	171
Synbranchiformes	0	0	0	0	1	0	1	0	0	5	13	19
Syngnathiformes	0	0	0	1	1	7	9	0	2	35	1	47
Tetraodontiformes	0	0	0	0	0	3	3	0	0	5	0	8
Subtotal (Actinopterygii)	**90**	**13**	**103**	**265**	**240**	**640**	**1,145**	**10**	**135**	**426**	**1,051**	**2,870**

Class SARCOPTERYGII

Order	EX	EW	Subtotal	CR	EN	VU	Subtotal	LR/cd	NT	DD	LC	TOTAL
Coelacanthiformes	0	0	0	1	0	1	2	0	0	0	0	2
Subtotal (Sarcopterygii)	0	0	0	1	0	1	2	0	0	0	0	2

Class ECHINOIDEA

Order	EX	EW	Subtotal	CR	EN	VU	Subtotal	LR/cd	NT	DD	LC	TOTAL
Echinoida	0	0	0	0	0	0	0	0	1	0	0	1
Subtotal (Echinoidea)	0	0	0	0	0	0	0	0	1	0	0	1

Class ARACHNIDA

Order	EX	EW	Subtotal	CR	EN	VU	Subtotal	LR/cd	NT	DD	LC	TOTAL
Araneae	0	0	0	2	5	9	16	0	2	8	3	29
Opiliones	0	0	0	0	0	1	1	0	0	0	0	1
Pseudoscorpionida	0	0	0	0	0	1	1	0	0	1	0	2
Subtotal (Arachnida)	0	0	0	2	5	11	18	0	2	9	3	32

Class CHILOPODA

Order	EX	EW	Subtotal	CR	EN	VU	Subtotal	LR/cd	NT	DD	LC	TOTAL
Scolopendromorpha	0	0	0	0	0	1	1	0	0	0	0	1
Subtotal (Chilopoda)	0	0	0	0	0	1	1	0	0	0	0	1

Class DIPLOPODA

Order	EX	EW	Subtotal	CR	EN	VU	Subtotal	LR/cd	NT	DD	LC	TOTAL
Spirostreptida	0	0	0	1	6	7	14	0	0	7	10	31
Subtotal (Diplopoda)	0	0	0	1	6	7	14	0	0	7	10	31

Class CRUSTACEA

Order	EX	EW	Subtotal	CR	EN	VU	Subtotal	LR/cd	NT	DD	LC	TOTAL
Amphipoda	2	0	2	7	6	56	69	0	0	0	0	71
Anaspidacea	0	0	0	0	0	4	4	0	0	0	0	4
Anomopoda	0	0	0	0	0	8	8	0	0	0	0	8
Anostraca	0	0	0	6	9	10	25	1	1	1	1	29
Calanoida	1	0	1	4	0	47	51	0	0	19	0	71
Conchostraca	0	0	0	0	0	4	4	0	0	0	0	4
Cyclopoida	1	0	1	1	0	6	7	5	0	0	0	13
Decapoda	1	0	1	52	102	211	365	0	18	638	428	1,450
Halocyprida	0	0	0	1	0	0	1	0	0	0	0	1
Harpacticoida	0	0	0	0	0	18	18	3	0	1	0	22
Isopoda	0	1	1	7	9	22	38	0	0	2	1	42
Mictacea	0	0	0	1	0	0	1	0	0	0	0	1
Misophrioida	0	0	0	2	0	0	2	0	0	0	0	2
Myodocopida	0	0	0	0	0	1	1	0	0	0	0	1
Mysidacea	0	0	0	2	0	0	2	0	0	0	0	2
Notostraca	0	0	0	0	1	0	1	0	0	0	0	1
Podocopida	2	0	2	1	0	8	9	0	0	0	0	11
Thoracica	0	0	0	0	0	0	0	0	0	2	0	2
Subtotal (Crustacea)	7	1	8	84	127	395	606	9	19	663	430	1,735

Class INSECTA

Order	EX	EW	Subtotal	CR	EN	VU	Subtotal	LR/cd	NT	DD	LC	TOTAL
Anoplura	0	0	0	1	0	0	1	0	0	0	0	1
Coleoptera	16	0	16	10	16	27	53	0	3	0	0	72
Dermaptera	0	0	0	1	0	0	1	0	0	0	0	1
Diptera	3	0	3	1	2	1	4	0	0	0	0	7
Ephemeroptera	2	0	2	0	0	1	1	0	0	0	0	3
Grylloblattaria	0	0	0	0	0	1	1	0	0	0	0	1
Homoptera	2	0	2	0	0	0	0	0	3	0	0	5
Hymenoptera	0	0	0	4	0	139	143	0	7	1	1	152
Lepidoptera	27	0	27	8	39	130	177	0	45	35	19	303
Mantodea	0	0	0	0	0	0	0	0	1	0	0	1
Odonata	2	0	2	36	67	73	176	0	34	90	327	629
Orthoptera	2	1	3	8	8	50	66	3	0	2	0	74
Phasmida	1	0	1	1	0	0	1	0	0	0	0	2
Plecoptera	1	0	1	0	0	2	2	0	0	1	0	4
Trichoptera	4	0	4	0	0	0	0	0	0	0	0	4
Subtotal (Insecta)	60	1	61	70	132	424	626	3	93	129	347	1,259

Class MEROSTOMATA

Order	EX	EW	Subtotal	CR	EN	VU	Subtotal	LR/cd	NT	DD	LC	TOTAL
Xiphosura	0	0	0	0	0	0	0	0	1	3	0	4
Subtotal (Merostomata)	0	0	0	0	0	0	0	0	1	3	0	4

Class ONYCHOPHORA

Order	EX	EW	Subtotal	CR	EN	VU	Subtotal	LR/cd	NT	DD	LC	TOTAL
Onychophora	0	0	0	3	2	4	9	0	1	1	0	11
Subtotal (Onychophora)	0	0	0	3	2	4	9	0	1	1	0	11

Class HIRUDINOIDEA

Order	EX	EW	Subtotal	CR	EN	VU	Subtotal	LR/cd	NT	DD	LC	TOTAL
Arhynchobdellae	0	0	0	0	0	0	0	0	1	0	0	1
Subtotal (Hirudinoidea)	0	0	0	0	0	0	0	0	1	0	0	1

Class OLIGOCHAETA

Order	EX	EW	Subtotal	CR	EN	VU	Subtotal	LR/cd	NT	DD	LC	TOTAL
Haplotaxida	1	0	1	1	0	4	5	0	1	0	0	7
Subtotal (Oligochaeta)	1	0	1	1	0	4	5	0	1	0	0	7

Class POLYCHAETA

Order	EX	EW	Subtotal	CR	EN	VU	Subtotal	LR/cd	NT	DD	LC	TOTAL
Eunicida	0	0	0	0	0	0	0	0	0	1	0	1
Nerillida	0	0	0	1	0	0	1	0	0	0	0	1
Subtotal (Polychaeta)	0	0	0	1	0	0	1	0	0	1	0	2

Class BIVALVIA

Order	EX	EW	Subtotal	CR	EN	VU	Subtotal	LR/cd	NT	DD	LC	TOTAL
Ostreoida	0	0	0	0	0	0	0	0	0	1	0	1
Unionoida	31	0	31	52	28	10	90	1	59	10	11	202
Veneroida	0	0	0	0	0	5	5	4	1	3	2	15
Subtotal (Bivalvia)	**31**	**0**	**31**	**52**	**28**	**15**	**95**	**5**	**60**	**14**	**13**	**218**

Class GASTROPODA

Order	EX	EW	Subtotal	CR	EN	VU	Subtotal	LR/cd	NT	DD	LC	TOTAL
Archaeogastropoda	2	0	2	7	2	3	12	0	1	25	1	41
Basommatophora	7	0	7	8	5	19	32	0	8	32	9	88
Mesogastropoda	57	3	60	56	76	204	336	3	38	182	43	662
Neogastropoda	0	0	0	0	0	4	4	0	3	22	0	29
Stylommatophora	191	11	202	145	113	241	499	11	136	296	30	1,174
Subtotal (Gastropoda)	**257**	**14**	**271**	**216**	**196**	**471**	**883**	**14**	**186**	**557**	**83**	**1,994**

Class ENOPLA

Order	EX	EW	Subtotal	CR	EN	VU	Subtotal	LR/cd	NT	DD	LC	TOTAL
Hoplonemertea	0	0	0	0	0	2	2	0	1	3	0	6
Subtotal (Enopla)	**0**	**0**	**0**	**0**	**0**	**2**	**2**	**0**	**1**	**3**	**0**	**6**

Class TURBELLARIA

Order	EX	EW	Subtotal	CR	EN	VU	Subtotal	LR/cd	NT	DD	LC	TOTAL
Tricladida	1	0	1	0	0	0	0	0	0	0	0	1
Subtotal (Turbellaria)	**1**	**0**	**1**	**0**	**0**	**0**	**0**	**0**	**0**	**0**	**0**	**1**

Class ANTHOZOA

Order	EX	EW	Subtotal	CR	EN	VU	Subtotal	LR/cd	NT	DD	LC	TOTAL
Actinaria	0	0	0	0	0	1	1	0	0	1	0	2
Gorgonacea	0	0	0	0	0	1	1	0	0	0	0	1
Helioporacea	0	0	0	0	0	1	1	0	0	0	0	1
Scleractinia	0	0	0	6	23	199	228	0	174	146	289	837
Stolonifera	0	0	0	0	0	0	0	0	1	0	0	1
Subtotal (Anthozoa)	**0**	**0**	**0**	**6**	**23**	**202**	**231**	**0**	**175**	**147**	**289**	**842**

Class HYDROZOA

Order	EX	EW	Subtotal	CR	EN	VU	Subtotal	LR/cd	NT	DD	LC	TOTAL
Milleporina	0	0	0	1	2	2	5	0	1	2	8	16
Subtotal (Hydrozoa)	**0**	**0**	**0**	**1**	**2**	**2**	**5**	**0**	**1**	**2**	**8**	**16**

	EX	EW	Subtotal	CR	EN	VU	Subtotal	LR/cd	NT	DD	LC	TOTAL
Total Fauna	717	37	754	1,665	2,488	4,309	8,462	45	2,448	4,830	16,226	32,765

Appendix 7. Number of species in each Red List Category in each major plant taxonomic group (Class, Family)

IUCN Red List Categories: EX - Extinct, **EW** - Extinct in the Wild, **CR** - Critically Endangered, **EN** - Endangered, **VU** - Vulnerable, **LR/cd** - Lower Risk/conservation dependent, **NT** - Near Threatened (includes **LR/nt** - Lower Risk/near threatened), **DD** - Data Deficient, **LC** - Least Concern (includes **LR/lc** - Lower Risk/least concern).

Class BRYOPSIDA

Family	EX	EW	Subtotal	CR	EN	VU	Subtotal	LR/cd	NT	DD	LC	TOTAL
Amblystegiaceae	0	0	0	0	1	2	3	0	0	0	0	3
Archidiaceae	0	0	0	0	1	0	1	0	0	0	0	1
Brachytheciaceae	1	0	1	0	0	0	0	0	0	0	0	1
Bryaceae	0	0	0	0	1	0	1	0	0	0	0	1
Bryoxiphiaceae	0	0	0	0	1	0	1	0	0	0	0	1
Daltoniaceae	0	0	0	1	0	0	1	0	0	0	0	1
Dicranaceae	0	0	0	0	1	0	1	0	0	0	0	1
Ditrichaceae	0	0	0	0	2	0	2	0	0	0	0	2
Echinodiaceae	0	0	0	0	0	2	2	0	0	0	0	2
Fabroniaceae	0	0	0	1	1	0	2	0	0	0	0	2
Fissidentaceae	0	0	0	1	0	0	1	0	0	0	0	1
Grimmiaceae	0	0	0	0	1	1	2	0	0	0	0	2
Hookeriaceae	0	0	0	1	1	0	2	0	0	0	0	2
Hypnobartlettiaceae	0	0	0	1	0	0	1	0	0	0	0	1
Leskeaceae	0	0	0	0	1	0	1	0	0	0	0	1
Neckeraceae	1	0	1	3	1	0	4	0	0	0	0	5
Orthotrichaceae	0	0	0	0	1	1	2	0	0	0	0	2
Pottiaceae	0	0	0	2	0	1	3	0	0	0	1	4
Pterobryaceae	0	0	0	1	1	0	2	0	0	0	0	2
Rhachitheciaceae	0	0	0	0	0	1	1	0	0	0	0	1
Sematophyllaceae	0	0	0	0	1	0	1	0	0	0	0	1
Sphagnaceae	0	0	0	0	0	2	2	0	0	0	0	2
Takakiaceae	0	0	0	0	0	1	1	0	0	0	0	1
Subtotal (Bryopsida)	**2**	**0**	**2**	**11**	**15**	**11**	**37**	**0**	**0**	**0**	**1**	**40**

Class ANTHOCEROTOPSIDA

Family	EX	EW	Subtotal	CR	EN	VU	Subtotal	LR/cd	NT	DD	LC	TOTAL
Anthocerotaceae	0	0	0	0	1	1	2	0	0	0	0	2
Subtotal (Anthocerotopsida)	**0**	**0**	**0**	**0**	**1**	**1**	**2**	**0**	**0**	**0**	**0**	**2**

Class MARCHANTIOPSIDA

Family	EX	EW	Subtotal	CR	EN	VU	Subtotal	LR/cd	NT	DD	LC	TOTAL
Aitchinsoniellaceae	0	0	0	0	1	0	1	0	0	0	0	1
Calypogeiaceae	0	0	0	0	0	1	1	0	0	0	0	1

Class MARCHANTIOPSIDA

Family	EX	EW	Subtotal	CR	EN	VU	Subtotal	LR/cd	NT	DD	LC	TOTAL
Cephaloziaceae	0	0	0	0	0	2	2	0	0	0	0	2
Cleveaceae	0	0	0	0	0	0	0	0	0	0	1	1
Exormothecaceae	0	0	0	0	1	0	1	0	0	0	0	1
Fossombroniaceae	0	0	0	0	0	1	1	0	0	0	0	1
Geocalycaceae	0	0	0	0	0	0	0	0	0	0	1	1
Gymnomitriaceae	0	0	0	0	0	0	0	0	0	0	1	1
Herbertaceae	0	0	0	0	0	0	0	0	0	0	1	1
Jubulaceae	0	0	0	0	0	0	0	0	0	0	1	1
Jungermanniaceae	0	0	0	0	2	4	6	0	0	0	1	7
Lejeuneaceae	0	0	0	6	7	3	16	0	0	0	2	18
Lepidoziaceae	0	0	0	2	0	0	2	0	0	0	1	3
Personiellaceae	0	0	0	0	0	1	1	0	0	0	0	1
Phycolepidoziaceae	0	0	0	1	0	0	1	0	0	0	0	1
Plagiochilaceae	0	0	0	1	0	0	1	0	0	0	0	1
Pleuroziaceae	0	0	0	0	1	0	1	0	0	0	0	1
Radulaceae	1	0	1	0	1	0	1	0	0	0	0	2
Ricciaceae	0	0	0	0	0	2	2	0	0	0	0	2
Scapaniaceae	0	0	0	0	0	1	1	0	0	0	0	1
Schistochilaceae	0	0	0	1	1	0	2	0	0	0	0	2
Sphaerocarpaceae	0	0	0	0	2	0	2	0	0	0	0	2
Vandiemeniaceae	0	0	0	1	0	0	1	0	0	0	0	1
Subtotal (Marchantiopsida)	**1**	**0**	**1**	**12**	**16**	**15**	**43**	**0**	**0**	**0**	**9**	**53**

Class LYCOPODIOPSIDA

Family	EX	EW	Subtotal	CR	EN	VU	Subtotal	LR/cd	NT	DD	LC	TOTAL
Lycopodiaceae	0	0	0	1	2	8	11	0	1	0	1	13
Subtotal (Lycopodiopsida)	**0**	**0**	**0**	**1**	**2**	**8**	**11**	**0**	**1**	**0**	**1**	**13**

Class SELLAGINELLOPSIDA

Family	EX	EW	Subtotal	CR	EN	VU	Subtotal	LR/cd	NT	DD	LC	TOTAL
Selaginellaceae	0	0	0	0	0	1	1	0	1	0	0	2
Subtotal (Sellaginellopsida)	**0**	**0**	**0**	**0**	**0**	**1**	**1**	**0**	**1**	**0**	**0**	**2**

Class ISOETOPSIDA

Family	EX	EW	Subtotal	CR	EN	VU	Subtotal	LR/cd	NT	DD	LC	TOTAL
Isoetaceae	0	0	0	2	0	1	3	0	0	0	0	3
Subtotal (Isoetopsida)	**0**	**0**	**0**	**2**	**0**	**1**	**3**	**0**	**0**	**0**	**0**	**3**

Class POLYPODIOPSIDA

Family	EX	EW	Subtotal	CR	EN	VU	Subtotal	LR/cd	NT	DD	LC	TOTAL
Adiantaceae	2	0	2	0	4	1	5	0	0	1	0	8
Aspleniaceae	0	0	0	0	1	1	2	0	3	2	1	8
Blechnaceae	0	0	0	0	2	3	5	0	0	5	0	10
Cyatheaceae	0	0	0	0	3	4	7	0	1	1	0	9
Dennstaedtiaceae	0	0	0	0	2	3	5	0	0	0	0	5

Class POLYPODIOPSIDA

Family	EX	EW	Subtotal	CR	EN	VU	Subtotal	LR/cd	NT	DD	LC	TOTAL
Dryopteridaceae	1	0	1	3	5	3	11	0	1	5	0	18
Grammitidaceae	0	0	0	0	3	3	6	0	1	1	0	8
Hymenophyllaceae	0	0	0	3	2	2	7	0	1	6	1	15
Lomariopsidaceae	0	0	0	18	5	12	35	0	1	15	0	51
Marattiaceae	0	0	0	0	0	0	0	0	1	0	0	1
Oleandraceae	0	0	0	0	1	0	1	0	0	0	0	1
Plagiogyriaceae	0	0	0	0	1	0	1	0	0	0	0	1
Polypodiaceae	0	0	0	2	2	4	8	0	1	6	2	17
Pteridaceae	0	0	0	1	0	0	1	0	0	0	0	1
Thelypteridaceae	0	0	0	1	3	12	16	0	1	0	1	18
Vittariaceae	0	0	0	0	0	1	1	0	0	1	0	2
Woodsiaceae	0	0	0	1	3	9	13	0	3	2	2	20
Subtotal (Polypodiopsida)	**3**	**0**	**3**	**29**	**37**	**58**	**124**	**0**	**14**	**45**	**7**	**193**

Class CONIFEROPSIDA

Family	EX	EW	Subtotal	CR	EN	VU	Subtotal	LR/cd	NT	DD	LC	TOTAL
Araucariaceae	0	0	0	3	3	12	18	11	5	0	6	40
Cephalotaxaceae	0	0	0	0	2	3	5	0	1	0	4	10
Cupressaceae	0	0	0	4	16	32	52	2	14	3	64	135
Phyllocladaceae	0	0	0	0	0	0	0	0	0	0	4	4
Pinaceae	0	0	0	6	12	24	42	4	21	4	154	225
Podocarpaceae	0	0	0	5	17	21	43	7	19	18	95	182
Sciadopityaceae	0	0	0	0	0	1	1	0	0	0	0	1
Taxaceae	0	0	0	3	4	4	11	1	3	1	7	23
Subtotal (Coniferopsida)	**0**	**0**	**0**	**21**	**54**	**97**	**172**	**25**	**63**	**26**	**334**	**620**

Class CYCADOPSIDA

Family	EX	EW	Subtotal	CR	EN	VU	Subtotal	LR/cd	NT	DD	LC	TOTAL
Cycadaceae	0	0	0	7	10	21	38	0	30	10	14	92
Stangeriaceae	0	0	0	0	0	0	0	0	1	0	2	3
Zamiaceae	0	4	4	38	30	44	112	0	36	8	34	194
Subtotal (Cycadopsida)	**0**	**4**	**4**	**45**	**40**	**65**	**150**	**0**	**67**	**18**	**50**	**289**

Class GINKGOOPSIDA

Family	EX	EW	Subtotal	CR	EN	VU	Subtotal	LR/cd	NT	DD	LC	TOTAL
Ginkgoaceae	0	0	0	0	1	0	1	0	0	0	0	1
Subtotal (Ginkoopsida)	**0**	**0**	**0**	**0**	**1**	**0**	**1**	**0**	**0**	**0**	**0**	**1**

Class MAGNOLIOPSIDA

Family	EX	EW	Subtotal	CR	EN	VU	Subtotal	LR/cd	NT	DD	LC	TOTAL
Acanthaceae	0	0	0	7	21	39	67	1	5	8	14	95
Aceraceae	0	0	0	2	1	2	5	0	3	0	0	8
Actinidiaceae	0	0	0	2	7	18	27	2	6	0	7	42
Adoxaceae	0	0	0	0	1	1	2	0	0	0	0	2
Aextoxicaceae	0	0	0	0	0	0	0	0	0	1	0	1

Class MAGNOLIOPSIDA												
Family	EX	EW	Subtotal	CR	EN	VU	Subtotal	LR/cd	NT	DD	LC	TOTAL
Aizoaceae	0	0	0	2	1	10	13	0	2	0	22	37
Alangiaceae	0	0	0	0	0	3	3	0	0	0	3	6
Amaranthaceae	2	0	2	2	3	15	20	0	1	2	4	29
Anacardiaceae	0	2	2	10	17	50	77	1	9	16	16	121
Ancistrocladaceae	0	0	0	0	0	1	1	0	0	0	0	1
Anisophylleaceae	0	0	0	0	0	13	13	0	0	0	4	17
Annonaceae	1	0	-	20	43	101	164	6	27	8	28	234
Apocynaceae	4	0	4	10	25	37	72	2	8	5	13	104
Aquifoliaceae	2	0	2	8	26	31	65	5	24	0	1	97
Araliaceae	0	0	0	27	30	78	135	5	10	12	14	176
Aristolochiaceae	0	0	0	3	4	10	17	0	0	0	0	17
Asclepiadaceae	0	0	0	14	10	34	58	0	2	4	20	84
Asteropeiaceae	0	0	0	1	4	1	6	0	1	0	1	8
Avicenniaceae	0	0	0	0	0	1	1	0	0	0	0	1
Balanopaceae	0	0	0	0	0	0	0	1	0	0	0	1
Balsaminaceae	0	0	0	0	6	3	9	0	0	1	1	11
Begoniaceae	1	0	-	3	11	30	44	0	4	0	1	50
Berberidaceae	0	0	0	2	3	23	28	0	2	13	2	45
Betulaceae	0	1	1	5	2	2	9	1	3	0	2	16
Bignoniaceae	0	0	0	3	9	22	34	1	1	0	3	39
Bombacaceae	0	0	0	3	15	18	36	1	4	6	0	47
Boraginaceae	0	0	0	11	7	30	48	0	6	6	12	72
Bretschneideraceae	0	0	0	0	1	0	1	0	0	0	0	1
Brunelliaceae	0	0	0	0	9	8	17	1	1	0	0	19
Buddlejaceae	0	0	0	1	1	3	5	0	1	0	0	6
Burseraceae	0	0	0	1	4	48	53	3	20	3	12	91
Buxaceae	0	0	0	1	2	2	5	0	2	0	0	7
Byblidaceae	0	0	0	1	0	0	1	0	0	0	4	5
Cactaceae	0	2	2	33	27	51	111	0	14	10	20	157
Callitrichaceae	0	0	0	0	0	0	0	0	0	0	1	1
Calyceraceae	0	0	0	0	0	1	1	0	0	0	0	1
Campanulaceae	10	4	14	31	58	34	123	0	8	1	5	151
Canellaceae	0	0	0	1	3	2	6	0	0	0	0	6
Capparaceae	0	0	0	5	4	5	14	0	1	7	1	23
Caprifoliaceae	0	0	0	5	1	4	10	1	2	2	1	16
Caricaceae	0	0	0	0	2	3	5	0	2	0	0	7
Caryocaraceae	0	0	0	0	3	2	5	0	0	0	0	5
Caryophyllaceae	0	0	0	13	6	5	24	0	0	0	5	29
Cecropiaceae	0	0	0	0	2	7	9	0	4	1	1	15
Celastraceae	0	0	0	11	12	39	62	2	8	1	14	87
Cephalotaceae	0	0	0	0	0	1	1	0	0	0	0	1
Cercidiphyllaceae	0	0	0	0	0	0	0	0	1	0	0	1
Chenopodiaceae	0	0	0	2	0	0	2	0	0	0	1	3
Chloranthaceae	0	0	0	0	2	2	4	0	0	0	0	4
Chrysobalanaceae	1	0	1	9	12	17	38	0	1	1	4	45

Class MAGNOLIOPSIDA

Family	EX	EW	Subtotal	CR	EN	VU	Subtotal	LR/cd	NT	DD	LC	TOTAL
Cistaceae	0	0	0	0	1	0	1	0	0	0	0	1
Clethraceae	0	0	0	1	0	2	3	1	2	1	1	8
Cobaeaceae	0	0	0	0	2	0	2	0	0	0	0	2
Cochlospermaceae	0	0	0	0	1	0	1	0	0	0	0	1
Combretaceae	0	1	1	2	11	16	29	0	6	2	0	38
Compositae	4	2	6	65	92	177	334	0	58	26	55	479
Connaraceae	0	0	0	2	2	3	7	0	2	1	0	10
Convolvulaceae	0	0	0	2	1	6	9	0	0	1	3	13
Cornaceae	0	0	0	2	0	8	10	0	0	0	1	11
Corylaceae	0	0	0	2	1	0	3	0	0	0	1	4
Crassulaceae	0	0	0	0	0	1	1	0	2	0	5	8
Cruciferae	0	0	0	7	4	12	23	0	2	0	1	26
Crypteroniaceae	0	0	0	0	0	1	1	0	0	0	0	1
Cucurbitaceae	0	0	0	0	0	4	4	0	0	0	1	5
Cunoniaceae	1	0	1	2	6	13	21	4	2	0	4	32
Datiscaceae	0	0	0	0	0	0	0	0	0	0	2	2
Degeneriaceae	0	0	0	0	0	1	1	0	1	0	0	2
Dichapetalaceae	0	0	0	0	0	11	11	0	1	0	0	12
Didiereaceae	0	0	0	0	0	0	0	0	1	0	0	1
Dilleniaceae	0	0	0	2	0	6	8	0	0	0	0	8
Dipentodontaceae	0	0	0	0	0	0	0	0	0	0	1	1
Dipsacaceae	0	0	0	0	1	1	2	0	1	0	0	3
Dipterocarpaceae	3	0	3	255	102	12	369	0	0	10	12	394
Dirachmaceae	0	0	0	0	1	1	2	0	0	0	0	2
Droseraceae	0	0	0	0	0	1	1	0	0	0	0	1
Ebenaceae	0	0	0	15	15	46	76	3	1	4	21	105
Elaeagnaceae	0	0	0	0	0	2	2	0	0	0	0	2
Elaeocarpaceae	0	0	0	4	3	23	30	8	2	1	2	43
Ericaceae	0	1	1	1	3	15	19	3	3	00	3	29
Erythroxylaceae	0	1	1	0	2	7	9	0	0	0	0	10
Escalloniaceae	0	0	0	0	0	0	0	1	1	0	0	2
Eucommiaceae	0	0	0	0	0	0	0	0	1	0	0	1
Eucryphiaceae	0	0	0	0	0	0	0	0	2	0	0	2
Euphorbiaceae	2	1	3	67	74	219	360	12	37	16	41	469
Eupteleaceae	0	0	0	0	0	0	0	0	0	0	1	1
Fagaceae	0	0	0	8	14	41	63	4	5	19	17	108
Flacourtiaceae	2	0	2	9	26	47	82	5	6	7	6	108
Frankeniaceae	0	0	0	0	0	1	1	0	0	0	0	1
Gentianaceae	0	0	0	1	10	12	23	0	2	0	8	33
Geraniaceae	0	0	0	2	1	6	9	0	0	0	0	9
Gesneriaceae	0	1	1	9	29	39	77	0	10	5	4	97
Goetzeaceae	0	0	0	0	1	0	1	0	0	0	0	1
Gomortegaceae	0	0	0	0	1	0	1	0	0	0	0	1
Goodeniaceae	0	0	0	0	1	3	4	0	0	0	1	5
Greyiaceae	0	0	0	0	0	0	0	0	0	1	0	1

Class MAGNOLIOPSIDA

Family	EX	EW	Subtotal	CR	EN	VU	Subtotal	LR/cd	NT	DD	LC	TOTAL
Grossulariaceae	0	0	0	2	0	2	4	0	0	0	1	5
Gunneraceae	0	0	0	0	0	1	1	0	0	0	0	1
Guttiferae	0	0	0	9	19	84	112	4	2	12	41	171
Hamamelidaceae	0	0	0	1	1	6	8	0	2	1	2	13
Hernandiaceae	1	0	1	2	2	2	6	0	4	2	1	14
Hippocastanaceae	0	0	0	0	0	1	1	0	0	0	0	1
Hoplestigmataceae	0	0	0	1	0	0	1	0	0	0	0	1
Huaceae	0	0	0	0	0	1	1	0	0	0	0	1
Humiriaceae	0	0	0	0	3	2	5	0	0	0	0	5
Icacinaceae	0	0	0	2	2	10	14	0	0	0	1	15
Illecebraceae	0	0	0	0	0	2	2	0	0	0	1	3
Illiciaceae	0	0	0	0	0	2	2	2	0	0	0	4
Irvingiaceae	0	0	0	0	0	0	0	0	1	0	1	2
Ixonanthaceae	0	0	0	0	0	2	2	0	0	0	0	2
Juglandaceae	0	0	0	0	4	9	13	0	2	0	4	19
Labiatae	0	0	0	8	6	18	32	0	4	2	10	48
Lacistemataceae	0	0	0	0	0	0	0	0	0	1	1	2
Lauraceae	0	0	0	24	50	125	199	12	21	11	34	277
Lecythidaceae	0	0	0	11	15	53	79	5	6	2	5	97
Leeaceae	0	0	0	0	0	0	0	0	1	0	0	1
Leguminosae	6	1	7	59	159	376	594	9	74	39	54	777
Leitneriaceae	0	0	0	0	0	0	0	0	1	0	0	1
Linaceae	0	0	0	1	0	2	3	0	0	0	0	3
Loasaceae	0	0	0	1	4	9	14	0	2	0	0	16
Loganiaceae	0	0	0	3	4	9	16	0	6	3	5	30
Loranthaceae	0	0	0	0	1	2	3	0	1	0	0	4
Lythraceae	0	0	0	2	2	5	9	0	0	0	3	12
Magnoliaceae	0	0	0	9	27	20	56	0	3	2	1	62
Malpighiaceae	0	0	0	2	5	9	16	0	0	4	1	21
Malvaceae	4	1	5	14	11	11	36	0	2	3	5	51
Marcgraviaceae	0	0	0	0	2	1	3	0	0	2	0	5
Medusagynaceae	0	0	0	1	0	0	1	0	0	0	0	1
Medusandraceae	0	0	0	0	1	1	2	0	0	0	0	2
Melanophyllaceae	0	0	0	2	1	1	4	0	1	0	2	7
Melastomataceae	0	0	0	25	96	148	269	1	28	8	20	326
Meliaceae	0	0	0	14	19	114	147	2	45	2	16	212
Melianthaceae	0	0	0	0	1	1	2	0	0	0	0	2
Meliosmaceae	0	0	0	0	2	2	4	0	0	1	0	5
Menispermaceae	0	0	0	2	3	3	8	0	2	0	0	10
Molluginaceae	0	0	0	0	0	0	0	0	0	0	1	1
Monimiaceae	0	0	0	7	5	7	19	0	3	0	2	24
Moraceae	0	0	0	7	15	24	46	0	3	1	25	75
Moringaceae	0	0	0	0	0	1	1	0	0	0	0	1
Myoporaceae	0	0	0	1	0	0	1	0	0	1	1	3
Myricaceae	0	0	0	1	1	1	3	0	0	0	1	4

▶ Class MAGNOLIOPSIDA

Family	EX	EW	Subtotal	CR	EN	VU	Subtotal	LR/cd	NT	DD	LC	TOTAL
Myristicaceae	0	0	0	4	8	142	154	5	27	7	32	225
Myrsinaceae	0	0	0	17	19	43	79	0	14	18	10	121
Myrtaceae	6	0	6	51	73	132	256	10	20	13	36	341
Nepenthaceae	0	0	0	5	11	35	51	7	3	8	16	85
Nyctaginaceae	0	0	0	1	6	2	9	0	2	1	3	15
Ochnaceae	0	0	0	2	2	10	14	0	2	3	3	22
Olacaceae	0	0	0	1	3	5	9	0	3	1	1	14
Oleaceae	0	0	0	7	5	8	20	2	2	2	1	27
Onagraceae	0	0	0	0	2	6	8	0	1	0	3	12
Opiliaceae	0	0	0	0	0	2	2	0	0	0	0	2
Oxalidaceae	0	0	0	1	3	3	7	0	1	0	4	12
Passifloraceae	0	0	0	0	11	9	20	0	2	1	5	28
Pedaliaceae	0	0	0	0	0	0	0	0	0	0	1	1
Piperaceae	0	0	0	34	47	20	101	0	7	7	2	117
Pittosporaceae	0	0	0	5	9	11	25	1	2	0	4	32
Plantaginaceae	0	0	0	0	0	1	1	0	0	0	0	1
Platanaceae	0	0	0	0	0	1	1	0	0	0	1	2
Plumbaginaceae	0	0	0	1	0	2	3	0	0	0	3	6
Podostemaceae	0	0	0	2	3	1	6	0	0	0	0	6
Polygalaceae	0	0	0	4	2	8	14	2	1	0	0	17
Polygonaceae	0	0	0	6	6	6	18	0	3	1	0	22
Portulacaceae	0	0	0	0	2	1	3	1	0	0	0	4
Primulaceae	0	1	1	0	0	0	0	0	0	0	0	1
Proteaceae	1	0	1	4	8	24	36	4	4	0	1	46
Quiinaceae	0	0	0	0	2	1	3	0	1	0	0	4
Rafflesiaceae	0	0	0	1	0	0	1	0	0	0	0	1
Ranunculaceae	0	0	0	4	1	0	5	0	1	0	0	6
Resedaceae	0	0	0	0	0	0	0	0	0	0	2	2
Rhamnaceae	1	0	1	6	5	12	23	1	3	2	3	33
Rhizophoraceae	0	0	0	3	3	6	12	0	2	2	2	18
Rhoipteleaceae	0	0	0	0	0	1	1	0	0	00	0	1
Rhynchocalycaceae	0	0	0	0	0	1	1	0	0	0	0	1
Rosaceae	0	0	0	15	15	50	80	2	7	12	17	118
Rubiaceae	5	0	5	56	80	234	370	3	35	33	18	464
Rutaceae	6	0	6	16	41	49	106	5	9	4	2	132
Salicaceae	0	0	0	2	2	4	8	0	1	0	2	11
Salvadoraceae	0	0	0	0	0	0	0	0	0	0	1	1
Santalaceae	1	0	1	0	2	3	5	0	1	0	1	8
Sapindaceae	2	0	2	11	24	62	97	3	8	6	6	122
Sapotaceae	6	0	6	28	64	146	238	15	44	5	21	329
Sarcolaenaceae	0	0	0	6	4	0	10	0	1	0	5	16
Sarraceniaceae	0	0	0	1	0	1	2	0	1	0	4	7
Scrophulariaceae	0	0	0	1	13	20	34	0	4	0	13	51
Scytopetalaceae	0	0	0	0	1	2	3	0	1	0	1	5
Simaroubaceae	0	0	0	0	2	9	11	0	0	0	3	14
Solanaceae	0	0	0	9	13	26	48	6	12	13	6	85
Staphyleaceae	0	0	0	0	0	3	3	0	0	0	0	3
Sterculiaceae	3	2	5	14	16	37	67	1	3	3	5	84

Class MAGNOLIOPSIDA

Family	EX	EW	Subtotal	CR	EN	VU	Subtotal	LR/cd	NT	DD	LC	TOTAL
Styracaceae	0	0	0	1	0	16	17	0	1	0	1	19
Symplocaceae	0	0	0	2	14	20	36	2	1	-	0	40
Tamaricaceae	0	0	0	0	0	0	0	0	0	0	1	1
Theaceae	0	1	1	16	10	51	77	7	5	1	1	92
Theophrastaceae	0	0	0	1	0	4	5	0	1	0	0	6
Thymelaeaceae	2	0	2	2	2	25	29	1	2	3	3	40
Ticodendraceae	0	0	0	0	0	1	1	0	0	0	0	1
Tiliaceae	0	0	0	2	7	18	27	6	2	6	1	42
Tropaeolaceae	0	0	0	1	1	6	8	0	0	1	0	9
Turneraceae	0	0	0	0	1	0	1	0	0	0	0	1
Ulmaceae	0	0	0	2	2	7	11	0	3	0	2	16
Umbelliferae	0	0	0	9	3	8	20	0	3	0	4	27
Urticaceae	0	0	0	7	3	5	15	0	7	1	0	23
Valerianaceae	1	0	1	2	1	3	6	0	1	0	2	10
Verbenaceae	0	0	0	10	10	40	60	1	5	2	8	76
Violaceae	0	0	0	4	10	11	25	0	6	0	1	32
Viscaceae	0	0	0	1	8	6	15	0	0	0	1	16
Vitaceae	0	0	0	0	0	0	0	0	0	0	6	6
Vochysiaceae	0	0	0	1	1	2	4	0	0	0	0	4
Winteraceae	0	0	0	0	2	3	5	0	0	0	0	5
Zygophyllaceae	0	0	0	2	3	0	5	2	0	2	2	11
Subtotal (Magnoliopsida)	**78**	**22**	**100**	**1,299**	**1,847**	**3,976**	**7,122**	**196**	**810**	**458**	**938**	**9,624**

Class LILIOPSIDA

Family	EX	EW	Subtotal	CR	EN	VU	Subtotal	LR/cd	NT	DD	LC	TOTAL
Alismataceae	0	0	0	0	0	1	1	0	0	0	0	1
Alliaceae	0	0	0	1	0	0	1	0	0	1	0	2
Aloaceae	0	0	0	3	3	3	9	0	3	2	7	21
Alstroemeriaceae	0	0	0	5	3	4	12	0	2	1	0	15
Amaryllidaceae	0	0	0	0	10	5	15	0	0	0	5	20
Anthericaceae	0	0	0	1	1	0	2	0	0	0	1	3
Aponogetonaceae	0	0	0	0	0	0	0	0	0	1	0	1
Araceae	0	0	0	12	15	40	67	0	14	77	10	168
Asparagaceae	0	0	0	0	0	0	0	0	0	1	0	1
Asphodelaceae	0	0	0	0	1	2	3	0	0	0	3	6
Asteliaceae	0	0	0	1	0	0	1	0	0	0	0	1
Bromeliaceae	0	0	0	7	46	65	118	0	21	2	10	151
Burmanniaceae	0	0	0	3	0	1	4	0	0	0	0	4
Colchicaceae	0	0	0	0	0	0	0	0	0	0	1	1
Commelinaceae	0	0	0	0	1	2	3	0	0	0	0	3
Costaceae	0	0	0	0	0	2	2	0	0	0	0	2
Cyclanthaceae	0	0	0	2	5	9	16	0	3	0	1	20
Cyperaceae	0	0	0	5	5	5	15	0	1	1	2	19
Dioscoreaceae	0	0	0	0	2	1	3	0	1	0	1	5
Dracaenaceae	1	0	1	0	7	5	12	0	0	1	0	14
Eriocaulaceae	0	0	0	0	0	4	4	0	0	0	1	5
Eriospermaceae	0	0	0	0	0	1	1	0	0	0	4	5

Class LILIOPSIDA

Family	EX	EW	Subtotal	CR	EN	VU	Subtotal	LR/cd	NT	DD	LC	TOTAL
Gramineae	1	0	1	3	14	25	42	0	10	11	13	77
Heliconiaceae	0	0	0	0	0	15	15	0	1	2	0	18
Hyacinthaceae	0	0	0	2	1	1	4	0	0	1	9	14
Iridaceae	0	0	0	0	0	1	1	0	0	0	5	6
Juncaceae	0	0	0	0	0	0	0	0	1	1	0	2
Lemnaceae	0	0	0	0	1	0	1	0	0	0	0	1
Marantaceae	0	0	0	2	9	9	20	0	6	0	0	26
Orchidaceae	0	0	0	30	71	46	147	0	5	0	1	153
Palmae	0	2	2	65	71	102	238	16	38	34	29	357
Pandanaceae	0	0	0	5	1	16	22	1	1	2	1	27
Triuridaceae	0	0	0	1	0	0	1	0	0	0	0	1
Xyridaceae	0	0	0	1	0	0	1	0	0	0	0	1
Zingiberaceae	0	0	0	0	0	1	1	0	2	0	1	4
Subtotal (Liliopsida)	**2**	**2**	**4**	**149**	**267**	**366**	**782**	**17**	**109**	**138**	**105**	**1,155**

Class CHLOROPHYCEAE

Family	EX	EW	Subtotal	CR	EN	VU	Subtotal	LR/cd	NT	DD	LC	TOTAL
Chaetophoraceae	0	0	0	0	0	0	0	0	0	1	0	1
Subtotal (Chlorophyceae)	**0**	**0**	**0**	**0**	**0**	**0**	**0**	**0**	**0**	**1**	**0**	**1**

Class ULVOPHYCEAE

Family	EX	EW	Subtotal	CR	EN	VU	Subtotal	LR/cd	NT	DD	LC	TOTAL
Cladophoraceae	0	0	0	0	0	0	0	0	0	1	0	1
Subtotal (Ulvophyceae)	**0**	**0**	**0**	**0**	**0**	**0**	**0**	**0**	**0**	**1**	**0**	**1**

Class FLORIDEOPHYCEAE

Family	EX	EW	Subtotal	CR	EN	VU	Subtotal	LR/cd	NT	DD	LC	TOTAL
Bonnemaisoniaceae	0	0	0	0	0	0	0	0	0	1	0	1
Ceramiaceae	0	0	0	0	0	0	0	0	0	7	1	8
Corallinaceae	0	0	0	0	0	0	0	0	0	11	0	11
Delesseriaceae	1	0	1	2	0	3	5	0	0	2	0	8
Galaxauraceae	0	0	0	1	0	0	1	0	0	1	0	2
Gigartinaceae	0	0	0	0	0	0	0	0	0	2	0	2
Gracilariaceae	0	0	0	1	0	0	1	0	0	1	0	2
Halymeniaceae	0	0	0	0	0	0	0	0	0	3	1	4
Hapalidiaceae	0	0	0	0	0	0	0	0	0	3	0	3
Kallymeniaceae	0	0	0	0	0	0	0	0	0	2	1	3
Phyllophoraceae	0	0	0	0	0	0	0	0	0	1	0	1
Rhizophyllidaceae	0	0	0	0	0	0	0	0	0	0	1	1
Rhodomelaceae	0	0	0	1	0	0	1	0	0	7	0	8
Rhodymeniaceae	0	0	0	0	0	0	0	0	0	2	0	2
Schizymeniaceae	0	0	0	1	0	0	1	0	0	0	0	1
Sebdeniaceae	0	0	0	0	0	0	0	0	0	1	0	1
Subtotal (Liliopsida)	**1**	**0**	**1**	**6**	**0**	**3**	**9**	**0**	**0**	**44**	**4**	**58**

	EX	EW	Subtotal	CR	EN	VU	Subtotal	LR/cd	NT	DD	LC	TOTAL
Total plants	**87**	**28**	**115**	**1,575**	**2,280**	**4,602**	**8,457**	**238**	**1,065**	**731**	**1,449**	**12,055**

Appendix 8. Number of threatened species in each major group of organisms in each country (Critically Endangered, Endangered and Vulnerable categories only)

The country and territory names used below are based on the short country names specified by the International Organization for Standardization (ISO) Maintenance Agency for ISO 3166 country codes (see http://www.iso.org/iso/country_codes/iso_3166_code_lists/english_country_names_and_code_elements.htm).

The figures for each country exclude species recorded as being introduced, vagrant or that have uncertain occurrences in that country. This note also applies to Appendices 9, 10 and 11.

AFRICA									
North Africa	**Mammals**	**Birds**	**Reptiles**	**Amphibians**	**Fishes**	**Molluscs**	**Other Inverts**	**Plants**	**Total**
Algeria	14	11	7	3	23	0	14	3	75
Egypt	17	10	11	0	24	0	46	2	110
Libyan Arab Jamahiriya	12	4	5	0	14	0	0	1	36
Morocco	18	10	10	2	31	0	9	2	82
Tunisia	14	8	4	1	20	0	7	0	54
Western Sahara	11	1	0	0	19	0	1	0	32
Sub-Saharan Africa	**Mammals**	**Birds**	**Reptiles**	**Amphibians**	**Fishes**	**Molluscs**	**Other Inverts**	**Plants**	**Total**
Angola	14	18	4	0	22	4	1	26	89
Benin	10	4	4	0	15	0	0	14	47
Botswana	6	7	0	0	2	0	0	0	15
Burkina Faso	8	5	1	0	0	0	0	2	16
Burundi	9	8	0	6	18	1	4	2	48
Cameroon	41	15	3	53	43	1	3	355	514
Cape Verde	3	4	1	0	18	0	0	2	28
Central African Republic	7	5	1	0	0	0	0	15	28
Chad	12	7	1	0	0	1	0	2	23
Comoros	5	8	2	0	7	0	62	5	89
Congo	11	3	2	0	15	1	3	35	70
Congo, The Democratic Republic of the	29	31	3	13	25	13	11	65	190
Côte d'Ivoire	24	14	4	13	19	1	0	105	180
Djibouti	8	7	0	0	14	0	50	2	81
Equatorial Guinea	18	5	4	4	13	0	0	63	107
Eritrea	9	9	6	0	14	0	50	3	91
Ethiopia	31	22	1	9	2	3	11	22	101
Gabon	13	5	3	3	21	0	0	108	153
Gambia	9	5	2	0	16	0	0	4	36
Ghana	17	8	4	10	17	0	1	117	174
Guinea	22	12	2	5	19	0	4	22	86
Guinea-Bissau	11	2	2	0	18	0	0	4	37
Kenya	27	27	5	7	71	16	55	103	311
Lesotho	2	5	0	0	1	0	2	1	11
Liberia	20	11	4	4	19	1	6	46	111
Madagascar	62	35	19	64	75	24	76	281	636
Malawi	6	12	0	5	101	9	7	14	154

Sub-Saharan Africa	Mammals	Birds	Reptiles	Amphibians	Fishes	Molluscs	Other Inverts	Plants	Total
Mali	11	6	1	0	1	0	0	6	25
Mauritania	14	8	3	0	23	0	1	0	49
Mauritius	6	11	7	0	11	27	69	88	219
Mayotte	1	3	2	0	3	0	59	0	68
Mozambique	11	21	5	3	45	4	54	46	189
Namibia	11	21	4	1	21	0	0	24	82
Niger	11	5	0	0	2	0	1	2	21
Nigeria	27	12	4	13	21	0	3	171	251
Réunion	5	6	0	0	6	14	58	15	104
Rwanda	19	10	0	8	9	0	3	3	52
Saint Helena	2	18	1	0	11	0	2	26	60
Sao Tomé and Principe	5	10	3	3	8	1	1	35	66
Senegal	15	8	6	0	28	0	0	7	64
Seychelles	5	10	10	6	14	2	63	45	155
Sierra Leone	16	10	3	2	16	0	0	47	94
Somalia	14	12	3	0	26	1	50	17	123
South Africa	23	35	19	21	65	24	137	74	398
Sudan	14	13	3	0	13	0	45	17	105
Swaziland	4	7	0	0	3	0	0	11	25
Tanzania, United Republic of	34	40	5	49	138	17	66	240	589
Togo	10	2	3	3	16	0	0	10	44
Uganda	21	18	0	6	54	10	12	38	159
Zambia	8	12	0	1	10	3	1	8	43
Zimbabwe	8	11	0	6	3	0	4	17	49

ANTARCTIC									
Antarctic	Mammals	Birds	Reptiles	Amphibians	Fishes	Molluscs	Other Inverts	Plants	Total
Antarctica	1	5	0	0	0	0	0	0	6
Bouvet Island	0	1	0	0	1	0	0	0	2
French Southern Territories (the)	3	13	2	0	3	0	0	0	21
Heard Island and McDonald Islands	0	10	0	0	1	0	0	0	11
South Georgia and the South Sandwich Islands	3	7	0	0	0	0	0	0	10

ASIA									
East Asia	Mammals	Birds	Reptiles	Amphibians	Fishes	Molluscs	Other Inverts	Plants	Total
China	74	85	30	90	70	1	20	446	816
Hong Kong	2	16	1	5	13	1	4	6	48
Japan	27	40	12	20	40	25	133	12	309
Korea, Democratic People's Republic of	9	20	0	1	8	0	2	3	43
Korea, Republic of	9	30	0	2	14	0	3	0	58
Macao	0	4	0	0	6	0	0	0	10
Mongolia	11	21	0	0	1	0	3	0	36
Taiwan, Province of China	10	22	8	10	37	1	120	78	286

North Asia	Mammals	Birds	Reptiles	Amphibians	Fishes	Molluscs	Other Inverts	Plants	Total
Belarus	4	4	0	0	1	0	8	0	17
Moldova	4	9	1	0	9	0	4	0	27
Russian Federation	33	51	6	0	32	1	28	7	158
Ukraine	11	12	2	0	20	0	14	1	60
South & Southeast Asia	**Mammals**	**Birds**	**Reptiles**	**Amphibians**	**Fishes**	**Molluscs**	**Other Inverts**	**Plants**	**Total**
Bangladesh	34	28	20	1	12	0	2	12	109
Bhutan	28	17	1	1	0	0	1	7	55
British Indian Ocean Territory	0	0	2	0	9	0	65	1	77
Brunei Darussalam	35	21	5	3	8	0	0	99	171
Cambodia	37	25	12	3	18	0	67	31	193
Disputed Territory (Spratly Islands)	0	0	0	0	1	0	0	0	1
India	96	76	25	65	40	2	109	246	659
Indonesia	183	115	27	33	111	3	229	386	1,087
Lao People's Democratic Republic	46	23	11	5	6	0	3	21	115
Malaysia	70	42	21	47	49	19	207	686	1,141
Maldives	2	0	3	0	12	0	38	0	55
Myanmar	45	41	22	0	17	1	63	38	227
Nepal	32	32	7	3	0	0	0	7	81
Philippines	39	67	9	48	60	3	199	216	641
Singapore	12	14	4	0	22	0	161	54	267
Sri Lanka	30	13	8	53	31	0	119	280	534
Thailand	57	44	22	4	50	1	179	86	443
Timor-Leste	4	5	1	0	5	0	0	0	15
Viet Nam	54	39	27	17	33	0	91	147	408
West & Central Asia	**Mammals**	**Birds**	**Reptiles**	**Amphibians**	**Fishes**	**Molluscs**	**Other Inverts**	**Plants**	**Total**
Afghanistan	11	13	1	1	3	0	1	2	32
Armenia	9	12	5	0	4	0	6	1	37
Azerbaijan	7	15	5	0	9	0	4	0	40
Bahrain	3	4	4	0	6	0	13	0	30
Cyprus	5	5	4	0	12	0	0	7	33
Georgia	10	10	7	1	12	0	9	0	49
Iran (Islamic Republic of)	16	20	9	4	21	0	19	1	90
Iraq	13	18	2	1	6	0	15	0	55
Israel	15	13	10	1	31	5	52	0	127
Jordan	13	8	5	0	14	0	49	0	89
Kazakhstan	16	21	2	1	13	0	4	16	73
Kuwait	6	8	2	0	10	0	13	0	39
Kyrgyzstan	6	12	2	0	3	0	3	14	40
Lebanon	10	6	6	0	15	0	3	0	40
Oman	9	9	4	0	20	0	26	6	74
Pakistan	23	27	10	0	22	0	15	2	99
Palestinian Territory, Occupied	3	7	4	1	1	0	1	0	17
Qatar	2	4	1	0	7	0	13	0	27
Saudi Arabia	9	14	2	0	16	0	53	3	97

West & Central Asia	Mammals	Birds	Reptiles	Amphibians	Fishes	Molluscs	Other Inverts	Plants	Total
Syrian Arab Republic	16	13	6	0	27	0	6	0	68
Tajikistan	8	9	1	0	8	0	2	14	42
Turkey	17	15	13	10	60	0	13	3	131
Turkmenistan	9	15	1	0	12	0	5	3	45
United Arab Emirates	7	8	2	0	9	0	16	0	42
Uzbekistan	11	15	2	0	8	0	1	15	52
Yemen	9	13	3	1	18	2	61	159	266

EUROPE

Europe	Mammals	Birds	Reptiles	Amphibians	Fishes	Molluscs	Other Inverts	Plants	Total
Albania	3	6	4	2	33	0	4	0	52
Andorra	2	0	1	0	2	1	3	0	9
Austria	4	9	1	0	9	22	21	4	70
Belgium	3	2	0	0	9	4	8	1	27
Bosnia and Herzegovina	4	6	2	1	27	0	10	1	51
Bulgaria	7	12	2	0	17	0	7	0	45
Croatia	7	11	2	2	46	0	15	1	84
Czech Republic	2	6	0	0	5	2	16	4	35
Denmark	2	2	0	0	13	1	10	3	31
Estonia	1	3	0	0	4	0	4	0	12
Faroe Islands	5	0	0	0	9	0	0	0	14
Finland	1	4	0	0	5	1	9	1	21
France	9	6	5	2	31	34	40	8	135
Germany	6	6	0	0	20	9	21	12	74
Gibraltar	5	3	0	0	10	2	0	0	20
Greece	10	11	5	5	62	1	13	11	118
Greenland	6	0	0	0	6	0	0	1	13
Guernsey	0	0	0	0	3	0	0	0	3
Holy See (Vatican City State)	1	0	0	0	1	0	0	0	2
Hungary	2	9	1	0	9	1	25	1	48
Iceland	5	0	0	0	12	0	0	0	17
Ireland	5	1	0	0	16	1	2	1	26
Isle of Man	0	0	0	0	3	0	0	0	3
Italy	7	8	5	8	33	16	42	19	138
Jersey	0	0	0	0	3	0	0	0	3
Latvia	1	4	0	0	6	1	9	0	21
Liechtenstein	0	0	0	0	0	0	4	0	4
Lithuania	3	4	0	0	6	0	6	0	19
Luxembourg	0	0	0	0	1	2	2	0	5
Macedonia, the former Yugoslav Republic of	5	10	2	0	14	0	5	0	36
Malta	3	3	0	0	13	3	0	3	25
Monaco	2	0	0	0	12	0	0	0	14
Montenegro	6	11	2	1	21	0	11	0	52
Netherlands	4	2	0	0	11	1	5	0	23
Norway	7	2	0	0	14	1	8	2	34

Europe	Mammals	Birds	Reptiles	Amphibians	Fishes	Molluscs	Other Inverts	Plants	Total
Poland	5	6	0	0	6	1	15	4	37
Portugal	11	8	2	1	38	67	16	16	159
Romania	7	12	2	0	16	0	22	1	60
San Marino	0	0	0	0	1	0	0	0	1
Serbia	6	11	0	1	8	0	16	1	43
Slovakia	3	7	1	0	7	6	13	2	39
Slovenia	4	4	1	2	24	0	42	0	77
Spain	16	15	18	6	52	27	35	49	218
Svalbard and Jan Mayen	1	0	0	0	2	0	0	0	3
Sweden	1	3	0	0	12	1	12	3	32
Switzerland	2	2	0	1	11	0	29	3	48
United Kingdom	5	2	0	0	34	2	8	14	65

NORTH & CENTRAL AMERICA

Mesoamerica	Mammals	Birds	Reptiles	Amphibians	Fishes	Molluscs	Other Inverts	Plants	Total
Belize	7	3	5	6	22	0	12	30	85
Costa Rica	8	17	8	59	19	0	28	11˙	250
El Salvador	5	3	7	10	7	0	6	26	64
Guatemala	16	11	13	80	16	2	7	83	228
Honduras	6	7	11	59	19	0	18	110	230
Mexico	100	54	95	211	114	5	57	26˙	897
Nicaragua	5	9	8	10	21	2	17	39	111
Panama	14	17	7	49	19	0	20	194	320

Caribbean Islands	Mammals	Birds	Reptiles	Amphibians	Fishes	Molluscs	Other Inverts	Plants	Total
Anguilla	1	0	3	0	14	0	10	3	31
Antigua and Barbuda	2	1	6	0	14	0	11	4	38
Aruba	3	1	2	0	15	0	1	0	22
Bahamas	7	5	6	0	20	0	11	5	54
Barbados	3	1	4	0	15	0	10	2	35
Bermuda	4	1	2	0	12	0	28	4	51
Cayman Islands	1	1	4	0	14	1	10	2	33
Cuba	14	17	8	49	28	0	15	163	294
Dominica	3	3	3	2	15	0	11	11	48
Dominican Republic	6	14	11	30	15	0	18	30	124
Grenada	3	1	4	1	15	0	10	3	37
Guadeloupe	5	1	5	3	14	1	15	7	51
Haiti	5	13	8	46	15	0	14	29	130
Jamaica	5	10	9	17	15	0	15	209	280
Martinique	2	2	6	2	10	1	0	8	31
Montserrat	3	2	2	1	14	0	11	3	36
Netherlands Antilles	4	1	6	0	15	0	11	2	39
Puerto Rico	3	8	9	14	13	0	1	53	101
Saint Barthélemy	1	0	2	0	1	0	0	2	6
Saint Kitts and Nevis	2	1	5	1	14	0	10	2	35
Saint Lucia	2	5	5	5	15	0	11	6	44
Saint Martin (French part)	1	0	2	0	1	0	0	2	6

Caribbean Islands	Mammals	Birds	Reptiles	Amphibians	Fishes	Molluscs	Other Inverts	Plants	Total
Saint Vincent and the Grenadines	2	2	3	1	16	0	10	4	38
Trinidad and Tobago	2	2	5	9	19	0	10	1	48
Turks and Caicos Islands	2	2	4	0	14	0	10	2	34
Virgin Islands, British	1	1	6	2	12	0	10	10	42
Virgin Islands, U.S.	2	1	4	2	11	0	0	11	31
North America	Mammals	Birds	Reptiles	Amphibians	Fishes	Molluscs	Other Inverts	Plants	Total
Canada	12	16	3	1	26	2	10	2	72
Saint Pierre and Miquelon	3	1	0	0	1	0	0	0	5
United States	37	74	32	56	164	273	312	244	1,192

SOUTH AMERICA									
South America	Mammals	Birds	Reptiles	Amphibians	Fishes	Molluscs	Other Inverts	Plants	Total
Argentina	35	49	5	29	31	0	10	44	203
Bolivia	19	29	2	39	0	0	1	71	161
Brazil	82	122	22	30	64	21	15	382	738
Chile	21	32	1	21	18	0	8	40	141
Colombia	52	86	15	214	31	0	31	223	652
Ecuador	43	69	11	171	15	48	12	1,839	2,208
Falkland Islands (Malvinas)	4	10	0	0	5	0	0	5	24
French Guiana	6	0	6	3	21	0	0	16	52
Guyana	8	3	5	7	22	0	1	22	68
Paraguay	8	27	2	0	0	0	0	10	47
Peru	53	93	6	96	10	0	3	275	536
Suriname	7	0	5	1	20	0	0	26	59
Uruguay	10	24	4	4	28	0	1	1	72
Venezuela	32	26	13	71	29	0	19	69	259

OCEANIA									
Oceania	Mammals	Birds	Reptiles	Amphibians	Fishes	Molluscs	Other Inverts	Plants	Total
American Samoa	1	8	2	0	8	5	52	1	77
Australia	57	49	38	48	84	175	282	55	788
Christmas Island	1	5	3	0	5	0	16	1	31
Cocos (Keeling) Islands	2	0	1	0	7	0	17	0	27
Cook Islands	1	15	1	0	7	0	25	1	50
Fiji	6	10	6	1	11	3	87	66	190
French Polynesia	1	32	1	0	13	29	26	47	149
Guam	2	12	2	0	9	6	0	4	35
Kiribati	1	5	1	0	7	1	72	0	87
Marshall Islands	2	5	1	0	10	1	66	0	85
Micronesia, Federated States of	6	9	3	0	13	4	104	5	144
Nauru	1	2	0	0	8	0	62	0	73
New Caledonia	9	14	2	0	17	11	84	218	355
New Zealand	8	69	12	4	14	5	10	21	143
Niue	2	8	1	0	7	0	23	0	41

Oceania	Mammals	Birds	Reptiles	Amphibians	Fishes	Molluscs	Other Inverts	Plants	Total
Norfolk Island	0	15	2	0	2	12	9	1	41
Northern Mariana Islands	5	14	1	0	9	4	47	5	85
Palau	4	2	2	0	12	5	97	4	126
Papua New Guinea	41	36	9	11	38	2	167	142	446
Pitcairn	2	10	0	0	6	5	10	7	40
Samoa	2	7	1	0	8	1	52	2	73
Solomon Islands	17	20	4	2	12	2	138	16	211
Tokelau	0	1	1	0	7	0	31	0	40
Tonga	2	4	2	0	9	2	33	4	56
Tuvalu	2	1	1	0	8	1	70	0	83
United States Minor Outlying Islands	0	11	1	0	12	0	44	0	68
Vanuatu	8	8	2	0	11	1	78	10	118
Wallis and Futuna	0	9	0	0	6	0	57	1	73

Appendix 9. Number of extinct, threatened and other species of animals in each Red List Category in each country

IUCN Red List Categories: EX - Extinct, **EW** - Extinct in the Wild, **CR** - Critically Endangered, **EN** - Endangered, **VU** - Vulnerable, **LR/cd** - Lower Risk/conservation dependent, **NT** - Near Threatened (includes **LR/nt** - Lower Risk/near threatened), **DD** - Data Deficient, **LC** - Least Concern (includes **LR/lc** - Lower Risk/least concern).

The country and territory names used below are based on the short country names specified by the International Organization for Standardization (ISO) Maintenance Agency for ISO 3166 country codes (see http://www.iso.org/iso/country_codes/iso_3166_code_lists/english_country_names_and_code_elements.htm).

AFRICA

North Africa	EX	EW	Subtotal	CR	EN	VU	Subtotal	LR/cd	NT	DD	LC	Total
Algeria	0	1	1	13	20	39	72	1	49	31	444	598
Egypt	0	1	1	8	14	86	108	2	134	53	671	969
Libyan Arab Jamahiriya	0	1	1	7	10	18	35	0	30	20	351	437
Morocco	1	1	2	15	20	45	80	1	62	43	489	677
Tunisia	0	1	1	11	16	27	54	0	40	22	388	505
Western Sahara	0	1	1	8	10	14	32	0	19	25	213	290

Sub-Saharan Africa	EX	EW	Subtotal	CR	EN	VU	Subtotal	LR/cd	NT	DD	LC	Total
Angola	0	0	0	9	19	35	63	0	50	114	1,384	1,611
Benin	0	0	0	5	8	20	33	0	31	30	698	792
Botswana	0	0	0	1	1	13	15	0	21	12	847	895
Burkina Faso	0	1	1	0	3	11	14	0	18	4	605	642
Burundi	0	0	0	2	15	29	46	0	37	18	899	1,000
Cameroon	0	0	0	27	61	71	159	0	59	80	1,334	1,632
Cape Verde	1	0	1	3	11	12	26	0	16	25	103	171
Central African Republic	0	0	0	1	2	10	13	0	25	24	993	1,055
Chad	0	1	1	3	3	15	21	0	22	7	649	700
Comoros	0	0	0	5	11	68	84	0	105	38	297	524
Congo	0	0	0	7	11	17	35	0	30	42	889	996
Congo, The Democratic Republic of the	0	0	0	10	45	70	125	0	71	152	1,909	2,257
Côte d'Ivoire	0	0	0	9	21	45	75	0	69	58	1,018	1,220
Djibouti	0	0	0	1	6	72	79	0	130	43	504	756
Equatorial Guinea	0	0	0	8	16	20	44	0	27	33	704	808
Eritrea	0	0	0	5	7	76	88	0	139	36	778	1,041
Ethiopia	0	0	0	5	27	47	79	0	41	27	1,106	1,253
Gabon	0	0	0	8	12	25	45	0	31	49	879	1,004
Gambia	0	0	0	4	8	20	32	0	30	27	651	740
Ghana	0	0	0	6	17	34	57	0	56	49	1,038	1,200
Guinea	0	0	0	8	21	35	64	0	61	51	921	1,097
Guinea-Bissau	0	0	0	4	9	20	33	0	34	25	599	691
Kenya	34	4	38	39	44	125	208	2	174	108	1,729	2,259
Lesotho	0	0	0	0	2	8	10	0	12	1	317	340
Liberia	0	0	0	10	18	37	65	0	53	49	781	948

Sub-Saharan Africa	EX	EW	Subtotal	CR	EN	VU	Subtotal	LR/cd	NT	DD	LC	Total
Madagascar	10	0	10	40	106	209	355	2	182	203	640	1,392
Malawi	0	0	0	2	25	113	140	0	36	39	1,238	1,453
Mali	0	1	1	2	4	13	19	0	23	6	717	766
Mauritania	0	1	1	11	10	28	49	0	40	35	555	680
Mauritius	40	0	40	10	26	95	131	2	116	49	242	580
Mayotte	5	0	5	1	5	62	68	0	96	25	267	461
Mozambique	0	0	0	8	28	107	143	2	174	101	1,407	1,827
Namibia	1	0	1	7	14	37	58	0	43	40	953	1,095
Niger	0	1	1	3	3	13	19	0	17	3	557	597
Nigeria	0	1	1	9	21	50	80	0	56	48	1,289	1,474
Réunion	16	0	16	3	13	73	89	0	100	40	229	474
Rwanda	1	0	1	6	14	29	49	0	30	8	806	894
Saint Helena	29	0	29	7	7	20	34	0	7	16	45	131
Sao Tomé and Principe	0	0	0	7	9	15	31	0	15	43	118	207
Senegal	0	1	1	9	14	34	57	0	46	44	781	929
Seychelles	4	0	4	6	19	85	110	2	119	34	296	565
Sierra Leone	0	0	0	4	15	28	47	0	56	48	845	996
Somalia	0	0	0	8	11	87	106	0	147	61	937	1,251
South Africa	5	0	5	43	87	194	324	3	159	133	1,322	1,946
Sudan	0	1	1	4	10	74	88	0	137	54	1,386	1,666
Swaziland	0	0	0	1	2	11	14	0	17	3	628	662
Tanzania, United Republic of	33	5	38	51	101	197	349	0	191	162	2,132	2,872
Togo	0	0	0	6	11	17	34	0	32	30	789	885
Uganda	34	4	38	28	37	56	121	0	56	39	1,534	1,788
Zambia	0	0	0	3	8	24	35	0	39	61	1,412	1,547
Zimbabwe	0	0	0	3	10	19	32	0	31	11	1,045	1,119

ANTARCTIC												
Antarctic	EX	EW	Subtotal	CR	EN	VU	Subtotal	LR/cd	NT	DD	LC	Total
Antarctica	0	0	0	0	1	5	6	0	6	4	32	48
Bouvet Island	0	0	0	0	0	2	2	0	1	2	16	21
French Southern Territories (the)	1	0	1	3	10	8	21	0	6	9	39	76
Heard Island and McDonald Islands	0	0	0	0	4	7	11	0	3	6	18	38
South Georgia and the South Sandwich Islands	0	0	0	0	5	5	10	0	6	7	91	114

ASIA												
East Asia	EX	EW	Subtotal	CR	EN	VU	Subtotal	LR/cd	NT	DD	LC	Total
China	4	1	5	54	111	205	370	1	184	350	1,797	2,707
Hong Kong	0	0	0	3	14	25	42	1	32	22	248	345
Japan	13	0	13	25	82	190	297	3	216	188	781	1,498
Korea, Democratic People's Republic of	1	0	1	3	11	26	40	0	23	17	382	463
Korea, Republic of	1	0	1	6	15	37	58	0	22	30	411	522
Macao	0	0	0	1	2	7	10	0	14	0	15	39
Mongolia	0	0	0	4	10	22	36	0	19	3	454	512
Taiwan, Province of China	0	0	0	8	36	164	208	2	205	97	676	1,188

North Asia	EX	EW	Subtotal	CR	EN	VU	Subtotal	LR/cd	NT	DD	LC	Total
Belarus	0	0	0	2	2	13	17	2	29	2	366	416
Moldova	1	0	1	2	7	18	27	0	32	7	353	420
Russian Federation	4	1	5	16	44	91	151	3	76	28	995	1,258
Ukraine	2	0	2	9	15	35	59	2	40	12	517	632
South & Southeast Asia	**EX**	**EW**	**Subtotal**	**CR**	**EN**	**VU**	**Subtotal**	**LR/cd**	**NT**	**DD**	**LC**	**Total**
Bangladesh	0	1	1	15	27	55	97	0	49	14	687	848
Bhutan	0	0	0	5	13	30	48	0	27	5	642	722
British Indian Ocean Territory	0	0	0	1	5	70	76	2	105	33	224	440
Brunei Darussalam	0	0	0	4	12	56	72	0	127	28	436	663
Cambodia	0	0	0	17	34	111	162	0	153	38	796	1,149
Disputed Territory (Spratly Islands)	0	0	0	0	1	0	1	0	0	5	5	11
India	1	0	1	51	105	257	413	2	252	231	1,631	2,530
Indonesia	1	0	1	63	157	481	701	4	513	455	2,008	3,682
Lao People's Democratic Republic	0	0	0	15	31	48	94	0	42	42	840	1,018
Malaysia	0	0	0	38	81	336	455	3	363	172	1,129	2,122
Maldives	0	0	0	1	5	49	55	2	92	34	251	434
Myanmar	0	0	0	18	39	132	189	3	223	104	1,344	1,863
Nepal	0	0	0	9	21	44	74	0	44	15	932	1,065
Philippines	0	0	0	42	65	318	425	5	284	192	807	1,713
Singapore	1	0	1	8	15	190	213	3	228	66	581	1,092
Sri Lanka	21	0	21	62	83	109	254	10	154	57	590	1,086
Thailand	1	1	2	25	67	265	357	2	318	172	1,349	2,200
Timor-Leste	0	0	0	2	6	7	15	0	22	23	259	319
Viet Nam	0	0	0	27	59	175	261	2	227	148	1,228	1,866
West & Central Asia	**EX**	**EW**	**Subtotal**	**CR**	**EN**	**VU**	**Subtotal**	**LR/cd**	**NT**	**DD**	**LC**	**Total**
Afghanistan	0	0	0	3	8	19	30	0	14	5	478	527
Armenia	0	0	0	3	9	24	36	1	31	8	367	443
Azerbaijan	0	1	1	3	13	24	40	2	32	10	461	546
Bahrain	0	0	0	2	3	25	30	0	52	8	287	377
Cyprus	0	0	0	3	11	12	26	0	27	12	308	373
Georgia	0	0	0	7	13	29	49	2	37	6	397	491
Iran (Islamic Republic of)	0	1	1	7	22	60	89	0	91	53	708	942
Iraq	1	0	1	3	12	40	55	0	59	8	477	600
Israel	4	1	5	17	28	82	127	0	134	45	640	951
Jordan	0	0	0	3	13	73	89	0	119	30	551	789
Kazakhstan	0	1	1	7	18	32	57	0	33	12	605	708
Kuwait	1	0	1	1	7	31	39	0	55	8	323	426
Kyrgyzstan	0	0	0	2	7	17	26	0	18	3	377	424
Lebanon	0	0	0	4	15	21	40	0	33	15	342	430
Oman	0	0	0	5	10	53	68	0	96	45	472	681
Pakistan	0	0	0	8	20	69	97	0	83	28	801	1,009
Palestinian Territory, Occupied	0	0	0	4	4	9	17	0	9	3	100	129
Qatar	0	0	0	1	1	25	27	0	55	7	284	373
Saudi Arabia	1	0	1	3	8	83	94	2	136	42	590	865
Syrian Arab Republic	1	0	1	7	26	35	68	0	32	23	373	497

West & Central Asia	EX	EW	Subtotal	CR	EN	VU	Subtotal	LR/cd	NT	DD	LC	Total
Tajikistan	0	0	0	3	10	15	28	0	15	3	368	414
Turkey	1	0	1	26	48	54	128	0	60	55	590	834
Turkmenistan	0	1	1	5	12	25	42	0	19	7	400	469
United Arab Emirates	0	0	0	1	7	34	42	0	59	24	370	495
Uzbekistan	0	0	0	5	11	21	37	0	20	5	395	457
Yemen	2	0	2	4	8	95	107	0	146	55	541	851

EUROPE

Europe	EX	EW	Subtotal	CR	EN	VU	Subtotal	LR/cd	NT	DD	LC	Total
Åland Islands	0	0	0	0	0	0	0	0	0	0	1	1
Albania	1	0	1	9	11	32	52	0	33	20	430	536
Andorra	0	0	0	1	1	7	9	0	7	0	193	209
Austria	4	0	4	11	17	38	66	3	39	17	463	592
Belgium	1	0	1	5	3	18	26	2	29	11	388	457
Bosnia and Herzegovina	0	0	0	5	11	34	50	0	26	4	430	510
Bulgaria	0	0	0	6	13	26	45	0	44	15	502	606
Croatia	1	0	1	15	21	47	83	0	38	10	477	609
Czech Republic	0	0	0	3	5	23	31	3	32	8	434	508
Denmark	0	0	0	4	6	18	28	3	27	12	392	462
Estonia	0	0	0	2	1	9	12	2	19	3	361	397
Faroe Islands	1	0	1	1	5	8	14	0	6	5	143	169
Finland	0	0	0	3	2	15	20	4	22	3	374	423
France	6	0	6	16	23	88	127	2	82	50	615	882
Germany	4	0	4	9	10	43	62	4	46	25	491	632
Gibraltar	0	0	0	5	6	9	20	0	21	15	204	260
Greece	1	0	1	22	32	53	107	0	62	35	504	709
Greenland	1	0	1	0	5	7	12	0	5	6	102	126
Guernsey	0	0	0	1	0	2	3	0	4	2	20	29
Holy See (Vatican City State)	0	0	0	1	0	1	2	0	4	0	19	25
Hungary	0	0	0	2	9	36	47	2	36	10	417	512
Iceland	1	0	1	3	6	8	17	0	8	11	140	177
Ireland	1	0	1	8	7	10	25	3	20	17	281	347
Isle of Man	0	0	0	1	0	2	3	0	3	1	18	25
Italy	1	0	1	17	20	82	119	2	68	31	555	776
Jersey	0	0	0	1	0	2	3	0	4	2	24	33
Latvia	0	0	0	2	3	16	21	4	25	4	382	436
Liechtenstein	0	0	0	0	0	4	4	1	13	0	258	276
Lithuania	0	0	0	3	1	15	19	3	23	5	357	407
Luxembourg	0	0	0	2	1	2	5	0	16	1	281	303
Macedonia, the former Yugoslav Republic of	2	0	2	3	5	28	36	0	25	13	454	530
Malta	0	0	0	4	7	11	22	0	25	17	224	288
Monaco	0	0	0	1	5	8	14	0	15	11	76	116
Montenegro	1	0	1	7	15	30	52	0	40	13	469	575
Netherlands	1	0	1	5	4	14	23	2	27	16	405	474
Norway	0	0	0	4	7	21	32	4	25	13	376	450

Europe	EX	EW	Subtotal	CR	EN	VU	Subtotal	LR/cd	NT	DD	LC	Total
Poland	1	0	1	4	5	24	33	4	39	6	450	533
Portugal	2	0	2	25	28	90	143	5	68	37	422	677
Romania	2	0	2	8	11	40	59	3	44	12	491	611
San Marino	0	0	0	1	0	0	1	0	5	0	52	58
Serbia	0	0	0	2	8	32	42	0	33	7	478	560
Slovakia	0	0	0	1	6	30	37	3	39	7	441	527
Slovenia	0	0	0	5	14	58	77	0	28	9	459	573
Spain	1	0	1	27	50	92	169	6	121	48	558	903
Svalbard and Jan Mayen	0	0	0	0	0	3	3	0	3	3	64	73
Sweden	0	0	0	4	4	21	29	4	30	10	411	484
Switzerland	6	0	6	3	3	39	45	4	39	13	446	553
United Kingdom	2	0	2	9	14	28	51	4	32	25	378	492

NORTH & CENTRAL AMERICA

Mesoamerica	EX	EW	Subtotal	CR	EN	VU	Subtotal	LR/cd	NT	DD	LC	Total
Belize	0	0	0	8	16	31	55	1	48	34	771	909
Costa Rica	3	0	3	30	49	60	139	0	71	72	1,205	1,490
El Salvador	0	0	0	8	11	19	38	0	42	26	676	782
Guatemala	1	0	1	39	49	57	145	1	73	57	966	1,243
Honduras	4	0	4	42	37	41	120	0	65	48	1,018	1,255
Mexico	22	6	28	179	222	235	636	3	158	299	1,998	3,122
Nicaragua	0	0	0	11	20	41	72	0	60	50	968	1,150
Panama	0	0	0	34	37	55	126	0	76	105	1,280	1,587
Caribbean Islands	**EX**	**EW**	**Subtotal**	**CR**	**EN**	**VU**	**Subtotal**	**LR/cd**	**NT**	**DD**	**LC**	**Total**
Anguilla	0	0	0	3	6	19	28	0	12	16	277	333
Antigua and Barbuda	0	0	0	6	6	22	34	0	17	19	287	357
Aruba	0	0	0	4	3	15	22	0	12	14	243	291
Bahamas	2	0	2	8	11	30	49	0	26	30	344	451
Barbados	0	0	0	6	5	22	33	0	15	22	286	356
Bermuda	0	0	0	27	7	13	47	0	17	16	180	260
Cayman Islands	1	0	1	6	6	19	31	0	16	20	296	364
Cuba	7	0	7	33	43	55	131	0	46	40	398	622
Dominica	1	0	1	5	8	24	37	0	13	19	287	357
Dominican Republic	9	0	9	20	33	41	94	0	34	23	314	474
Grenada	0	0	0	6	8	20	34	0	14	22	189	259
Guadeloupe	7	0	7	6	8	30	44	0	21	18	316	406
Haiti	10	0	10	39	24	38	101	0	33	23	321	488
Jamaica	6	0	6	19	18	34	71	0	31	34	257	399
Martinique	8	0	8	4	3	16	23	0	12	11	234	288
Montserrat	0	0	0	6	6	21	33	0	12	6	269	320
Netherlands Antilles	0	0	0	5	9	23	37	0	15	24	309	385
Puerto Rico	2	0	2	13	11	24	48	0	23	15	303	391
Saint Barthélemy	0	0	0	0	2	2	4	0	0	0	0	4
Saint Kitts and Nevis	0	0	0	6	6	21	33	0	14	18	282	347
Saint Lucia	1	0	1	6	10	22	38	0	16	16	286	357
Saint Martin (French part)	0	0	0	0	2	2	4	0	0	0	0	4

Caribbean Islands	EX	EW	Subtotal	CR	EN	VU	Subtotal	LR/cd	NT	DD	LC	Total
Saint Vincent and the Grenadines	1	0	1	6	7	21	34	0	15	19	286	355
Trinidad and Tobago	0	0	0	10	9	28	47	1	27	30	607	712
Turks and Caicos Islands	0	0	0	6	8	18	32	0	17	11	302	362
Virgin Islands, British	0	0	0	8	6	18	32	0	17	14	211	274
Virgin Islands, U.S.	2	0	2	4	4	12	20	0	17	13	175	227
North America	EX	EW	Subtotal	CR	EN	VU	Subtotal	LR/cd	NT	DD	LC	Total
Canada	6	0	6	7	16	47	70	0	35	32	783	926
Saint Pierre and Miquelon	0	0	0	0	3	2	5	0	4	6	205	220
United States	229	4	233	193	200	555	948	7	278	294	1,634	3,394

SOUTH AMERICA												
South America	EX	EW	Subtotal	CR	EN	VU	Subtotal	LR/cd	NT	DD	LC	Total
Argentina	2	3	5	17	46	96	159	0	120	123	1,234	1,641
Bolivia	0	0	0	9	24	57	90	2	72	51	1,823	2,038
Brazil	9	1	10	65	98	193	356	9	181	404	2,493	3,453
Chile	0	0	0	17	21	63	101	0	70	138	521	830
Colombia	3	0	3	79	133	217	429	2	160	271	2,433	3,298
Ecuador	6	0	6	78	108	183	369	2	129	172	1,966	2,644
Falkland Islands (Malvinas)	1	0	1	0	7	12	19	0	13	16	129	178
French Guiana	0	0	0	6	5	25	36	1	31	44	989	1,101
Guyana	0	0	0	6	8	32	46	2	35	59	1,099	1,241
Paraguay	0	3	3	5	9	23	37	0	49	18	842	949
Peru	3	0	3	33	84	144	261	2	124	231	2,197	2,818
Suriname	0	0	0	6	5	22	33	0	31	42	989	1,095
Uruguay	0	0	0	7	23	41	71	0	40	37	480	628
Venezuela	2	0	2	34	54	102	190	1	76	181	1,803	2,253

OCEANIA												
Oceania	EX	EW	Subtotal	CR	EN	VU	Subtotal	LR/cd	NT	DD	LC	Total
American Samoa	1	0	1	3	9	64	76	3	87	33	217	417
Australia	40	0	40	66	148	519	733	3	373	224	1,325	2,698
Christmas Island	0	0	0	2	3	25	30	0	24	16	128	198
Cocos (Keeling) Islands	0	0	0	0	3	24	27	0	27	13	118	190
Cook Islands	15	0	15	1	5	43	49	1	53	23	158	299
Fiji	1	0	1	6	8	110	124	3	149	45	308	630
French Polynesia	68	11	79	26	20	56	102	2	62	27	184	456
Guam	3	1	4	6	12	13	31	2	11	64	102	214
Kiribati	0	0	0	0	5	82	87	3	118	33	238	479
Marshall Islands	0	0	0	0	4	81	85	3	119	36	251	494
Micronesia, Federated States of	2	0	2	6	10	123	139	3	140	97	328	709
Nauru	0	0	0	0	1	72	73	0	110	24	206	413
New Caledonia	4	0	4	7	17	113	137	3	146	47	330	667
New Zealand	19	0	19	14	31	77	122	0	46	59	175	421
Niue	0	0	0	1	4	36	41	0	48	23	161	273
Norfolk Island	10	0	10	1	8	31	40	0	24	4	72	150

Oceania	EX	EW	Subtotal	CR	EN	VU	Subtotal	LR/cd	NT	DD	LC	Total
Northern Mariana Islands	1	0	1	6	15	59	80	3	84	43	265	476
Palau	1	0	1	4	9	109	122	4	141	95	346	709
Papua New Guinea	1	0	1	21	33	250	304	3	250	269	1,208	2,035
Pitcairn	0	0	0	0	8	25	33	1	24	10	66	134
Samoa	0	0	0	2	6	63	71	3	87	37	221	419
Solomon Islands	2	0	2	7	19	169	195	3	182	70	470	922
Tokelau	0	0	0	0	2	38	40	2	55	15	153	265
Tonga	1	0	1	0	6	46	52	3	65	28	202	351
Tuvalu	0	0	0	0	3	80	83	3	117	24	213	440
United States Minor Outlying Islands	1	0	1	4	7	57	68	1	71	23	237	401
Vanuatu	1	0	1	1	7	100	108	3	132	33	295	572
Wallis and Futuna	0	0	0	1	2	69	72	0	93	22	218	405

Appendix 10. Number of extinct, threatened and other species of plants in each Red List Category in each country

IUCN Red List Categories: EX - Extinct, **EW** - Extinct in the Wild, **CR** - Critically Endangered, **EN** - Endangered, **VU** - Vulnerable, **LR/cd** - Lower Risk/conservation dependent, **NT** - Near Threatened (includes **LR/nt** - Lower Risk/near threatened), **DD** - Data Deficient, **LC** - Least Concern (includes **LR/lc** - Lower Risk/least concern).

The country and territory names used below are based on the short country names specified by the International Organization for Standardization (ISO) Maintenance Agency for ISO 3166 country codes (see http://www.iso.org/iso/country_codes/iso_3166_code_lists/english_country_names_and_code_elements.htm).

AFRICA

North Africa	EX	EW	Subtotal	CR	EN	VU	Subtotal	LR/cd	NT	DD	LC	Total
Algeria	0	0	0	1	1	1	3	0	1	0	8	12
Egypt	0	0	0	1	1	0	2	0	0	0	1	3
Libyan Arab Jamahiriya	0	0	0	0	0	1	1	0	0	0	3	4
Morocco	0	0	0	0	1	1	2	0	3	0	8	13
Tunisia	0	0	0	0	0	0	0	0	0	0	4	4

Sub-Saharan Africa	EX	EW	Subtotal	CR	EN	VU	Subtotal	LR/cd	NT	DD	LC	Total
Angola	0	0	0	0	2	24	26	0	6	1	6	39
Benin	0	0	0	0	0	14	14	0	2	0	2	18
Botswana	0	0	0	0	0	0	0	0	3	0	3	6
Burkina Faso	0	0	0	0	0	2	2	0	1	0	1	4
Burundi	0	0	0	0	0	2	2	1	1	0	2	6
Cameroon	1	0	1	66	69	220	355	2	37	4	7	406
Cape Verde	0	0	0	0	0	2	2	0	0	0	0	2
Central African Republic	0	0	0	1	2	12	15	1	3	3	2	24
Chad	0	0	0	0	0	2	2	0	1	0	0	3
Comoros	0	0	0	2	2	1	5	0	0	0	3	8
Congo	0	0	0	1	7	27	35	1	4	0	3	43
Congo, The Democratic Republic of the	0	0	0	0	12	53	65	3	14	2	13	97
Côte d'Ivoire	2	0	2	2	18	85	105	1	10	1	5	124
Djibouti	0	0	0	0	1	1	2	0	1	0	0	3
Equatorial Guinea	0	0	0	3	13	47	63	0	13	0	3	79
Eritrea	0	0	0	0	0	3	3	0	2	0	0	5
Ethiopia	0	0	0	0	1	21	22	1	30	1	4	58
Gabon	0	0	0	4	14	90	108	1	16	3	4	132
Gambia	0	0	0	0	0	4	4	0	0	0	1	5
Ghana	0	0	0	3	19	95	117	1	10	1	5	134
Guinea	0	0	0	0	0	22	22	1	3	0	3	29
Guinea-Bissau	0	0	0	0	0	4	4	0	0	0	1	5
Kenya	0	0	0	5	14	84	103	1	26	1	15	146
Lesotho	0	0	0	0	0	1	1	0	0	0	1	2
Liberia	0	0	0	0	4	42	46	0	2	0	4	52
Madagascar	0	0	0	62	102	117	281	0	30	12	39	362
Malawi	0	0	0	0	3	11	14	1	3	0	6	24

Sub-Saharan Africa	EX	EW	Subtotal	CR	EN	VU	Subtotal	LR/cd	NT	DD	LC	Total
Mali	0	0	0	0	2	4	6	0	1	0	1	8
Mauritius	2	0	2	65	14	9	88	0	0	0	0	90
Mayotte	0	0	0	0	0	0	0	0	0	0	1	1
Mozambique	0	0	0	2	6	38	46	0	10	23	14	93
Namibia	0	0	0	2	2	20	24	0	12	3	122	161
Niger	0	0	0	0	0	2	2	0	0	0	0	2
Nigeria	0	0	0	16	18	137	171	2	14	1	6	194
Réunion	0	0	0	9	5	1	15	0	0	0	0	15
Rwanda	0	0	0	0	0	3	3	1	2	0	4	10
Saint Helena	7	2	9	11	8	7	26	0	3	5	7	50
Sao Tomé and Principe	0	0	0	0	2	33	35	0	10	0	1	46
Senegal	0	0	0	0	0	7	7	0	2	0	1	10
Seychelles	0	0	0	7	4	34	45	1	3	0	1	50
Sierra Leone	0	0	0	1	4	42	47	1	4	0	3	55
Somalia	0	0	0	0	3	14	17	0	42	0	2	61
South Africa	0	3	3	16	16	42	74	4	21	9	21	132
Sudan	0	0	0	1	1	15	17	1	5	2	4	29
Swaziland	0	1	1	2	2	7	11	1	2	2	8	25
Tanzania, United Republic of	2	0	2	8	33	199	240	2	17	7	17	285
Togo	0	0	0	0	0	10	10	0	2	0	1	13
Uganda	0	0	0	2	4	32	38	1	9	1	9	58
Zambia	0	0	0	0	0	8	8	0	4	0	7	19
Zimbabwe	0	0	0	0	6	11	17	0	8	0	7	32

ASIA												
East Asia	EX	EW	Subtotal	CR	EN	VU	Subtotal	LR/cd	NT	DD	LC	Total
China	3	1	4	74	174	198	446	5	51	19	116	641
Hong Kong	0	0	0	2	1	3	6	0	1	0	1	8
Japan	0	0	0	0	3	9	12	0	6	6	31	55
Korea, Democratic People's Republic of	0	0	0	0	0	3	3	0	1	1	14	19
Korea, Republic of	0	0	0	0	0	0	0	0	2	1	12	15
Mongolia	0	0	0	0	0	0	0	0	0	2	11	13
Taiwan, Province of China	0	1	1	12	39	27	78	1	9	7	23	119
North Asia	EX	EW	Subtotal	CR	EN	VU	Subtotal	LR/cd	NT	DD	LC	Total
Moldova	0	0	0	0	0	0	0	0	0	0	3	3
Russian Federation	0	0	0	0	2	5	7	0	4	3	35	49
Ukraine	0	0	0	0	0	1	1	0	0	0	14	15
South & Southeast Asia	EX	EW	Subtotal	CR	EN	VU	Subtotal	LR/cd	NT	DD	LC	Total
Bangladesh	0	0	0	4	2	6	12	0	1	0	5	18
Bhutan	0	0	0	1	2	4	7	0	3	1	17	28
British Indian Ocean Territory	0	0	0	0	0	1	1	0	0	0	2	3
Brunei Darussalam	0	0	0	38	23	38	99	4	16	2	48	169
Cambodia	0	0	0	9	13	9	31	0	5	4	33	73
India	7	2	9	45	112	89	246	1	22	18	70	366
Indonesia	1	1	2	113	69	204	386	9	80	38	171	686

South & Southeast Asia	EX	EW	Subtotal	CR	EN	VU	Subtotal	LR/cd	NT	DD	LC	Total
Lao People's Democratic Republic	0	0	0	5	7	9	21	0	5	5	24	55
Malaysia	2	-	3	186	99	401	686	113	70	27	281	1,180
Myanmar	0	0	0	13	12	13	38	0	10	8	54	110
Nepal	0	0	0	0	2	5	7	0	2	1	24	34
Philippines	0	0	0	52	34	130	216	3	24	12	66	321
Singapore	0	0	0	11	13	30	54	8	20	2	109	193
Sri Lanka	1	0	1	78	73	129	280	5	1	3	15	305
Thailand	0	0	0	29	21	36	86	3	25	13	75	202
Timor-Leste	0	0	0	0	0	0	0	0	0	0	2	2
Viet Nam	0	0	0	25	39	83	147	1	32	14	65	259
West & Central Asia	EX	EW	Subtotal	CR	EN	VU	Subtotal	LR/cd	NT	DD	LC	Total
Afghanistan	0	0	0	0	1	1	2	0	4	2	14	22
Armenia	0	0	0	0	0	1	1	0	1	0	11	13
Azerbaijan	0	0	0	0	0	0	0	0	2	0	12	14
Cyprus	0	0	0	6	0	1	7	0	0	0	7	14
Georgia	0	0	0	0	0	0	0	1	3	0	11	15
Iran (Islamic Republic of)	0	0	0	0	0	1	1	0	5	1	11	18
Iraq	0	0	0	0	0	0	0	0	0	0	3	3
Israel	0	0	0	0	0	0	0	0	0	0	1	1
Jordan	0	0	0	0	0	0	0	0	0	0	1	1
Kazakhstan	0	0	0	7	7	2	16	0	2	11	23	52
Kyrgyzstan	0	0	0	6	6	2	14	0	3	5	22	44
Lebanon	0	0	0	0	0	0	0	0	0	0	11	11
Oman	0	0	0	0	1	5	6	0	5	0	1	12
Pakistan	0	0	0	0	0	2	2	0	3	3	11	19
Palestinian Territory, Occupied	0	0	0	0	0	0	0	0	0	0	1	1
Saudi Arabia	0	0	0	0	2	1	3	0	1	0	2	6
Syrian Arab Republic	0	0	0	0	0	0	0	0	0	0	9	9
Tajikistan	0	0	0	7	3	4	14	0	4	2	14	34
Turkey	0	0	0	0	1	2	3	1	10	1	25	40
Turkmenistan	0	0	0	0	1	2	3	0	3	0	6	12
Uzbekistan	0	0	0	4	7	4	15	0	3	3	19	40
Yemen	3	0	3	3	28	128	159	0	23	28	113	326

EUROPE												
Europe	EX	EW	Subtotal	CR	EN	VU	Subtotal	LR/cd	NT	DD	LC	Total
Albania	0	0	0	0	0	0	0	0	1	0	17	18
Andorra	0	0	0	0	0	0	0	0	0	0	6	6
Austria	0	0	0	0	2	2	4	0	0	0	11	15
Belgium	0	0	0	0	0	1	1	0	0	0	3	4
Bosnia and Herzegovina	0	0	0	0	0	1	1	0	0	1	2	4
Bulgaria	0	0	0	0	0	0	0	0	1	0	14	15
Croatia	0	0	0	0	0	1	1	0	0	2	2	5
Czech Republic	0	0	0	0	1	3	4	0	0	0	10	14
Denmark	0	0	0	0	0	3	3	0	0	0	3	6

Europe	EX	EW	Subtotal	CR	EN	VU	Subtotal	LR/cd	NT	DD	LC	Total
Estonia	0	0	0	0	0	0	0	0	0	0	3	3
Faroe Islands	0	0	0	0	0	0	0	0	0	0	1	1
Finland	0	0	0	0	0	1	1	0	0	0	5	6
France	0	0	0	6	0	2	8	0	0	0	17	25
Germany	0	0	0	2	3	7	12	0	0	0	10	22
Gibraltar	0	0	0	0	0	0	0	0	0	0	5	5
Greece	0	0	0	9	0	2	11	0	5	1	20	37
Greenland	0	0	0	0	0	1	1	0	0	0	1	2
Guernsey	0	0	0	0	0	0	0	0	0	0	1	1
Hungary	0	0	0	0	0	1	1	0	0	0	6	7
Iceland	0	0	0	0	0	0	0	0	0	0	1	1
Ireland	0	0	0	0	0	1	1	0	0	0	3	4
Isle of Man	0	0	0	0	0	0	0	0	0	0	1	1
Italy	1	0	1	16	2	1	19	0	0	0	17	37
Jersey	0	0	0	0	0	0	0	0	0	0	1	1
Latvia	0	0	0	0	0	0	0	0	0	0	3	3
Liechtenstein	0	0	0	0	0	0	0	0	0	0	2	2
Luxembourg	0	0	0	0	0	0	0	0	0	0	1	1
Macedonia, the former Yugoslav Republic of	0	0	0	0	0	0	0	0	0	0	2	2
Malta	0	0	0	3	0	0	3	0	1	0	4	8
Monaco	0	0	0	0	0	0	0	0	0	0	6	6
Montenegro	0	0	0	0	0	0	0	0	1	0	13	14
Netherlands	0	0	0	0	0	0	0	0	0	0	3	3
Norway	0	0	0	0	0	2	2	0	0	0	6	8
Poland	0	1	1	0	1	3	4	0	0	0	10	15
Portugal	0	0	0	2	6	8	16	3	4	0	17	40
Romania	0	0	0	0	0	1	1	0	0	0	12	13
San Marino	0	0	0	0	0	0	0	0	0	0	1	1
Serbia	0	0	0	0	0	1	1	0	1	0	11	13
Slovakia	0	0	0	1	0	1	2	0	0	0	2	4
Spain	1	1	2	21	17	11	49	6	7	0	20	84
Sweden	0	0	0	0	0	3	3	0	0	0	5	8
Switzerland	0	0	0	0	1	2	3	0	0	0	10	13
United Kingdom	0	0	0	5	2	7	14	0	1	0	7	22

NORTH & CENTRAL AMERICA												
Mesoamerica	EX	EW	Subtotal	CR	EN	VU	Subtotal	LR/cd	NT	DD	LC	Total
Belize	0	0	0	1	11	18	30	0	5	0	3	38
Costa Rica	0	0	0	4	33	74	111	3	37	7	18	176
El Salvador	0	0	0	1	6	19	26	2	4	6	6	44
Guatemala	0	0	0	5	29	49	83	4	11	7	17	122
Honduras	0	1	1	42	38	30	110	2	5	1	12	131
Mexico	0	2	2	40	75	146	261	8	25	18	88	402
Nicaragua	0	0	0	3	16	20	39	2	17	1	7	66
Panama	0	0	0	19	71	104	194	2	47	40	25	308

Caribbean Islands	EX	EW	Subtotal	CR	EN	VU	Subtotal	LR/cd	NT	DD	LC	Total
Anguilla	0	0	0	0	2	1	3	0	0	0	0	3
Antigua and Barbuda	0	0	0	0	3	1	4	0	0	0	0	4
Bahamas	0	0	0	0	3	2	5	0	2	3	1	11
Barbados	0	0	0	0	1	1	2	0	0	0	0	2
Bermuda	0	0	0	1	1	2	4	0	0	0	0	4
Cayman Islands	0	0	0	0	1	1	2	0	1	0	0	3
Cuba	4	-	5	23	57	83	163	0	3	4	6	181
Dominica	0	0	0	1	4	6	11	0	0	2	0	13
Dominican Republic	0	0	0	2	8	20	30	0	2	2	1	35
Grenada	0	0	0	0	2	1	3	0	0	0	0	3
Guadeloupe	0	0	0	0	4	3	7	0	0	1	2	10
Haiti	0	0	0	5	6	18	29	0	1	2	0	32
Jamaica	2	0	2	40	53	116	209	0	73	5	1	290
Martinique	0	0	0	0	4	4	8	0	0	1	2	11
Montserrat	0	0	0	0	2	1	3	0	0	0	1	4
Netherlands Antilles	0	0	0	0	2	0	2	0	0	0	0	2
Puerto Rico	0	0	0	22	16	15	53	0	2	3	1	59
Saint Barthélemy	0	0	0	0	2	0	2	0	0	0	0	2
Saint Kitts and Nevis	0	0	0	0	1	1	2	0	0	1	1	4
Saint Lucia	0	0	0	0	2	4	6	0	0	1	1	8
Saint Martin (French part)	0	0	0	0	2	0	2	0	0	0	0	2
Saint Vincent and the Grenadines	0	0	0	0	2	2	4	0	0	1	0	5
Trinidad and Tobago	0	0	0	0	1	0	1	0	2	1	3	7
Turks and Caicos Islands	0	0	0	0	2	0	2	0	0	1	1	4
Virgin Islands, British	0	0	0	6	4	0	10	0	0	0	0	10
Virgin Islands, U.S	0	0	0	3	7	1	11	0	0	0	0	11
North America	**EX**	**EW**	**Subtotal**	**CR**	**EN**	**VU**	**Subtotal**	**LR/cd**	**NT**	**DD**	**LC**	**Total**
Canada	1	0	1	0	1	1	2	0	1	0	36	40
United States	23	7	30	104	65	75	244	4	23	3	83	387

SOUTH AMERICA												
South America	**EX**	**EW**	**Subtotal**	**CR**	**EN**	**VU**	**Subtotal**	**LR/cd**	**NT**	**DD**	**LC**	**Total**
Argentina	0	1	1	1	11	32	44	1	13	11	21	91
Bolivia	1	0	1	4	10	57	71	2	10	9	17	110
Brazil	5	1	6	46	117	219	382	22	69	34	86	599
Chile	1	1	2	15	4	21	40	1	10	3	3	59
Colombia	3	0	3	31	85	107	223	4	42	11	48	331
Ecuador	1	0	1	246	668	925	1,839	1	263	285	153	2,542
Falkland Islands (Malvinas)	0	0	0	0	1	4	5	0	1	0	6	12
French Guiana	0	0	0	3	2	11	16	1	2	0	16	35
Guyana	0	0	0	1	3	18	22	1	8	2	23	56
Paraguay	0	0	0	1	5	4	10	1	5	7	14	37
Peru	1	0	1	9	15	251	275	4	37	18	38	373
Suriname	0	0	0	1	2	23	26	0	3	5	13	47
Uruguay	0	0	0	0	1	0	1	0	1	3	2	7
Venezuela	0	0	0	3	7	59	69	2	70	0	50	191

OCEANIA											
Oceania	EX	EW	Subtotal	CR	EN	VU	Subtotal	LR/cd	NT	DD	LC	Total
American Samoa	0	0	0	0	0	1	1	0	1	0	3	5
Australia	1	0	1	4	14	37	55	7	20	0	88	171
Christmas Island	0	0	0	0	0	1	1	0	0	0	0	1
Cocos (Keeling) Islands	0	0	0	0	0	0	0	0	0	0	2	2
Cook Islands	0	0	0	0	0	1	1	0	0	0	0	1
Fiji	1	0	1	21	13	32	66	0	19	3	43	132
French Polynesia	6	0	6	26	4	17	47	0	18	34	50	155
Guam	0	0	0	1	1	2	4	0	0	1	1	6
Micronesia, Federated States of	0	0	0	0	1	4	5	0	0	0	2	7
New Caledonia	3	0	3	28	65	125	218	37	4	0	17	279
New Zealand	0	0	0	2	7	12	21	3	15	1	16	56
Norfolk Island	1	0	1	0	0	1	1	0	0	0	0	2
Northern Mariana Islands	0	0	0	2	1	2	5	0	0	0	0	5
Palau	0	0	0	0	1	3	4	0	1	0	0	5
Papua New Guinea	0	0	0	14	15	113	142	0	33	19	70	264
Pitcairn	0	0	0	1	1	5	7	0	0	0	2	9
Samoa	0	0	0	1	1	0	2	0	1	0	2	5
Solomon Islands	0	0	0	0	1	15	16	0	10	14	20	60
Tonga	0	0	0	1	0	3	4	0	0	0	1	5
Tuvalu	0	0	0	0	0	0	0	0	0	0	2	2
Vanuatu	0	0	0	1	2	7	10	1	5	0	8	24
Wallis and Futuna	0	0	0	0	0	1	1	0	1	0	0	2

Appendix 11. Number of endemic and threatened endemic species per country for completely assessed taxonomic groups (mammals, birds, amphibians, freshwater crabs, reef-forming corals, conifers, cycads)

The country and territory names used below are based on the short country names specified by the International Organization for Standardization (ISO) Maintenance Agency for ISO 3166 country codes (see http://www.iso.org/iso/country_codes/iso_3166_code_lists/english_country_names_and_code_elements.htm).

AFRICA														
North Africa	**Mammals**		**Birds**		**Amphibians**		**FW Crabs**		**Reef-forming Corals**		**Conifers**		**Cycads**	
	Total endemics	Threatened endemics	Total endemics	Threatened endemics	Total endemics	Threatened endemics	Total endemics	Threatened endemics	Total endemics	Threatened endemics	Total endemics	Threatened endemics	Total endemics	Threatened endemics
Algeria	2	0	1	1	1	1	0	0	0	0	1	1	0	0
Egypt	4	0	0	0	1	0	0	0	0	0	0	0	0	0
Libyan Arab Jamahiriya	3	0	0	0	0	0	0	0	0	0	0	0	0	0
Morocco	5	2	0	0	1	0	0	0	0	0	0	0	0	0
Tunisia	1	0	0	0	0	0	0	0	0	0	0	0	0	0
Western Sahara	0	0	0	0	0	0	0	0	0	0	0	0	0	0
Sub-Saharan Africa	**Mammals**		**Birds**		**Amphibians**		**FW Crabs**		**Reef-forming Corals**		**Conifers**		**Cycads**	
	Total endemics	Threatened endemics	Total endemics	Threatened endemics	Total endemics	Threatened endemics	Total endemics	Threatened endemics	Total endemics	Threatened endemics	Total endemics	Threatened endemics	Total endemics	Threatened endemics
Angola	8	0	10	6	19	0	3	0	0	0	0	0	0	0
Benin	0	0	0	0	0	0	0	0	0	0	0	0	0	0
Botswana	0	0	0	0	0	0	0	0	0	0	0	0	0	0
Burkina Faso	0	0	0	0	0	0	0	0	0	0	0	0	0	0
Burundi	0	0	0	0	2	0	0	0	0	0	0	0	0	0
Cameroon	15	14	7	6	51	33	5	2	0	0	0	0	0	0
Cape Verde	0	0	4	2	0	0	0	0	0	0	0	0	0	0
Central African Republic	2	0	0	0	2	0	1	0	0	0	0	0	0	0
Chad	0	0	0	0	1	0	0	0	0	0	0	0	0	0
Comoros	4	2	8	6	0	0	0	0	0	0	0	0	0	0
Congo	3	0	0	0	0	0	0	0	0	0	0	0	0	0
Congo, The Democratic Republic of the	25	4	10	6	41	3	12	1	0	0	0	0	3	1
Côte d'Ivoire	1	1	0	0	3	1	0	0	0	0	0	0	0	0
Djibouti	0	0	1	1	0	0	0	0	0	0	0	0	0	0
Equatorial Guinea	2	2	3	2	1	0	0	0	0	0	0	0	0	0
Eritrea	0	0	0	0	1	0	0	0	0	0	0	0	0	0
Ethiopia	32	18	16	11	24	9	2	1	0	0	0	0	0	0
Gabon	2	1	0	0	3	1	0	0	0	0	0	0	0	0
Gambia	0	0	0	0	0	0	0	0	0	0	0	0	0	0
Ghana	1	0	0	0	4	1	1	1	0	0	0	0	0	0

Sub-Saharan Africa	Mammals		Birds		Amphibians		FW Crabs		Reef-forming Corals		Conifers		Cycads	
	Total endemics	Threatened endemics	Total endemics	Threatened endemics	Total endemics	Threatened endemics	Total endemics	Threatened endemics	Total endemics	Threatened endemics	Total endemics	Threatened endemics	Total endemics	Threatened endemics
Guinea	2	2	0	0	4	0	2	1	0	0	0	0	0	0
Guinea-Bissau	0	0	0	0	0	0	0	0	0	0	0	0	0	0
Kenya	12	7	7	6	13	4	6	0	0	0	0	0	2	1
Lesotho	0	0	0	0	0	0	0	0	0	0	0	0	0	0
Liberia	0	0	1	1	0	0	3	3	0	0	0	0	0	0
Madagascar	181	56	104	25	241	64	14	2	3	2	6	4	0	0
Malawi	2	0	1	1	3	3	2	2	0	0	1	1	0	0
Mali	1	0	0	0	2	0	0	0	0	0	0	0	0	0
Mauritania	0	0	0	0	0	0	0	0	0	0	0	0	0	0
Mauritius	1	1	24	8	0	0	0	0	0	0	0	0	0	0
Mayotte	0	0	1	1	0	0	0	0	0	0	0	0	0	0
Mozambique	2	1	1	0	0	0	0	0	0	0	0	0	3	2
Namibia	1	0	1	0	3	0	0	0	0	0	0	0	0	0
Niger	0	0	0	0	0	0	0	0	0	0	0	0	0	0
Nigeria	4	2	4	2	1	1	3	1	0	0	0	0	0	0
Réunion	0	0	13	2	0	0	0	0	0	0	0	0	0	0
Rwanda	1	1	0	0	1	0	1	0	0	0	0	0	0	0
Saint Helena	0	0	16	6	0	0	0	0	0	0	0	0	0	0
Sao Tomé and Principe	5	3	26	9	7	3	2	0	0	0	1	1	0	0
Senegal	0	0	0	0	0	0	1	0	0	0	0	0	0	0
Seychelles	2	2	14	7	11	6	1	0	0	0	0	0	0	0
Sierra Leone	0	0	0	0	2	1	2	0	0	0	0	0	0	0
Somalia	8	0	10	5	3	0	0	0	0	0	0	0	0	0
South Africa	29	10	15	3	41	16	7	1	0	0	3	2	29	18
Sudan	8	0	1	0	1	0	0	0	0	0	0	0	0	0
Swaziland	0	0	0	0	0	0	0	0	0	0	0	0	1	0
Tanzania, United Republic of	22	18	21	15	70	47	5	3	0	0	0	0	3	2
Togo	0	0	0	0	1	0	0	0	0	0	0	0	0	0
Uganda	1	0	1	0	0	0	4	1	0	0	0	0	3	3
Zambia	5	2	2	1	2	0	1	0	0	0	0	0	0	0
Zimbabwe	0	0	0	0	0	0	0	0	0	0	0	0	1	1

ANTARCTIC

Antarctic	Mammals		Birds		Amphibians		FW Crabs		Reef-forming Corals		Conifers		Cycads	
	Total endemics	Threatened endemics	Total endemics	Threatened endemics	Total endemics	Threatened endemics	Total endemics	Threatened endemics	Total endemics	Threatened endemics	Total endemics	Threatened endemics	Total endemics	Threatened endemics
Antarctica	0	0	0	0	0	0	0	0	0	0	0	0	0	0
Bouvet Island	0	0	0	0	0	0	0	0	0	0	0	0	0	0
French Southern Territories (the)	0	0	1	0	0	0	0	0	0	0	0	0	0	0
Heard Island and McDonald Islands	0	0	0	0	0	0	0	0	0	0	0	0	0	0
South Georgia and the South Sandwich Islands	0	0	1	0	0	0	0	0	0	0	0	0	0	0

ASIA

East Asia	Mammals		Birds		Amphibians		FW Crabs		Reef-forming Corals		Conifers		Cycads	
	Total endemics	Threatened endemics	Total endemics	Threatened endemics	Total endemics	Threatened endemics	Total endemics	Threatened endemics	Total endemics	Threatened endemics	Total endemics	Threatened endemics	Total endemics	Threatened endemics
China	82	17	51	17	173	76	216	4	0	0	56	26	12	8
Hong Kong	0	0	0	0	3	2	4	2	0	0	0	0	0	0
Japan	40	16	15	6	44	18	21	2	4	2	22	5	0	0
Korea, Democratic People's Republic of	0	0	0	0	0	0	0	0	0	0	0	0	0	0
Korea, Republic of	0	0	0	0	3	1	0	0	0	0	0	0	0	0
Macao	0	0	0	0	0	0	0	0	0	0	0	0	0	0
Mongolia	0	0	0	0	0	0	0	0	0	0	0	0	0	0
Taiwan, Province of China	14	0	13	1	16	8	36	11	0	0	10	6	1	1

North Asia	Mammals		Birds		Amphibians		FW Crabs		Reef-forming Corals		Conifers		Cycads	
	Total endemics	Threatened endemics	Total endemics	Threatened endemics	Total endemics	Threatened endemics	Total endemics	Threatened endemics	Total endemics	Threatened endemics	Total endemics	Threatened endemics	Total endemics	Threatened endemics
Belarus	0	0	0	0	0	0	0	0	0	0	0	0	0	0
Moldova	0	0	0	0	0	0	0	0	0	0	0	0	0	0
Russian Federation	21	1	1	0	0	0	0	0	0	0	2	0	0	0
Ukraine	2	2	0	0	0	0	0	0	0	0	0	0	0	0

South & Southeast Asia	Mammals		Birds		Amphibians		FW Crabs		Reef-forming Corals		Conifers		Cycads	
	Total endemics	Threatened endemics	Total endemics	Threatened endemics	Total endemics	Threatened endemics	Total endemics	Threatened endemics	Total endemics	Threatened endemics	Total endemics	Threatened endemics	Total endemics	Threatened endemics
Bangladesh	0	0	0	0	1	0	0	0	0	0	0	0	0	0
Bhutan	0	0	0	0	1	0	1	0	0	0	0	0	0	0
British Indian Ocean Territory	0	0	0	0	0	0	0	0	0	0	0	0	0	0
Brunei Darussalam	0	0	0	0	0	0	1	0	0	0	0	0	0	0
Cambodia	1	0	1	0	3	0	1	0	0	0	0	0	0	0
Disputed Territory	0	0	0	0	0	0	0	0	0	0	0	0	0	0
India	43	29	50	15	153	60	63	3	0	0	2	1	3	1
Indonesia	258	114	369	67	170	22	76	11	4	2	6	1	2	0
Lao People's Democratic Republic	6	3	0	0	6	0	13	3	0	0	0	0	0	0
Malaysia	19	3	6	0	44	22	78	28	0	0	14	11	0	0
Maldives	0	0	0	0	0	0	0	0	0	0	0	0	0	0
Myanmar	4	0	4	1	9	0	26	1	0	0	1	0	0	0
Nepal	2	1	0	0	3	0	0	0	0	0	0	0	0	0
Philippines	110	24	194	56	79	48	42	4	0	0	3	2	5	1
Singapore	1	0	0	0	1	0	3	3	0	0	0	0	0	0
Sri Lanka	17	15	21	7	89	52	50	40	1	0	0	0	1	0
Thailand	4	1	1	1	14	1	86	17	0	0	0	0	3	3
Timor-Leste	1	0	0	0	0	0	0	0	0	0	0	0	0	0
Viet Nam	9	3	9	5	33	4	36	4	0	0	4	4	12	10

West & Central Asia	Mammals		Birds		Amphibians		FW Crabs		Reef-forming Corals		Conifers		Cycads	
	Total endemics	Threatened endemics	Total endemics	Threatened endemics	Total endemics	Threatened endemics	Total endemics	Threatened endemics	Total endemics	Threatened endemics	Total endemics	Threatened endemics	Total endemics	Threatened endemics
Afghanistan	0	0	0	0	1	1	1	0	0	0	0	0	0	0
Armenia	2	2	0	0	0	0	0	0	0	0	0	0	0	0

West & Central Asia	Mammals		Birds		Amphibians		FW Crabs		Reef-forming Corals		Conifers		Cycads	
	Total endemics	Threatened endemics	Total endemics	Threatened endemics	Total endemics	Threatened endemics	Total endemics	Threatened endemics	Total endemics	Threatened endemics	Total endemics	Threatened endemics	Total endemics	Threatened endemics
Azerbaijan	0	0	0	0	0	0	0	0	0	0	0	0	0	0
Bahrain	0	0	0	0	0	0	0	0	0	0	0	0	0	0
Cyprus	1	0	0	0	0	0	0	0	0	0	1	1	0	0
Georgia	0	0	0	0	0	0	0	0	0	0	0	0	0	0
Iran, Islamic Republic of	5	0	0	0	4	2	3	0	0	0	0	0	0	0
Iraq	0	0	0	0	0	0	0	0	0	0	0	0	0	0
Israel	1	0	0	0	1	0	0	0	0	0	0	0	0	0
Jordan	0	0	0	0	0	0	0	0	0	0	0	0	0	0
Kazakhstan	4	0	0	0	0	0	0	0	0	0	0	0	0	0
Kuwait	0	0	0	0	0	0	0	0	0	0	0	0	0	0
Kyrgyzstan	1	0	0	0	0	0	0	0	0	0	0	0	0	0
Lebanon	0	0	0	0	0	0	0	0	0	0	0	0	0	0
Oman	1	0	0	0	0	0	0	0	2	0	0	0	0	0
Pakistan	2	1	0	0	0	0	1	0	0	0	0	0	0	0
Palestinian Territory, Occupied	0	0	0	0	0	0	0	0	0	0	0	0	0	0
Qatar	0	0	0	0	0	0	0	0	0	0	0	0	0	0
Saudi Arabia	1	0	0	0	0	0	0	0	1	0	0	0	0	0
Syrian Arab Republic	1	0	0	0	0	0	0	0	0	0	0	0	0	0
Tajikistan	0	0	0	0	0	0	0	0	0	0	0	0	0	0
Turkey	6	0	0	0	9	7	2	1	0	0	0	0	0	0
Turkmenistan	0	0	0	0	0	0	0	0	0	0	0	0	0	0
United Arab Emirates	0	0	0	0	0	0	0	0	0	0	0	0	0	0
Uzbekistan	0	0	0	0	0	0	0	0	0	0	0	0	0	0
Yemen	2	0	7	1	1	1	3	0	0	0	0	0	0	0

EUROPE

Europe	Mammals		Birds		Amphibians		FW Crabs		Reef-forming Corals		Conifers		Cycads	
	Total endemics	Threatened endemics	Total endemics	Threatened endemics	Total endemics	Threatened endemics	Total endemics	Threatened endemics	Total endemics	Threatened endemics	Total endemics	Threatened endemics	Total endemics	Threatened endemics
Åland Islands	0	0	0	0	0	0	0	0	0	0	0	0	0	0
Albania	0	0	0	0	0	0	0	0	0	0	0	0	0	0
Andorra	0	0	0	0	0	0	0	0	0	0	0	0	0	0
Austria	0	0	0	0	0	0	0	0	0	0	0	0	0	0
Belgium	0	0	0	0	0	0	0	0	0	0	0	0	0	0
Bosnia and Herzegovina	0	0	0	0	0	0	0	0	0	0	1	1	0	0
Bulgaria	0	0	0	0	0	0	0	0	0	0	0	0	0	0
Croatia	0	0	0	0	0	0	0	0	0	0	0	0	0	0
Czech Republic	0	0	0	0	0	0	0	0	0	0	0	0	0	0
Denmark	0	0	0	0	0	0	0	0	0	0	0	0	0	0
Estonia	0	0	0	0	0	0	0	0	0	0	0	0	0	0
Faroe Islands	0	0	0	0	0	0	0	0	0	0	0	0	0	0
Finland	0	0	0	0	0	0	0	0	0	0	0	0	0	0
France	0	0	0	0	3	0	0	0	0	0	0	0	0	0
Germany	0	0	0	0	0	0	0	0	0	0	0	0	0	0

Europe	Mammals		Birds		Amphibians		FW Crabs		Reef-forming Corals		Con fers		Cycads	
	Total endemics	Threatened endemics	Total endemics	Threatened endemics	Total endemics	Threatened endemics	Total endemics	Threatened endemics	Total endemics	Threatened endemics	Total endemics	Threatened endemics	Total endemics	Threatened endemics
Gibraltar	0	0	0	0	0	0	0	0	0	0	0	0	0	0
Greece	2	1	0	0	3	3	0	0	0	0	1	0	0	0
Greenland	0	0	0	0	0	0	0	0	0	0	0	0	0	0
Guernsey	0	0	0	0	0	0	0	0	0	0	0	0	0	0
Holy See (Vatican City State)	0	0	0	0	0	0	0	0	0	0	0	0	0	0
Hungary	0	0	0	0	0	0	0	0	0	0	0	0	0	0
Iceland	0	0	0	0	0	0	0	0	0	0	0	0	0	0
Ireland	0	0	0	0	0	0	0	0	0	0	0	0	0	0
Isle of Man	0	0	0	0	0	0	0	0	0	0	0	0	0	0
Italy	4	1	0	0	13	5	0	0	0	0	1	1	0	0
Jersey	0	0	0	0	0	0	0	0	0	0	0	0	0	0
Latvia	0	0	0	0	0	0	0	0	0	0	0	0	0	0
Liechtenstein	0	0	0	0	0	0	0	0	0	0	0	0	0	0
Lithuania	0	0	0	0	0	0	0	0	0	0	0	0	0	0
Luxembourg	0	0	0	0	0	0	0	0	0	0	0	0	0	0
Macedonia, the former Yugoslav Republic of	0	0	0	0	0	0	0	0	0	0	0	0	0	0
Malta	0	0	0	0	0	0	0	0	0	0	0	0	0	0
Monaco	0	0	0	0	0	0	0	0	0	0	0	0	0	0
Montenegro	0	0	0	0	0	0	0	0	0	0	0	0	0	0
Netherlands	0	0	0	0	0	0	0	0	0	0	0	0	0	0
Norway	0	0	0	0	0	0	0	0	0	0	0	0	0	0
Poland	1	0	0	0	0	0	0	0	0	0	0	0	0	0
Portugal	1	1	2	1	0	0	0	0	0	0	1	1	0	0
Romania	0	0	0	0	0	0	0	0	0	0	0	0	0	0
San Marino	0	0	0	0	0	0	0	0	0	0	0	0	0	0
Serbia	0	0	0	0	0	0	0	0	0	0	0	0	0	0
Slovakia	0	0	0	0	0	0	0	0	0	0	0	0	0	0
Slovenia	0	0	0	0	0	0	0	0	0	0	0	0	0	0
Spain	3	3	5	2	4	3	0	0	0	0	1	0	0	0
Svalbard and Jan Mayen	0	0	0	0	0	0	0	0	0	0	0	0	0	0
Sweden	0	0	0	0	0	0	0	0	0	0	0	0	0	0
Switzerland	0	0	0	0	0	0	0	0	0	0	0	0	0	0
United Kingdom	0	0	1	0	0	0	0	0	0	0	0	0	0	0

NORTH & CENTRAL AMERICA

Mesoamerica	Mammals		Birds		Amphibians		FW Crabs		Reef-forming Corals		Conifers		Cycads	
	Total endemics	Threatened endemics	Total endemics	Threatened endemics	Total endemics	Threatened endemics	Total endemics	Threatened endemics	Total endemics	Threatened endemics	Total endemics	Threatened endemics	Total endemics	Threatened endemics
Belize	0	0	0	0	0	0	2	0	0	0	0	0	1	1
Costa Rica	5	0	6	5	39	23	9	1	0	0	3	1	0	0
El Salvador	0	0	0	0	0	0	3	1	0	0	0	0	0	0
Guatemala	3	1	1	0	32	24	6	1	0	0	0	0	2	1
Honduras	3	1	1	1	41	37	1	1	0	0	0	0	1	1
Mexico	157	80	86	21	245	171	54	6	2	1	30	10	36	32
Nicaragua	2	0	0	0	3	3	3	0	0	0	0	0	0	0
Panama	13	4	8	3	26	12	11	0	1	1	0	0	5	3

Caribbean Islands	Mammals		Birds		Amphibians		FW Crabs		Reef-forming Corals		Conifers		Cycads	
	Total endemics	Threatened endemics	Total endemics	Threatened endemics	Total endemics	Threatened endemics	Total endemics	Threatened endemics	Total endemics	Threatened endemics	Total endemics	Threatened endemics	Total endemics	Threatened endemics
Anguilla	0	0	0	0	0	0	0	0	0	0	0	0	0	0
Antigua and Barbuda	0	0	1	0	0	0	0	0	0	0	0	0	0	0
Aruba	0	0	0	0	0	0	0	0	0	0	0	0	0	0
Bahamas	2	1	2	0	1	0	0	0	0	0	0	0	1	0
Barbados	0	0	1	0	0	0	0	0	0	0	0	0	0	0
Bermuda	0	0	0	0	0	0	0	0	0	0	1	1	0	0
Cayman Islands	0	0	1	0	0	0	0	0	0	0	0	0	0	0
Cuba	21	12	21	8	59	49	2	0	0	0	4	2	3	1
Dominica	0	0	3	2	1	1	0	0	0	0	0	0	0	0
Dominican Republic	1	0	0	0	10	9	0	0	0	0	1	1	0	0
Grenada	0	0	1	1	1	1	0	0	0	0	0	0	0	0
Guadeloupe	1	1	3	0	2	2	0	0	0	0	0	0	0	0
Haiti	2	0	0	0	26	26	0	0	0	0	0	0	0	0
Jamaica	7	3	32	6	21	17	0	0	0	0	2	1	0	0
Martinique	1	0	2	1	1	1	0	0	0	0	0	0	0	0
Montserrat	0	0	1	1	0	0	0	0	0	0	0	0	0	0
Netherlands Antilles	1	0	0	0	0	0	0	0	0	0	0	0	0	0
Puerto Rico	1	0	8	4	14	13	0	0	0	0	0	0	1	1
Saint Barthélemy	0	0	0	0	0	0	0	0	0	0	0	0	0	0
Saint Kitts and Nevis	0	0	0	0	0	0	0	0	0	0	0	0	0	0
Saint Lucia	1	0	5	3	0	0	0	0	0	0	0	0	0	0
Saint Martin (French part)	0	0	0	0	0	0	0	0	0	0	0	0	0	0
Saint Vincent and the Grenadines	1	0	2	2	1	1	0	0	0	0	0	0	0	0
Trinidad and Tobago	1	0	1	1	7	7	1	0	0	0	1	0	0	0
Turks and Caicos Islands	0	0	0	0	0	0	0	0	0	0	0	0	0	0
Virgin Islands, British	0	0	0	0	0	0	0	0	0	0	0	0	0	0
Virgin Islands, U.S.	0	0	0	0	1	1	0	0	0	0	0	0	0	0

North America	Mammals		Birds		Amphibians		FW Crabs		Reef-forming Corals		Conifers		Cycads	
	Total endemics	Threatened endemics	Total endemics	Threatened endemics	Total endemics	Threatened endemics	Total endemics	Threatened endemics	Total endemics	Threatened endemics	Total endemics	Threatened endemics	Total endemics	Threatened endemics
Canada	5	1	0	0	0	0	0	0	0	0	0	0	0	0
Saint Pierre and Miquelon	0	0	0	0	0	0	0	0	0	0	0	0	0	0
United States	106	20	62	32	178	49	0	0	9	3	39	12	0	0

SOUTH AMERICA

South America	Mammals		Birds		Amphibians		FW Crabs		Reef-forming Corals		Conifers		Cycads	
	Total endemics	Threatened endemics	Total endemics	Threatened endemics	Total endemics	Threatened endemics	Total endemics	Threatened endemics	Total endemics	Threatened endemics	Total endemics	Threatened endemics	Total endemics	Threatened endemics
Argentina	82	13	12	0	37	21	0	0	0	0	0	0	0	0
Bolivia	22	4	15	5	63	32	1	0	0	0	2	0	1	0
Brazil	183	55	197	71	496	26	13	1	8	0	4	0	0	0
Chile	17	5	11	3	29	12	0	0	0	0	1	1	0	0
Colombia	37	9	65	40	333	158	77	10	0	0	0	0	6	6
Ecuador	29	11	32	17	155	100	13	2	0	0	0	0	1	1

South America	Mammals		Birds		Amphibians		FW Crabs		Reef-forming Corals		Conifers		Cycads	
	Total endemics	Threatened endemics	Total endemics	Threatened endemics	Total endemics	Threatened endemics	Total endemics	Threatened endemics	Total endemics	Threatened endemics	Total endemics	Threatened endemics	Total endemics	Threatened endemics
Falkland Islands (Malvinas)	1	0	1	1	0	0	0	0	0	0	0	0	0	0
French Guiana	1	0	1	0	6	2	0	0	0	0	0	0	0	0
Guyana	0	0	0	0	19	4	2	0	0	0	0	0	0	0
Paraguay	2	1	1	0	1	0	0	0	0	0	0	0	0	0
Peru	55	19	106	36	217	69	8	0	0	0	0	0	2	2
Suriname	0	0	0	0	4	0	1	0	0	0	0	0	0	0
Uruguay	1	1	0	0	1	1	0	0	0	0	0	0	0	0
Venezuela	19	6	38	14	139	62	19	3	0	0	5	1	0	0

OCEANIA

Oceania	Mammals		Birds		Amphibians		FW Crabs		Reef-forming Corals		Conifers		Cycads	
	Total endemics	Threatened endemics	Total endemics	Threatened endemics	Total endemics	Threatened endemics	Total endemics	Threatened endemics	Total endemics	Threatened endemics	Total endemics	Threatened endemics	Total endemics	Threatened endemics
American Samoa	0	0	0	0	0	0	0	0	0	0	0	0	0	0
Australia	243	50	299	24	206	46	7	2	5	0	37	10	69	18
Christmas Island	1	1	3	3	0	0	0	0	0	0	0	0	0	0
Cocos (Keeling) Islands	0	0	0	0	0	0	0	0	0	0	0	0	0	0
Cook Islands	0	0	7	5	0	0	0	0	0	0	0	0	0	0
Fiji	1	1	27	8	2	1	1	0	0	0	4	3	0	0
French Polynesia	0	0	34	16	0	0	0	0	0	0	0	0	0	0
Guam	1	0	1	0	0	0	0	0	0	0	0	0	0	0
Kiribati	0	0	1	0	0	0	0	0	0	0	0	0	0	0
Marshall Islands	0	0	0	0	0	0	0	0	0	0	0	0	0	0
Micronesia, Federated States of	4	3	17	4	0	0	0	0	0	0	0	0	0	0
Nauru	0	0	1	1	0	0	0	0	0	0	0	0	0	0
New Caledonia	6	6	25	7	0	0	0	0	0	0	44	17	0	0
New Zealand	4	4	74	41	4	4	0	0	0	0	18	1	0	0
Niue	0	0	0	0	0	0	0	0	0	0	0	0	0	0
Norfolk Island	0	0	6	4	0	0	0	0	0	0	1	1	0	0
Northern Mariana Islands	0	0	3	3	0	0	0	0	0	0	0	0	0	0
Palau	2	0	10	0	1	0	0	0	0	0	0	0	0	0
Papua New Guinea	70	23	83	17	174	9	12	0	0	0	4	0	2	0
Pitcairn	0	0	5	5	0	0	0	0	0	0	0	0	0	0
Samoa	0	0	9	4	0	0	0	0	0	0	0	0	0	0
Solomon Islands	19	13	45	8	1	0	0	0	0	0	1	0	0	0
Tokelau	0	0	0	0	0	0	0	0	0	0	0	0	0	0
Tonga	0	0	2	1	0	0	0	0	0	0	1	1	0	0
Tuvalu	0	0	0	0	0	0	0	0	0	0	0	0	0	0
United States Minor Outlying Islands	0	0	0	0	0	0	0	0	1	1	0	0	0	0
Vanuatu	2	2	10	5	0	0	0	0	0	0	1	1	0	0
Wallis and Futuna	0	0	0	0	0	0	0	0	0	0	0	0	0	0

Appendix 12. Species changing IUCN Red List Category for genuine reasons

IUCN Red List Categories: **EX** - Extinct, **EW** - Extinct in the Wild, **CR** - Critically Endangered, **EN** - Endangered, **VU** - Vulnerable, **LR/cd** - Lower Risk/conservation dependent, **NT** - Near Threatened (includes **LR/nt** - Lower Risk/near threatened), **DD** - Data Deficient, **LC** - Least Concern (includes **LR/lc** - Lower Risk/least concern).

Species may move between categories for a variety of reasons, including genuine improvement or deterioration in status, new information being available about the species that was not known at the time of the previous assessment, taxonomic changes, or mistakes being made in previous assessments (e.g., incorrect information used previously, missapplication of the IUCN Red List Criteria, etc.). To help Red List users interpret the changes between the current and the previous Red List, a summary of species that have changed category **for genuine reasons only** in 2008 is provided in the table below. These genuine re-categorizations are used to calculate the Red List Index (see chapters 1 and 2). Note, however, that the list of genuine changes used in the RLI calculation for a particular group may not match exactly the list of genuine changes published here. For example, a species may have had the same category published for multiple Red List assessments, but information has subsequently come to light showing that it ought to have qualified for a lower or higher category in one or more previous assessments owing to a genuine change in status. This retrospective recalculation is included in the RLI calculation but is not reflected in the published history of Red List assessments.

MAMMALS			
Scientific name	**Common name**	**2007 IUCN Red List Category**	**2008 IUCN Red List Category**
Genuine Improvements			
Gulo gulo	Wolverine	VU	NT
Mustela nigripes	Black-footed Ferret	EW	EN
Arctocephalus philippii	Juan Fernández Fur Seal	VU	NT
Arctocephalus townsendi	Guadalupe Fur Seal	VU	NT
Balaena mysticetus	Bowhead Whale	LR/cd	LC
Eubalaena australis	Southern Right Whale	LR/cd	LC
Megaptera novaeangliae	Humpback Whale	VU	LC
Bison bonasus	European Bison	EN	VU
Capra pyrenaica	Spanish Ibex	LR/nt	LC
Capra walie	Walia Ibex	CR	EN
Procapra przewalskii	Przewalski's Gazelle	CR	EN
Rupicapra pyrenaica	Pyrenean Chamoix	LR/cd	LC
Eschrichtius robustus	Gray Whale	LR/cd	LC
Pteropus dasymallus	Ryukyu Flying Fox	EN	NT
Pteropus molossinus	Pohnpei Flying Fox	CR	VU
Pteropus samoensis	Samoan Flying Fox	VU	NT
Corynorhinus rafinesquii	Rafinesque's Big-eared Bat	VU	LC
Corynorhinus townsendii	Townsend's Big-eared Bat	VU	LC
Myotis emarginatus	Geoffroy's Bat	VU	LC
Myotis grisescens	Gray Myotis	EN	NT
Dasyurus geoffroii	Western Quoll	VU	NT
Dendrolagus lumholtzi	Lumholtz's Tree Kangaroo	LR/nt	LC
Macropus irma	Western Brush Wallaby	LR/nt	LC
Bettongia lesueur	Burrowing Bettong	VU	NT
Pseudochirulus cinereus	Daintree River Ringtail Possum	LR/nt	LC
Pseudochirulus herbertensis	Herbert River Ringtail Possum	LR/nt	LC
Equus ferus	Wild Horse	EW	CR
Equus zebra	Mountain Zebra	EN	VU
Rhinoceros unicornis	Indian Rhinoceros	EN	VU
Leontopithecus chrysopygus	Black Lion Tamarin	CR	EN

Genuine Improvements			
Loxodonta africana	African Elephant	VU	NT
Castor fiber	Eurasian Beaver	NT	LC
Leporillus conditor	Greater Stick-nest Rat	EN	VU
Pseudomys fieldi	Shark Bay Mouse	CR	VU
Pseudomys occidentalis	Western Mouse	EN	LC
Spermophilus suslicus	Speckled Ground Squirrel	VU	NT
Spermophilus washingtoni	Washington Ground Squirrel	VU	NT

Genuine deteriorations			
Panthera pardus	Leopard	LC	NT
Prionailurus planiceps	Flat-headed Cat	VU	EN
Prionailurus viverrinus	Fishing Cat	VU	EN
Spilogale pygmaea	Pygmy Spotted Skunk	LR/lc	VU
Aonyx cinerea	Asian Small-Clawed Otter	NT	VU
Mustela altaica	Altai Weasel	LR/lc	NT
Vormela peregusna	Marbled Polecat	LR/lc	VU
Arctocephalus galapagoensis	Galápagos Fur Seal	VU	EN
Neophoca cinerea	Australian Sea Lion	LR/lc	EN
Zalophus wollebaeki	Galápagos Sea Lion	VU	EN
Monachus schauinslandi	Hawaiian Monk Seal	EN	CR
Hemigalus derbyanus	Banded Civet	LR/lc	VU
Paradoxurus zeylonensis	Golden Palm Civet	LR/lc	VU
Cephalophus jentinki	Jentink's Duiker	VU	EN
Kobus megaceros	Nile Lechwe	LR/nt	EN
Axis kuhlii	Bawean Deer	EN	CR
Hydropotes inermis	Chinese Water Deer	LR/nt	VU
Rucervus eldii	Eld's Deer	VU	EN
Rusa timorensis	Javan Rusa	LR/lc	VU
Rusa unicolor	Sambar	LR/lc	VU
Sus barbatus	Bearded Pig	LR/lc	VU
Sus celebensis	Sulawesi Warty Pig	LR/lc	NT
Tayassu pecari	White-Lipped Peccary	LR/lc	NT
Amorphochilus schnablii	Smoky Bat	VU	EN
Coelops robinsoni	Malayan Tailess Leaf-nosed Bat	LR/nt	VU
Tadarida johorensis	Northern Free-tailed Bat	LR/nt	VU
Tadarida mops	Malayan Free-tailed Bat	LR/lc	NT
Nycteris tragata	Malayan Slit-faced Bat	LR/lc	NT
Lonchorhina fernandezi	Fernandez's Sword-Nosed Bat	VU	EN
Lonchorhina orinocensis	Orinoco Sword-Nosed Bat	LR/nt	VU
Platyrrhinus chocoensis	Choco Broad-Nosed Bat	VU	EN
Acerodon mackloti	Sunda Fruit Bat	LR/lc	VU
Eidolon helvum	Straw-Coloured Fruit Bat	LC	NT
Megaerops wetmorei	White-collared Fruit Bat	LR/lc	VU
Pteropus caniceps	North Moluccan Flying Fox	LR/lc	NT
Pteropus lylei	Lyle's Flying Fox	LR/lc	VU
Pteropus melanopogon	Black-bearded Flying Fox	LR/lc	EN
Pteropus niger	Mauritan Flying Fox	VU	EN

Genuine deteriorations			
Pteropus temminckii	Temminck's Flying Fox	LR/nt	VU
Pteropus vampyrus	Large Flying-fox	LR/lc	NT
Pteropus woodfordi	Dwarf Flying Fox	LR/lc	VU
Rousettus bidens	Manado Rousette	LR/nt	VU
Styloctenium wallacei	Stripe-faced Fruit Bat	LR/nt	VU
Rhinolophus cognatus	Andaman Horseshoe Bat	VU	EN
Rhinolophus sedulus	Lesser Wooly Horseshoe Bat	LR/lc	NT
Corynorhinus mexicanus	Mexican Big-eared Bat	LR/lc	NT
Hesperoptenus tomesi	Large False Serotine	LR/lc	VU
Kerivoula pellucida	Clear-winged Woolly Bat	LR/lc	NT
Murina aenea	Bronze Tube-nosed Bat	LR/nt	VU
Murina rozendaali	Gilded Tube-nosed Bat	LR/nt	VU
Phoniscus atrox	Groove-toothed Trumpet-eared Bat	LR/lc	NT
Rhogeessa minutilla	Tiny Yellow Bat	LR/nt	VU
Dasyurus hallucatus	Northern Quoll	LR/nt	EN
Pseudantechinus bilarni	Sandstone Pseudantechinus	LR/lc	NT
Sarcophilus harrisii	Tasmanian Devil	LR/lc	EN
Myrmecobius fasciatus	Numbat	VU	EN
Burramys parvus	Mountain Pygmy Possum	EN	CR
Dendrolagus mbaiso	Dingiso	VU	CR
Dendrolagus scottae	Tenkile	EN	CR
Dorcopsis luctuosa	Grey Dorcopsis	LR/lc	VU
Dorcopsulus vanheurni	Small Dorcopsis	LR/lc	NT
Petaurus abidi	Northern Glider	VU	CR
Phalanger lullulae	Woodlark Cuscus	LR/lc	EN
Phalanger matanim	Telefomin Cuscus	EN	CR
Spilocuscus kraemeri	Admiralty Cuscus	LR/lc	NT
Spilocuscus rufoniger	Black-spotted Cuscus	EN	CR
Bettongia penicillata	Woylie	LR/cd	CR
Pseudochirops coronatus	Reclusive Ringtail	LR/lc	VU
Ochotona iliensis	Ili Pika	VU	EN
Equus hemionus	Asiatic Wild Ass	VU	EN
Manis javanica	Sunda Pangolin	LR/nt	EN
Manis pentadactyla	Chinese Pangolin	LR/nt	EN
Saguinus niger	Black-handed Tamarin	LC	VU
Macaca hecki	Heck's Macaque	LR/nt	VU
Macaca nigra	Celebes Crested Macaque	EN	CR
Macaca tonkeana	Tonkean Macaque	LR/nt	VU
Presbytis melalophos	Sumatran Surili	LR/nt	EN
Presbytis thomasi	Thomas's Langur	LR/nt	VU
Pygathrix cinerea	Grey-shanked Douc Langur	EN	CR
Trachypithecus obscurus	Dusky Leaf-monkey	LR/lc	NT
Hylobates agilis	Agile Gibbon	LR/nt	EN
Hylobates albibarbis	Bornean White-bearded Gibbon	LR/nt	EN
Hylobates klossii	Kloss's Gibbon	VU	EN
Hylobates lar	Lar Gibbon	LR/nt	EN

Genuine deteriorations			
Hylobates muelleri	Müller's Bornean Gibbon	LR/nt	EN
Nomascus concolor	Black Crested Gibbon	EN	CR
Symphalangus syndactylus	Siamang	LR/nt	EN
Nycticebus coucang	Greater Slow Loris	LR/lc	VU
Tarsius bancanus	Horsfield's Tarsier	LR/lc	VU
Abrothrix illuteus	Gray Grass Mouse	LR/lc	NT
Abrothrix sanborni	Sanborn's Grass Mouse	LR/lc	NT
Akodon latebricola	Ecuadorean Grass Mouse	LR/lc	VU
Akodon surdus	Silent Grass Mouse	LR/lc	VU
Arborimus longicaudus	Red Tree Vole	LR/lc	NT
Arvicola sapidus	Southern Water Vole	LR/nt	VU
Calomys hummelincki	Hummelinck's Vesper Mouse	LR/lc	VU
Mesocricetus brandti	Brandt's Hamster	LR/lc	NT
Microtus oaxacensis	Tarabundi Vole	LR/nt	EN
Microtus quasiater	Japalapan Pine Vole	LR/lc	NT
Microtus umbrosus	Zepoal Tepec Vole	LR/lc	EN
Neotoma palatina	Bolano's Woodrat	LR/nt	VU
Reithrodontomys microdon	Small-toothed Harvest Mouse	LR/nt	VU
Reithrodontomys spectabilis	Cozumel Harvest Mouse	EN	CR
Sigmodon alleni	Allen's Cotton Rat	LR/lc	VU
Thomasomys hylophilus	Woodland Oldfield Mouse	LR/lc	EN
Wilfredomys oenax	Greater Wilfred's Mouse	LR/lc	EN
Xenomys nelsoni	Magedelena Wood Rat	LR/nt	EN
Ctenomys australis	Southern Tuco-tuco	LR/lc	EN
Allactaga vinogradovi	Vinogradov's Jerboa	LR/lc	NT
Pappogeomys alcorni	Alcorn's Pocket Gopher	VU	CR
Chaetodipus goldmani	Goldman's Pocket Mouse	LR/lc	VU
Dipodomys nitratoides	Fresno Kangaroo Rat	LR/nt	VU
Gerbillus gleadowi	Little Hairy-footed Gerbil	LR/lc	NT
Maxomys rajah	Rajah Sundaic Maxomys	LR/lc	VU
Maxomys whiteheadi	Whitehead's Sundaic Maxomys	LR/lc	VU
Nesokia bunnii	Bunn's Short-Tailed Bandicoot Rat	LR/nt	EN
Niviventer cremoriventer	Sundaic Arboreal Niviventer	LR/lc	VU
Pseudomys fumeus	Smoky Mouse	VU	EN
Pseudomys novaehollandiae	New Holland Mouse	LR/lc	VU
Rattus richardsoni	Glacier Rat	LR/lc	VU
Solomys ponceleti	Poncelet's Giant Rat	EN	CR
Solomys salebrosus	Bougainville Giant Rat	LR/nt	EN
Solomys sapientis	Isabel Giant Rat	VU	EN
Uromys neobritannicus	Bismarck Giant Rat	LR/lc	EN
Zyzomys maini	Arnhem Land Rock Rat	LR/lc	NT
Octodon bridgesi	Bridges's Degu	LR/lc	VU
Octodon pacificus		VU	CR
Callosciurus adamsi	Ear-spot Squirrel	LR/lc	VU
Callosciurus melanogaster	Mentawai Squirrel	LR/lc	VU
Marmota sibirica	Mongolian Marmot	LR/lc	EN

Genuine deteriorations			
Marmota vancouverensis	Vancouver Island Marmot	EN	CR
Petaurista nobilis	Bhutan Giant Flying Squirrel	LR/nt	VU
Petinomys genibarbis	Whiskered Flying Squirrel	LR/lc	VU
Petinomys lugens	Siberut Flying Squirrel	LR/nt	EN
Petinomys setosus	Temminck's Flying Squirrel	LR/lc	VU
Petinomys vordermanni	Vordermann's Flying Squirrel	LR/lc	VU
Pteromyscus pulverulentus	Smoky Flying Squirrel	LR/nt	EN
Ratufa bicolor	Black Giant Squirrel	LR/lc	NT
Rhinosciurus laticaudatus	Shrew-faced Squirrel	LR/lc	NT
Rubrisciurus rubriventer	Sulawesi Giant Squirrel	LR/lc	VU
Spermophilus perotensis	Perote Ground Squirrel	LR/nt	EN
Spermophilus xanthoprymnus	Asia Minor Ground Squirrel	LR/lc	NT
Spalax arenarius	Sandy Mole Rate	VU	EN
Spalax zemni	Podolian Mole Rat	LR/lc	VU

BIRDS			
Scientific name	**Common name**	**2007 IUCN Red List Category**	**2008 IUCN Red List Category**
Genuine improvements			
Ducula galeata	Marquesan Imperial-pigeon	CR	EN
Apteryx owenii	Little Spotted Kiwi	VU	NT
Genuine deteriorations			
Aythya baeri	Baer's Pochard	VU	EN
Sterna nereis	Fairy Tern	LC	VU
Eurynorhynchus pygmeus	Spoon-billed Sandpiper	EN	CR
Numenius arquata	Eurasian Curlew	LC	NT
Ducula finschii	Finsch's Imperial-pigeon	LC	NT
Ducula rubricera	Red-knobbed Imperial-pigeon	LC	NT
Reinwardtoena browni	Pied Cuckoo-dove	LC	NT
Centropus violaceus	Violaceous Coucal	LC	NT
Accipiter princeps	New Britain Goshawk	NT	VU
Pauxi pauxi	Helmeted Curassow	VU	EN
Tetrao mlokosiewiczi	Caucasian Grouse	DD	NT
Atrichornis rufescens	Rufous Scrub-bird	NT	VU
Coracina newtoni	Réunion Cuckooshrike	EN	CR
Corvus kubaryi	Mariana Crow	EN	CR
Loxops caeruleirostris	Akekee	EN	CR
Stipiturus mallee	Mallee Emuwren	VU	EN
Melidectes whitemanensis	Bismarck Melidectes	LC	NT
Mimus trifasciatus	Floreana Mockingbird	EN	CR
Megalurulus grosvenori	Bismarck Thicketbird	DD	VU
Sylvia undata	Dartford Warbler	LC	NT
Diomedea dabbenena	Tristan Albatross	EN	CR
Cacatua ophthalmica	Blue-eyed Cockatoo	LC	VU
Ninox odiosa	Russet Hawk-owl	LC	VU
Tyto aurantia	Bismarck Masked-owl	DD	VU

REPTILES

Scientific name	Common name	2007 IUCN Red List Category	2008 IUCN Red List Category
Genuine deteriorations			
Crocodylus rhombifer	Cuban Crocodile	EN	CR
Erymnochelys madagascariensis	Madagascar Big-headed Turtle	EN	CR
Astrochelys radiata	Radiated Tortoise	VU	CR
Astrochelys yniphora	Ploughshare Tortoise	EN	CR
Pyxis arachnoides	Spider Tortoise	VU	CR
Pyxis planicauda	Flat-tailed Tortoise	EN	CR

AMPHIBIANS

Scientific name	Common name	2007 IUCN Red List Category	2008 IUCN Red List Category
Genuine improvement			
Thorius macdougalli		EN	VU
Genuine deteriorations			
Incilius holdridgei		CR	EX
Centrolene buckleyi		NT	VU
Craugastor escoces		CR	EX
Parvimolge townsendi		EN	CR
Pseudoeurycea gigantea		EN	CR
Pseudoeurycea juarezi		EN	CR
Thorius munificus		EN	CR

INVERTEBRATES

Scientific name	Common name	2007 IUCN Red List Category	2008 IUCN Red List Category
Genuine deterioration			
Hemiphlebia mirabilis	Ancient Greenling	VU	EN